Emerging Paradigms in Pharmaceutical Research

Series Editor
Mukesh Nandave, School of Pharmaceutical Sciences
Delhi Pharmaceutical Sciences and Research University
New Delhi, Delhi, India

This series presents the dynamic advancements in pharmaceutical science and provides a comprehensive resource for researchers, academicians, and professionals. Each book in the series focuses on a critical area of pharmaceutical sciences, integrating foundational principles with the latest research and applications.

Research in pharmaceutical sciences is evolving rapidly, with groundbreaking innovations and interdisciplinary approaches reshaping drug discovery, development, and delivery. The objective of this series is to bridge the gap between fundamental research and practical applications, offering readers a platform to explore cutting-edge knowledge, novel methodologies, and emerging technologies in pharmaceutical sciences. By incorporating diverse perspectives from experts worldwide, the series seeks to foster collaboration and innovation.

This series targeted at a diverse audience, including:

- Academic researchers in pharmaceutical sciences, pharmacology, and medicinal chemistry.
- Industry professionals in drug development, formulation, and regulatory affairs.
- Postgraduate and doctoral students seeking advanced knowledge in specialized areas.
- Policymakers and regulatory authorities involved in healthcare and drug development.

Arpita Roy • Soumya Pandit
Mukesh Nandave
Editors

Mycoremediation of Xenobiotics

Harnessing Fungi for Environmental Detoxification

Springer

Editors
Arpita Roy
Department of Biotechnology, Sharda School of Bioscience and Technology
Sharda University
Greater Noida, Uttar Pradesh, India

Soumya Pandit
Department of Life Sciences, Sharda School of Bioscience and Technology
Sharda University
Greater Noida, Uttar Pradesh, India

Mukesh Nandave
Pharmacology and Pharmaceutical Biotechnology
Delhi Pharmaceutical Sciences and Research University
New Delhi, Delhi, India

ISSN 3091-3179 ISSN 3091-3187 (electronic)
Emerging Paradigms in Pharmaceutical Research
ISBN 978-981-95-5742-4 ISBN 978-981-95-5743-1 (eBook)
https://doi.org/10.1007/978-981-95-5743-1

This Springer imprint is published by the registered company Springer Nature Singapore Pte Ltd.
The registered company address is: 152 Beach Road, #21-01/04 Gateway East, Singapore 189721, Singapore

Preface

In the current development, environmental pollution is one of the major threats all over the world. Due to rapid urbanization and industrialization, various types of pollutants are coming up at an alarming rate, which leads to potential damage to the quality ecosystems. In the current situation, remediation of pollutants is one of the major challenges which our societies are facing around the globe. Increase in human population is also one of the major factors for pollution. This book entitled *Mycoremediation of Xenobiotics: Harnessing Fungi for Environmental Detoxification* provides broad-spectrum potential alternatives for remediation of pollutants. Mycoremediation is a viable, economical, and environmentally responsible method of cleaning up xenobiotic-contaminated settings. In addition to presenting mycoremediation as a workable answer to one of the most important environmental issues of our day, this book will address the urgent need for a complete knowledge of the technique. This book will be an invaluable resource for environmental scientists, politicians, and industry experts looking to promote or use mycoremediation methods, since it highlights state-of-the-art research and effective case studies. Additionally, by bridging the gap between mycology, environmental science, and biotechnology, it will promote multidisciplinary cooperation.

We are enormously thankful to all the authors who have contributed in this volume and to the Springer team for their substantial teamwork for printing up this volume.

Greater Noida, Uttar Pradesh, India — Arpita Roy
Soumya Pandit
New Delhi, Delhi, India — Mukesh Nandave

Contents

About the Editors

Arpita Roy, PhD is working as an Assistant Professor in the School of Bioscience and Technology (Department of Biotechnology) at Sharda University, Greater Noida. She obtained her MTech (Industrial Biotechnology) and PhD (Biotechnology) from Delhi Technological University. Her areas of interest include Plant Biotechnology, Environmental Biotechnology, Green nanoparticle synthesis, and Microbiology. She has published more than 120 papers in peer-reviewed journals (h-index 48; i-index 144; Total citation >8500), 14 patents, and 30 book chapters, and has edited 12 books. She has qualified GATE, CSIR-UGC-NET, and ICAR-ASRB NET. She has been enlisted in Top 2% scientists in the world for four consecutive years (Stanford University report published—2022, 2023, 2024, and 2025). She has edited several special issues in different journals. She is a recipient of "Commendable Research Excellence Award DTU 2019, 2020 and 2021" from Delhi Technological University, Delhi. She has presented more than 15 papers in national and international conferences.

Soumya Pandit is an Associate Professor at Sharda University, Greater Noida, Delhi NCR, India. He pursued his doctoral studies from Indian Institute of Technology, Kharagpur, and completed his postdoctoral research work (PBC fellow) at the Department of Desalination and Water Treatment, The Zuckerberg Institute for Water Research (ZIWR), Ben-Gurion University of the Negev. He is a recipient of North West University, South Arica Post-doctoral fellowship. He has authored 150 research papers in peer-reviewed journals and 22 book chapters and published 15 Indian patents. He holds a BTech and an MTech in Biotechnology. His current research focuses on microbial electrochemical systems for bioenergy harvesting, biohythane, bacterial biofilm and biofouling study, etc. He serves as a reviewer for *Frontiers in Environmental Microbiology*, editorial board member in *SCIREA Journal of Biology*, and the *Journal of Korean Society of Environmental Engineers* (JKSEE).

Mukesh Nandave is Associate Dean Research, Head of School of Pharmaceutical Sciences, and Head of Department of Pharmacology at Delhi Pharmaceutical Sciences and Research University, Govt. of NCT of Delhi, New Delhi. He earned his PhD in Pharmacology from AIIMS, Delhi, and received his Post-Doctoral training from the Division of Cardiothoracic Surgery, the Ohio State University Medical Center, Columbus, USA. Since more than 19 years he has been investigating the role of nutraceuticals, herbomineral formulations, plant extracts, and phytoconstituents in Myocardial ischemia and reperfusion injury, Diabetes, Obesity, Pain management. He has published more than 75 papers in peer-reviewed national and international journals. His lab received more than 2 crore total funding from the Government as well as industries, including Pharmazz, Dabur, Charak, Madhavbaug, Sandu, etc. He has received numerous awards, including G. Achari Gold Medal by Indian Pharmacological Society, Association of Pharmaceutical Teachers of India (APTI) Young Pharmacy Teacher of the Year Award, Early Investigator Award by International Society for Heart Research, Prof. Duggirala Visweswaram and Prof. Sreemantula Satyanarayana Prize, Best Research Output of the Year for 2014–2015, 2013–2014, and 2012–2013 Award of SVKM's NMIMS University, and the Indus Foundation's Award for Research Excellence. He is a life member of various professional bodies, including International Society for Heart Research (ISHR), International Academy of Cardiovascular Sciences (IACS), and Indian Pharmacological Society (IPS). He is also Secretary General of International Academy of Cardiovascular Sciences (IACS), India Section.

Overview of Environmental Contamination and Xenobiotics

1

Hiruni Wijesinghe, Waruni Isurika, Gobika Thiripuranathar, Keerthi Aththanayake, and Sagarika Ekanayake

Abstract

Environmental contamination by xenobiotics has become a defining challenge of the Anthropocene. Xenobiotics, including synthetic organic compounds, heavy metals, pharmaceuticals, and persistent pollutants, are foreign to natural biological systems and often resist degradation. These compounds share key characteristics, such as persistence, toxicity, bioaccumulation, and biomagnification, which amplify their ecological and human health impacts. Major anthropogenic sources include industrial emissions, agricultural practices, urban waste, and transportation, while air, water, soil, and sediments serve as critical compartments for contaminant distribution. The ecological consequences of xenobiotic pollution are wide-ranging, encompassing biodiversity loss, disruption of ecosystem functions, eutrophication, soil degradation, and deterioration of water quality. Human health risks associated with xenobiotics are equally concerning, spanning carcinogenicity, teratogenicity, endocrine disruption, neurotoxicity, and immunotoxicity. Despite the availability of conventional remediation techniques, including incineration, landfilling, soil washing, and pump-and-treat systems, these approaches often present significant limitations, such as high operational costs, elevated energy demands, and additional environmental disturbances. Consequently, there is a growing emphasis on sustainable and environmentally compatible remediation strategies. Emerging approaches, particularly bioremediation and mycoremediation, offer promising, cost-effective, and eco-

H. Wijesinghe · W. Isurika · G. Thiripuranathar (✉) · K. Aththanayake
College of Chemical Sciences, Institute of Chemistry Ceylon,
Welikada, Rajagiriya, Sri Lanka
e-mail: tgobika@ichemc.edu.lk

S. Ekanayake
Department of Biochemistry, Faculty of Medical Sciences, University of Sri Jayewardenepura, Gangodawila, Nugegoda, Sri Lanka

A. Roy et al. (eds.), *Mycoremediation of Xenobiotics*, Emerging Paradigms in Pharmaceutical Research, https://doi.org/10.1007/978-981-95-5743-1_1

friendly alternatives capable of degrading or transforming persistent pollutants while minimizing ecological harm. Together, these insights underscore the urgency of addressing xenobiotic pollution and highlight the potential for innovative remediation strategies to restore environmental quality. By integrating scientific innovation, regulatory frameworks, and community engagement, it is possible to mitigate the risks posed by xenobiotics, safeguard human health, and promote the sustainable management of ecosystems in the Anthropocene.

Keywords
Xenobiotics · Environmental contamination · Bioaccumulation · Ecotoxicology · Mycoremediation

1.1 Introduction

1.1.1 Environmental Contamination: A Legacy and a Modern Crisis

The human-dominated era is marked by the large-scale human impact on the planet, especially alterations to the environment. Environmental contamination refers to chemical, physical, or biological agents in air, water, soil, or biota at concentrations or in locations that are out of place and may pose risks to ecosystems and human health. Contamination could be from natural processes, such as naturally occurring heavy metals and volcanic ash and from human activities such as industrial discharge, agriculture, waste disposal, and urban runoff (Ojija 2024).

Although the terms contamination and pollution are often used in the same way, contamination refers to the presence of an unwanted substance in the environment. Pollution, on the other hand, means contamination that causes, or likely will cause, harmful effects on living things or ecosystems (Dinka 2018). In other words, all pollutants are contaminants, but not every contaminant counts as a pollutant unless it creates negative impacts. Bioavailability and toxicity define that difference (Manisalidis et al. 2020). Monitoring and managing contamination require understanding sources, pathways, persistence, and biological effects of pollutants.

Primary sources of environmental contamination include industrial effluents from chemical manufacturing, mining, and metallurgy. They also include agricultural inputs and runoff, such as the usage of pesticides, herbicides, and fertilizers. Municipal wastewater and sewage containing pharmaceutical and personal-care product residues represent another source of contamination (Suleiman and Ibrahim 2024). Furthermore, atmospheric deposition from combustion includes heavy metals and poly aromatic hydrocarbons (PAHs), which play a significant role in contamination. Other sources are plastic and microplastic fragmentation and improper disposal of hazardous wastes. In addition, emerging technologies and products, such as nanomaterials, novel flame retardants, and per- and polyfluoroalkyl substances (PFAS), add newly recognized sources and pathways, increasing the diversity of contaminants in soils, surfaces, groundwater, and biota (Singh et al. 2023).

1.1.2 The Global Pervasiveness of Contamination

Environmental contamination has become a global issue, not limited to industrial areas and cities. Pollutants travel between continents and oceans through atmospheric circulation, ocean currents and global trade. They reach even the most distant, seemingly untouched parts of the planet (U.S. Environmental Protection Agency (EPA)). The strongest proof of this global reach is the presence of persistent organic pollutants (POPs) in the ecosystems of the Arctic and Antarctic (Persistent Organic Pollutants: A Global Issue, A Global Response | US EPA 2025). Chemicals such as polychlorinated biphenyls (PCBs) and dichlorodiphenyltrichloroethane (DDT), which were never used in these polar areas, have traveled thousands of kilometers through the air and sea (Castañeda et al. 2025). They accumulate in the fatty tissues of animals, such as polar bears, seals, and seabirds (Fromberg et al. 2011).

Furthermore, microplastics, which are tiny particles formed due to the breakdown of larger plastic items, have also become a common pollutant worldwide. They have been discovered in the deepest parts of the ocean, like the Mariana Trench, in alpine snow and ice, and in the air breathed and the water consumed by living beings (Li et al. 2023). This widespread contamination is elevated by the huge amount of chemicals produced and released. The usage of agricultural chemicals also significantly increased, with global pesticide application estimated to have reached 3.5 million tons by 2020 (Tang et al. 2021). Over 350,000 different chemicals and mixtures are registered for production and use around the world, creating an enormous long lasting xenobiotic reservoir in the environment (Sobral et al. 2025). This ongoing chemical presence on the planet ensures that no ecosystem remains completely isolated from human influence. It highlights the urgent need for worldwide cooperation and sustainable solutions to address this shared issue.

1.1.3 Xenobiotics: Strangers to Life

Derived from the Greek words xenos, meaning "foreign" or "strange," and bios, meaning "life," the term "xenobiotic" describes any substance, chemical, or ingredient found in a living organism that is not naturally occurring or not usually found in that organism or biological system (Singh et al. 2023). A variety of compounds, such as synthetic substances made through human activities, such as pesticides, pharmaceuticals, industrial solvents, and plasticizers, come under xenobiotics. These are not natural compounds as they have no natural biological analogues. When introduced into living organisms, they can disrupt the normal biochemical pathways, as those organisms had overlooked the evolutionary advantage of accommodating these compounds (Sobral et al. 2025).

The idea of xenobiotics is subjective and includes more than just synthetic chemicals. Naturally occurring substances can also be viewed as xenobiotics when they enter organisms that do not usually synthesize or encounter them. For example, plant-derived alkaloids are natural compounds, but they act as xenobiotics in insects and mammals, since these organisms are unable to break them down easily. This shows that xenobiotics are classified based on the physiology and ecology of organisms, not just the origin of the compound. In environmental science, xenobiotic is often used with a narrower meaning, mainly referring to synthetic chemicals that are foreign to the entire biosphere (Singh et al. 2023). These pollutants, like PCBs, heavy metals, and microplastics, create significant challenges for ecosystems. Unlike natural compounds, which organisms have adapted to process over time, xenobiotics often resist natural breakdown. Their persistence in the environment leads to bioaccumulation and biomagnification, affecting not just individual organisms but entire ecological networks (Mohapatra et al. 2025). Additionally, introducing xenobiotics negatively affects the balance of microbial communities, which are vital for nutrient cycling and ecosystem stability.

1.2 Key Characteristics of Xenobiotics

Xenobiotics demonstrate various characteristics that influence their behavior in the environment, the biological effects, and overall risk. A key trait of many xenobiotics is their persistence or recalcitrance (Biradar et al. 2024). Unlike natural organic molecules, in which microbes are capable of breaking down and recycling, many synthetic compounds have chemical structures that are new in evolutionary terms. Traits such as extensive halogenation, including carbon-chlorine or carbon-fluorine bonds, complex branching, and aromatic rings, make these molecules hard for microbial enzymes to recognize and break down. As a result, their resistance to microbial, photochemical, and hydrolytic breakdown leads to long environmental half-lives, allowing persistence for years, decades, or even centuries, creating long-term contamination risks (Mukherjee et al. 2024).

The biological effects of xenobiotics depend on how well they interact with living organisms. This begins with their bioavailability. A substance may be in soil or sediment, and its bioavailability could be low if it is tightly bound to mineral particles. Factors, like solubility, particle size, and environmental pH, influence the bioavailability of chemicals and determine the amount of chemicals that could enter the biological systems (Iyevhobu et al. 2025).

Bioaccumulation is the process where a chemical is taken up by an organism at a rate faster than it is eliminated through metabolism or excretion (Dinka 2018). This leads to a net increase of the chemical in the tissues of an organism over time. Lipophilic compounds, which do not dissolve well in water, tend to accumulate in the fatty tissues of organisms, where their concentration may exceed that of the surrounding environment. Such accumulation has serious biological and ecological effects.

Biomagnification, also called trophic magnification, is the gradual increase in concentration of a bioaccumulated substance as it moves up the food chain. When a predator eats prey that contains a low concentration of a persistent lipophilic xenobiotic, the contaminant transfers and resides in the predator's fatty tissues. As this process continues up the food chain the concentration grows (Li et al. 2023). This means that even small amounts of a pollutant in the environment reach dangerously high levels in top predators, including birds of prey, marine mammals, and humans.

The toxicity of xenobiotics comes from their natural ability to cause harmful biological effects, which could present as acute toxicity (damage from a single, high-dose exposure) or chronic toxicity (harm from repeated, low-dose exposure over time). These effects may range from immediate cell death to subtle, long-lasting disruptions in physiology, reproduction, or development. A crucial idea in toxicology is that the original xenobiotic compound is not always in its most dangerous form. Through metabolic processes in organisms or environmental breakdown, parent compounds may change into metabolites or byproducts that are more harmful than the original. This process, known as "toxicant activation" or "metabolic activation," raises significant concerns regarding many pollutants, as partial breakdown could unknowingly increase the overall risk (Sobral et al. 2025).

These traits often drive their industrial use, creating a tension between their intended benefits and their environmental impact. This conflict is a main focus of modern ecotoxicology. It highlights the need for a "safe-by-design" approach to chemical development, where environmental factors are considered at the molecular level. For example, the strong chemical stability of per- and polyfluoroalkyl substances (PFAS) makes them useful for purposes such as nonstick coatings and firefighting foams. Conversely, this same stability leads to their classification as "forever chemicals" as they remain in the environment indefinitely (Brunn et al. 2023). Similarly, the long-lasting nature of organochlorine pesticides such as DDT was once appreciated for extended insect control, but this persistence allows for their long-range transport and bioaccumulation (Esther et al. 2022).

1.3 Sources and Environmental Pathways of Xenobiotic Release

Contaminants enter the environment through both human-made and natural sources (Fig. 1.1). Knowledge of where these substances originate from and how they are transported is key to understanding their effects on ecosystems and human health, as well as creating strategies to reduce them.

1.3.1 Major Anthropogenic Sources

Human activities are the primary drivers of xenobiotic contamination, releasing a diverse array of pollutants into the environment. These sources are broadly categorized as shown in the following table (Table 1.1).

Industrial sectors, such as chemical manufacturing, power generation, metallurgy, electronics and mining, release priority pollutants to air and water, including heavy metals (e.g., Hg, Pb, Cd), VOCs and SVOCs, PAHs and legacy (DDT, PCBs, etc.), and emerging persistent organic pollutants. Stack emissions and fugitive releases contribute to atmospheric burdens that later deposit to land and water

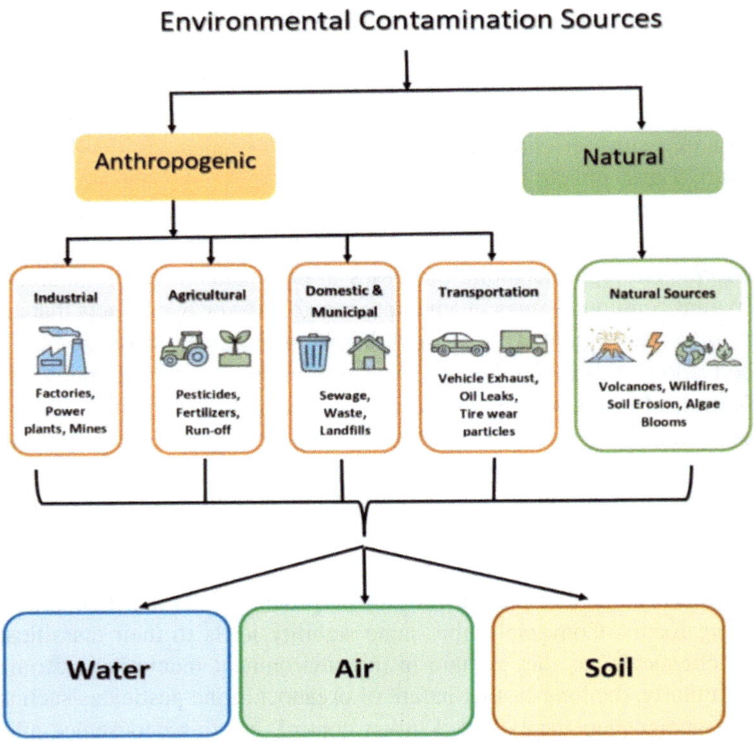

Fig. 1.1 Environmental contamination sources and pathways

Table 1.1 Major anthropogenic sources of xenobiotics

Source	Examples of xenobiotics released
Industrial emissions/ discharges	PAHs, heavy metals (lead, cadmium, mercury), volatile organic compounds (VOCs)
Agricultural runoff	Pesticides (atrazine, glyphosate), herbicides, nitrate and phosphate fertilizers
Urban and domestic waste	Pharmaceuticals and personal care products (PPCPs), detergents, microplastics, endocrine-disrupting compounds
Transportation	Vehicle emissions (PAHs, NOx, CO), oil spills, tire wear particles
Waste incineration	Dioxins, furans, mercury, other heavy metals

(Berríos-Rolón et al. 2025). Effluents and process waters convey metals and organics to receiving waters when treatment is inadequate (Naidu et al. 2021). Mining and ore processing generate acid mine drainage with low pH and elevated levels of sulfate and dissolved metals that migrate into streams and aquifers (Singh et al. 2024).

Plantations and croplands release pesticides, herbicides, veterinary medications, and excess minerals, such as phosphates and nitrates, into rivers, lakes, and coastal areas through diffuse runoff. In surface waters close to intensive agricultural areas, pesticide residues and their transformation products are commonly found. The transport of these chemicals to streams is further accelerated by subsurface pipes (Wang et al. 2023). The development of hypoxic conditions brought on by nutrient-rich runoff causes eutrophication, which modifies contaminant cycling and increases toxicity hazards (Khalili and Moridi 2025).

On the other hand, municipal solid-waste landfills generate leachate containing diverse xenobiotics, notably PFAS and PPCPs, that may reach groundwater or surface waters if not fully captured and treated (Berríos-Rolón et al. 2025). Wastewater treatment plants (WWTPs) are key control points but are not designed to completely remove many trace organics; thus, effluents and biosolids act as sources to receiving waters and soils (Sun et al. 2019).

In addition to combustion-derived PAHs and metals, non-exhaust particulates like tire and road-wear particles that serve as carriers of additives and sorbed pollutants are released by road, rail, aviation, and maritime transportation (Berríos-Rolón et al. 2025). Background loading results from regular bilge discharges and small spills that happen over time. On the other hand, major marine oil spills release complex mixtures of hydrocarbons all at once. This sudden release causes both immediate and long-term harm to the environment (Patterson et al. 2023).

Furthermore, municipal solid-waste and hazardous-waste incineration release trace amounts of dioxins, furans, and metals if combustion and air-pollution-control conditions are not optimum. Residual ash may contain metals that need safe disposal (Jia et al. 2018).

1.3.2 Environmental Compartments and Transport Pathways

The dispersion, retention, and transformation of contaminants are controlled by a complex web of physical and chemical processes that interact with the environment. The ways that pollutants move and accumulate depend on many compartments, including air, water, soil, and sediments. For instance, the atmosphere can help chemicals in the air move rapidly over great distances, and soils and sediments frequently serve as reservoirs, holding onto pollutants for a long time. In order to move materials across landscapes and link terrestrial and aquatic ecosystems, water systems combine inputs from several sources. Pollutants are also influenced by biological interactions, especially in food webs, which have a cascade effect on ecological networks. These interactions can have a ripple effect on ecological networks. Analysis of these compartments and pathways provides important information about contaminant behavior, supporting monitoring, management, and remediation techniques (Singh et al. 2023).

The atmosphere is a key medium for transporting volatile and semi-volatile contaminants, which travel locally or across long distances either in gaseous form or attached to particles. They are deposited onto land and water through wet processes, such as precipitation or dry settling of dust. Persistent compounds, such as PAHs, PCBs, and DDT, reach remote regions, including the polar areas (Wania and Shunthirasingham 2020). Transport is shaped by meteorological factors such as wind, turbulence, and temperature inversions as well as by contaminant properties like volatility and photoreactivity. The "grasshopper effect" describes repeated cycles of volatilization and deposition that enable pollutants to move far from their sources (Berríos-Rolón et al. 2025). For example, mercury from coal combustion can form reactive gaseous species, while PAHs from vehicles may transform into more toxic derivatives and spread over hundreds of kilometers (Degrendele et al. 2020). While urban and industrial emissions release nitrogen oxides, sulfur dioxide, PAHs, and heavy metals, biomass burning and wildfires also contribute large amounts of organic compounds and particulates.

Aquatic environments similarly act as major pathways for contamination through inputs from agriculture, industry, urban runoff, wastewater, and landfills. Hydrophobic substances, such as PCBs, PAHs, and DDT, attach to particles and accumulate in sediments, while soluble pollutants, including nutrients, heavy metals, and pharmaceuticals, disperse widely. Their movement is shaped by rainfall, river flow, groundwater, and seasonal events, with persistent substances often concentrating in riverbeds and coastal zones. Many of these pollutants bioaccumulate in aquatic organisms and magnify through food webs, i.e., methylmercury in fish. Human modifications to waterways, such as dams or channelization, further alter these transport dynamics and often intensify ecological impacts (Naidu et al. 2021).

Soils and sediments function as major reservoirs of contaminants, acting both as sinks and secondary sources. Pollutants enter soils through deposition, irrigation, biosolids, pesticide application, and atmospheric intake, where persistent compounds and heavy metals accumulate over time. Their behavior depends on factors like pH, redox potential, microbial activity, and organic matter, which influence degradation or transformation (Wang et al. 2023). Sediments in rivers, lakes, and coastal zones similarly store hydrophobic pollutants and metals, removing them from the water column but creating long-term ecological risks. Industrial effluents, agricultural erosion, urban runoff, and mining are key sources and disturbances, such as dredging, storms, or chemical shifts, that remobilize toxins (Ojija 2024).

Furthermore, contaminants enter living beings when they eat, breathe, or absorb them through their skin. Chemicals, such as PCBs, DDT, methylmercury, brominated flame retardants, and PFAS, tend to accumulate in fatty tissues, particularly in top predators, birds, and humans (Lamb et al. 2022). For example, methylmercury collects in fish like tuna and swordfish, while DDT has caused eggshell thinning in birds, such as eagles and falcons (Thilagam and Gopalakrishnan 2022). Microplastics carry other pollutants and transfer them from small creatures, such as zooplankton, to larger organisms like fish and marine mammals (Bajt 2021). When these contaminants move through food webs, they harm reproduction, growth, immune systems, and survival, sometimes causing widespread effects in ecosystems (Dinka 2018).

1.4 Classification of Xenobiotic Contaminants

The vast array of xenobiotic compounds is broadly categorized into organic and inorganic classes. Each class possesses distinct sources, environmental behaviors, and toxicological profiles. The following table (Table 1.2) provides a summary of these major classes of xenobiotics, which are discussed in detail in the subsequent sections.

1.4.1 Organic Xenobiotics: A Diverse and Persistent Threat

Persistent organic pollutants (POPs) are harmful organic compounds recognized under the Stockholm Convention for their dangerous effects. There are 12 POPs listed under this convention. They are intentionally made endrin, DDT, chlordane, mirex, toxaphene, heptachlor, aldrin, dieldrin, and unintentional byproducts, like

Table 1.2 Major classes of xenobiotic contaminants

Xenobiotic class	Examples	Primary sources	Key environmental & health impacts
Organic			
Persistent organic pollutants (POPs)	DDT, PCBs, dioxins	Industrial byproducts, legacy pesticides	Bioaccumulation, endocrine disruption, carcinogenic, long-range transport
Polycyclic aromatic hydrocarbons (PAHs)	Benzo[a]pyrene	Incomplete combustion of organic matter	Carcinogenic, mutagenic, toxic to aquatic life
Petroleum hydrocarbons	Benzene, toluene (BTEX)	Oil spills, fuel leaks, urban runoff	Neurotoxic, carcinogenic, ecosystem damage
Pharmaceuticals & personal care products (PPCPs)	Antibiotics, hormones, triclosan	Wastewater effluent	Endocrine disruption, antibiotic resistance, aquatic toxicity
Per- and polyfluoroalkyl substances (PFAS)	PFOA, PFOS	Firefighting foam, nonstick coatings	Extreme persistence ("forever chemicals"), immunotoxicity, carcinogenic
Volatile organic compounds (VOCs)	Formaldehyde, benzene	Solvents, paints, building materials	Smog formation, indoor air quality issues, respiratory irritant, carcinogenic
Inorganic			
Heavy metals	Lead (Pb), mercury (Hg), cadmium (Cd)	Mining, smelting, industrial discharge	Neurotoxicity (Pb, Hg), nephrotoxicity (Cd), carcinogenic, nondegradable
Other inorganic pollutants	Cyanides, asbestos, nitrates/phosphates	Mining, insulation, agricultural runoff	Acute toxicity (cyanide), mesothelioma (asbestos), eutrophication (nutrients)

PCBs, dioxins, hexachlorobenzene, and furans (Persistent Organic Pollutants: A Global Issue, A Global Response | US EPA 2025). Due to the ubiquitous nature of POPs, they can even be found in pristine areas. Even after many years of bans or restrictions in developed countries, their persistence continues to pose global threats. Developing countries face greater risks because of weaker regulations and poor disposal practices, making it harder to manage these pollutants (Dinka 2018).

Polycyclic aromatic hydrocarbons (PAHs) consist of over 100 chemicals mainly produced during the incomplete burning of organic materials. Naphthalene, phenanthrene, fluoranthene, pyrene, chrysene, benzo[a]pyrene are some examples of PAHs. PAHs often appear as complex mixtures, making detection and cleanup difficult. Their toxicity is well-known, with compounds like benzo[a]pyrene categorized as mutagens and carcinogens. Long-term exposure may lead to oxidative stress, DNA damage, and higher risks of heart disease (Bukowska et al. 2022).

Petroleum hydrocarbons are complex mixtures found in crude oil and refined products, like gasoline, diesel, and heating oil. Environmental contamination happens due to spills, leaks from storage tanks and pipelines, and runoff in cities. As a major event, the Deepwater Horizon oil spill has caused significant harm to marine ecosystems (Beyer et al. 2016). These hydrocarbons contain volatile organic compounds, like benzene and heavier PAHs, which are toxic to various organisms. Contamination in soil, groundwater, and marine environments can last for decades, requiring advanced cleanup methods, like bioremediation and chemical dispersants. It has been shown by recent studies that microbial consortia can effectively break down complex hydrocarbons (Mukherjee et al. 2024).

PPCPs cover a wide range of chemicals used in medicine, cosmetics, and hygiene products, including antibiotics, anti-inflammatories like ibuprofen, antidepressants, hormones, and ingredients in soaps and sunscreens like triclosan. While PPCPs are not inherently persistent, their ongoing use results in "pseudo-persistence" in water environments (Ali et al. 2025). PPCPs remain biologically active even at low concentrations. New analytical methods, including high-resolution mass spectrometry, have improved the detection of PPCPs. However, innovative treatment technologies, such as advanced oxidation processes and membrane filtration, are necessary to tackle this growing problem (Ali et al. 2025).

PFAS are synthetic compounds that have been employed in consumer goods and industrial settings since the 1940s due to their ability to withstand heat, water, and oil. Common applications include nonstick cookware, stain-resistant fabrics, firefighting foams, and food packaging (Mohapatra et al. 2025). Exposure to PFAS has been connected to harmful health effects, such as immune system suppression, developmental problems, and increased chances of certain cancers. Reducing PFAS contamination needs improved filtration methods, regulatory efforts, and safer alternatives to reduce their presence in the environment.

Organic chemicals that have a high vapor pressure at room temperature are classified as volatile organic compounds (VOCs). They evaporate easily from solids or liquids into gas. This broad group includes many different compounds, such as benzene, toluene, ethylbenzene, and xylene (known together as BTEX), as well as formaldehyde, methylene chloride, acetone, and perchloroethylene (Paolin and

Strlič 2024). There are serious environmental and health effects of VOCs. As they aid in forming ground-level ozone or smog, VOCs are regulated outdoors. This happens through chemical reactions involving VOCs and nitrogen oxides (NOx) when sunlight is present. This pollution can harm respiratory health and worsen climate change. VOC levels inside can be ten times higher than those outside. This often results from limited ventilation and the presence of materials that emit VOCs. This higher exposure, especially through inhalation, brings significant health risks (Volatile Organic Compounds' Impact on Indoor Air Quality | US EPA 2025).

1.4.2 Inorganic Xenobiotics: The Indestructible Pollutants

Unlike organic pollutants, which can break down into simpler, harmless compounds, heavy metals remain in their original form and do not decompose. The most significant heavy metals concerning environmental and biological impacts include Pb, Hg, Cd, As, and Cr. Each metal has its own risks.

Exposure to Pb causes severe brain damage, especially in children, resulting in developmental delays and problems with thinking skills (Bjørklund et al. 2024). Hg is one of the top-most hazardous substances, especially harmful to the nervous system. Methylmercury is the most hazardous organic form of Hg, while monomethyl mercury, dimethyl mercury, and phenyl methyl mercury are also common forms (Dinka 2018). A striking example of methylmercury exposure is the Minamata Bay incident in Japan, where consumption of contaminated fish and shellfish caused severe neurological disorders, birth defects, and deaths (Yorifuji et al. 2023). Cd is another hazardous metal that can damage the kidneys and weaken bones. Itai-itai disease could result following prolonged exposure to Cd (Yang et al. 2025). Although classified as a metalloid, As has many toxic traits similar to heavy metals. It is linked with skin problems, developmental issues, and various types of cancer (Ali et al. 2020). Chromium, especially in its hexavalent form (Cr (VI)), is known to cause cancer and respiratory problems, along with skin irritation (Mukherjee et al. 2024).

Besides heavy metals, other inorganic xenobiotics also threaten human health and the environment. One such group is cyanides, which are highly toxic compounds used in mining and chemical production. It exists in several forms, including the volatile gas hydrogen cyanide (HCN), water-soluble salts like sodium cyanide (NaCN) and potassium cyanide (KCN), and naturally occurring cyanogenic glycosides found in plants such as cassava and bitter almonds. Other forms include metal-cyanide complexes and cyanogen chloride. Cyanide disrupts cellular respiration by binding to cytochrome c oxidase, blocking aerobic metabolism, and leading to histotoxic hypoxia. This results in rapid ATP depletion and accumulation of lactic acid. Cyanide's lethality is exacerbated by its inhibition of various enzymes, causing reactive oxygen species accumulation and cellular damage (Nath and Bebarta 2023). Cyanides in water systems cause immediate harm and death to fish and other aquatic life. Prolonged exposure also harms human health, leading to brain and heart issues (Bhattacharya and Flora 2009). Asbestos is another serious health hazard. Asbestos is a group of naturally occurring fibrous silicate minerals that have been widely used in various industrial and commercial applications due to their desirable properties, such as heat resistance, durability, and insulating capabilities. Amphibole asbestos and Chrysotile are the two main types of asbestos (Macher et al. 2025). When asbestos fibers become airborne and are inhaled, they can cause lung diseases such as asbestosis, lung cancer, and mesothelioma (Jaurand et al. 2021). These diseases can take decades to show symptoms, which makes public health responses difficult. Despite strict regulations and bans on the use of asbestos in many countries, it remains a critical issue, particularly in older buildings and products where asbestos may still be present (Macher et al. 2025). Nutrients like nitrates and phosphates also become pollutants when their levels exceed natural concentrations (Ali et al. 2020).

1.5 Environmental and Health Impacts of Xenobiotics

1.5.1 Ecological Impacts

Heavy metals such as Cd, Pb, and Hg, along with POPs like PCBs and dioxins, reduce the quantity of nitrifying bacteria, nitrogen-fixing microbes, and decomposers in soils and sediments (Jouquet et al. 2022). For example, in rivers polluted by industry, metal accumulation has been linked to reduced benthic microbial diversity, which in turn affects nutrient cycling and the breakdown of organic matter (Rubin-Blum et al. 2022). Beyond microorganisms, xenobiotics also impact keystone species. For example, pesticide runoff has been shown to reduce populations of amphibians, such as frogs and salamanders, which are important for controlling insect populations (Öberg 2023). Similarly, neonicotinoid insecticides have been linked to declines in bee populations. They disrupt pollination and plant reproduction, affecting multiple levels of the food chain and altering predator-prey relationships, competition, and overall ecosystem stability (Mohapatra et al. 2025).

The population and number of species decrease due to direct toxicity to plants and animals as well as damage to their habitats. Amphibians, freshwater fish, and invertebrates are especially sensitive to xenobiotics, such as atrazine, glyphosate, and pharmaceuticals, which affect their development, reproduction, and survival in aquatic environments (Hasan et al. 2021). As pesticides and heavy metals build up in soils and vegetation, ecosystems on land encounter comparable difficulties. Prolonged exposure to these substances reduces the diversity of soil fauna, including earthworms and arthropods that are essential to the nutrient cycle. For instance, fish and bird populations in the Amazon basin have declined as a result of ongoing Hg pollution (Okorondu et al. 2022). These cases demonstrate that persistent xenobiotics pose a threat to biodiversity at multiple trophic levels. As a result, biodiversity loss reduces ecosystem resilience and increases habitat vulnerability to climate change and other environmental stressors.

Overnutrition, which is frequently present with pollution, results in eutrophication, which lowers oxygen levels, produces algal blooms, and kills more aquatic life. In addition to carrying nitrogen and phosphate fertilizers, farm runoff typically contains pesticide residues, exacerbating the issue (Ali et al. 2020). An obvious example is the Gulf of Mexico, where nutrient runoff from the Mississippi River Basin introduces industrial pollutants and herbicides, resulting in seasonal dead zones (Xu et al. 2022). Algal blooms damage delicate fish and invertebrates while blocking sunlight, lowering water oxygen levels and changing which species may live. They benefit hardy organisms like cyanobacteria. Pollutants that persist for a long time accumulate in sediments or algae and then pass through the food chain (Thiagarajan and Devarajan 2025).

Soil degradation describes the gradual decline of soil health and productivity, often driven by intensive farming, industrial pollutants, and poor land management. Healthy soils typically support crops, cycle nutrients, and sustain diverse communities of organisms; however, exposure to xenobiotics disrupts this balance. The stability of ecosystems and food security are eventually threatened by toxins that diminish microbial life, decrease fertility, and degrade soils over time (Pan et al. 2021). Microbes, fungi, and earthworms that are vital for decomposition and nutrient turnover are inhibited by xenobiotics, which also interfere with the bacteria and invertebrates that drive the nitrogen cycle. Herbicides, petroleum hydrocarbons, and heavy metals are especially dangerous. For example, Cd-contaminated soils in some regions of China slow down the decomposition of organic matter due to decreased microbial biomass and enzyme activity (Wang et al. 2023). Likewise, glyphosate modifies the microbial makeup of soil and restricts the availability of nitrogen and phosphorus for plants (Biradar et al. 2024). As xenobiotics accumulate, they degrade soil structure, reduce organic matter, and increase acidity or salinity, ultimately lowering crop yields and undermining the long-term health of ecosystems.

Water quality degradation is a prominent outcome of xenobiotic pollution since water bodies act as both sinks and transport routes for contaminants. Freshwater and marine systems are critical for drinking, agriculture, fisheries, and biodiversity. However, they remain highly susceptible to chemical inputs from industrial, agricultural, and urban sources. These contaminants impair aquatic life by reducing reproductive success, growth, and survival among sensitive species. Persistent chemicals make water unsafe for human consumption and disrupt aquatic biodiversity (Naidu et al. 2021). For example, in the Great Lakes, the accumulation of PCBs in fish has resulted in consumption advisories due to their neurotoxic and endocrine-disrupting effects on humans (Martyniuk et al. 2020). The presence of antibiotics and endocrine-active pharmaceuticals in surface waters can disrupt fish reproduction and microbial communities, thereby altering the functions of ecosystems.

1.5.2 Human Health Impacts

Human exposure to these contaminants occurs primarily through ingestion of contaminated food and water, inhalation of polluted air, and dermal contact with soil or occupational chemicals. The resulting health effects depend on chemical properties, concentration, exposure duration, and individual susceptibility and may range from acute poisoning to chronic disorders (Lavezzi and Ramos-Molina 2023). Persistent and bioaccumulative compounds, including heavy metals, pesticides, and industrial solvents, interfere with cellular, organ, and systemic functions, resulting in carcinogenic, mutagenic, teratogenic, neurotoxic, immunotoxic, and endocrine-disrupting effects.

1.5.2.1 Carcinogenic and Mutagenic Effects

Chronic exposure to chemicals such as benzene, As, and PAHs increases the risk of leukemia, liver, lung, and skin cancers (Grandjean and Landrigan 2014; Rahman et al. 2023). POPs such as PCBs, dioxins, and certain organochlorine pesticides induce liver, breast, and thyroid cancers through endocrine disruption, oxidative stress, and epigenetic modifications. Occupational exposure to aromatic amines, formaldehyde, and heavy metals, such as Cd and Cr, has also been linked to bladder, lung, and prostate cancers. Additionally, lifestyle-related exposures such as consumption of charred or smoked foods containing PAHs contribute to dietary carcinogen intake. The mechanisms of xenobiotic-induced carcinogenesis include direct DNA adduct formation, mutations in oncogenes or tumor suppressor genes, disruption of cell cycle regulation, and interference with apoptosis, collectively increasing the probability of tumor development over time (Chi et al. 2021).

Xenobiotics, such as PAHs, nitrosamines, benzene, and heavy metals like Cd, As, and Cr, induce genetic mutations through multiple mechanisms. PAHs, found in vehicle exhaust, tobacco smoke, and charred foods, form bulky DNA adducts, disrupting normal base pairing and triggering errors during replication. Nitrosamines commonly present in processed meats and some industrial chemicals, alkylate DNA, causing point mutations and chromosomal aberrations. Benzene generates reactive metabolites that induce DNA strand breaks, oxidative stress, and chromosomal aneuploidy. Heavy metals, such as Cd and As, interfere with DNA repair enzymes and inhibit mismatch repair pathways and generate ROS, leading to oxidative damage and increased mutation frequency. Cr (VI) used in leather tanning and metal plating forms DNA-protein cross-links, resulting in genomic instability. These genetic alterations may accumulate silently over time, contributing to cancer development, heritable genetic disorders, reproductive abnormalities, and developmental defects in offspring. For instance, chronic exposure to benzene is strongly linked to leukemia, while Cd exposure has been associated with increased risk of prostate and kidney cancers (Guo et al. 2021). Occupational exposure scenarios, such as in industrial workers handling solvents, dyes, and metal compounds, as well as environmental exposures like contaminated drinking water or polluted air, significantly elevate mutagenic risks. The mutagenic potential of xenobiotics is further influenced by individual susceptibility factors, including genetic polymorphisms in detoxification enzymes (e.g., cytochrome P450 variants), age, nutritional status, and concurrent exposure to other chemicals. Consequently, even low-dose, long-term exposure can lead to cumulative DNA damage, increasing the likelihood of cancer and reproductive dysfunction later in life (Choukrallah et al. 2018; Rahman et al. 2023).

1.5.2.2 Teratogenic Effects

Among the many health consequences of xenobiotic exposure, reproduction and early development are especially concerning. The developing fetus is uniquely vulnerable, since its organs and systems are still forming, and even minor derangements have lifelong consequences. Unlike adults who may metabolize or excrete certain toxins, embryos and fetuses have limited defense mechanisms, making them highly sensitive to chemical stressors.

Exposure to xenobiotics during critical periods of fetal development results in birth defects, growth retardation, and neurodevelopmental delays. Chemicals, such as dioxins, organophosphate pesticides, PCBs, and heavy metals, like Hg and Pb, can disrupt embryonic cellular signaling, organogenesis, and neural development (Peller et al. 2021). These disruptions may occur even at low doses, particularly during sensitive windows of gestation. Epidemiological studies have linked maternal exposure to organophosphate pesticides and PCBs with neural tube defects, limb malformations, congenital heart defects, and impaired cognitive development in children (Dinka 2018). Mechanistically, xenobiotics may induce oxidative stress, alter gene expression, or trigger apoptosis in developing tissues, contributing to teratogenic outcomes.

1.5.2.3 Endocrine Disruption

The endocrine system is a complex network of glands and hormones that regulates essential biological functions, including growth, reproduction, metabolism, and stress responses. Hormones act at very low concentrations and in highly specific pathways; thus, the system is extremely sensitive to interferences. Endocrine disruption occurs when external chemicals, known as endocrine-disrupting chemicals (EDCs), interfere with the normal production, release, transport, metabolism, or action of hormones. Such disruptions lead to long-lasting and sometimes irreversible health effects, particularly when exposures occur during critical developmental windows (Martyniuk et al. 2020).

EDCs, including bisphenol A (BPA), phthalates, PFAS, and certain pesticides, such as atrazine and DDT, mimic, block, or modulate natural hormone activity, thereby disrupting the endocrine system (Gore et al. 2015). These disruptions affect reproductive, thyroid, adrenal, and metabolic functions. Observed effects include reduced fertility, altered timing of puberty, decreased sperm quality, hormone-dependent cancers (e.g., breast, prostate, and testicular cancers), impaired glucose metabolism, thyroid dysfunction, and increased risk of obesity and metabolic syndrome (Bertram et al. 2022).

Epidemiological studies have demonstrated that prenatal exposure to BPA and phthalates is associated with altered fetal growth, early onset of puberty in girls, and reduced anogenital distance in boys (Bertram et al. 2022). PFAS exposure has been linked to decreased semen quality and delayed menarche (Maxwell et al. 2024). Pesticides, such as atrazine, have been shown to cause feminization in amphibians and are associated with reproductive hormone imbalances in humans (Dinka 2018). EDCs bind hormone receptors, disrupt receptor signaling, alter gene expression, or induce epigenetic modifications that persist throughout life. Both prenatal and adult exposures are important, as even low-dose chronic exposure may result in long-term hormonal deregulation. Furthermore, mixtures of EDCs may produce additive or synergistic effects, thereby increasing risks to human health (Lin and Zhou 2025).

1.5.2.4 Neurotoxicity

The nervous system is one of the most sensitive targets of chemical exposure as it controls essential functions ranging from cognition and behavior to sensory processing and motor coordination. Even small disruptions in neural signaling or brain development have lasting impacts, particularly when exposures occur during early life stages. Neurotoxicity refers to the adverse effects of chemical substances on the structure or function of the nervous system. Xenobiotics represent a significant group of neurotoxicants capable of damaging both the central and peripheral nervous systems through diverse mechanisms.

Xenobiotics, including heavy metals such as Pb and Hg, organophosphate pesticides, industrial solvents like toluene, and certain brominated flame retardants (PBDEs), profoundly impair neural health (Grandjean and Landrigan 2014; Moreira et al. 2023; Sosnowska-Sienkiewicz et al. 2021). These chemicals exert neurotoxic effects through multiple pathways, including disruption of neurotransmitter signaling, induction of oxidative stress, mitochondrial dysfunction, and interference with calcium homeostasis, and triggering of neuronal apoptosis. Children and developing fetuses are particularly susceptible, as early-life exposure can interfere with critical neurodevelopmental processes such as synaptogenesis, myelination, and neuronal migration (Patel and Kumar 2025).

Epidemiological studies have linked prenatal and early childhood exposure to Pb with decreased IQ, attention deficits, learning disabilities, and increased risk of behavioral disorders, including ADHD and conduct disorders (McFarland et al. 2022). Methylmercury exposure through maternal consumption of contaminated seafood has been associated with cognitive deficits, motor impairment, and language delays in children. Organophosphate pesticide exposure has been correlated with reduced executive function and increased risk of developmental disorders.

In adults, chronic exposure to neurotoxic xenobiotics could contribute to neurodegenerative diseases, such as Parkinson's disease, Alzheimer's disease, and peripheral neuropathies (Capucciati et al. 2021). Solvents like toluene and n-hexane cause white matter degeneration and sensory-motor deficits, while PBDEs and PCBs are linked to subtle cognitive impairments, mood disorders, and memory deficits. Mechanistically, these compounds induce oxidative stress, impair mitochondrial function, and disrupt calcium signaling, leading to apoptosis and neuroinflammation. Occupational exposures, industrial emissions, contaminated food and water, and indoor pollutants represent key sources of neurotoxic xenobiotics.

1.5.2.5 Immunotoxicity

The immune system plays a central role in protecting the body against infections, eliminating abnormal cells, and maintaining overall health through a balance of innate and adaptive responses. Due to its complexity and sensitivity, the immune system is highly vulnerable to disruption by environmental contaminants. Even subtle disturbances in immune regulation can increase the risk of disease, compromise vaccine effectiveness, or trigger harmful inflammatory reactions.

Xenobiotics could impair both innate and adaptive immune responses, thus increasing susceptibility to infections, reducing vaccine efficacy, and promoting autoimmune or inflammatory disorders (Dimitrov et al. 2017). Chemicals such as dioxins, PCBs, PFAS, and heavy metals like As and Cd modulate cytokine production, T- and B-cell function, and antibody formation (Lewis et al. 2022). Chronic exposure has been associated with heightened vulnerability to respiratory infections, impaired response to vaccination, and increased incidence of autoimmune diseases such as systemic lupus erythematosus.

Xenobiotics induce oxidative stress, disrupt immune cell signaling, interfere with receptor-mediated regulation, and cause long-term epigenetic changes. Prolonged or repeated exposure leads to persistent immune dysregulation and increases the risk of infectious and noninfectious diseases (Zhai et al. 2023).

1.5.2.6 Organ-Specific Toxicity

Many xenobiotics target specific organs, leading to functional impairment and structural damage. Cd and Hg primarily affect the kidneys, causing tubular damage, proteinuria, and reduced renal function. Hepatotoxicity results from exposure to carbon tetrachloride, acetaminophen overdose, aflatoxins, and certain industrial solvents, leading to liver inflammation, fibrosis, or necrosis. Airborne pollutants, such as particulate matter, ozone, and VOCs, induce pulmonary inflammation, exacerbate asthma, and contribute to chronic obstructive pulmonary disease (COPD) (Aghapour et al. 2022). Some xenobiotics also affect the cardiovascular system, causing arrhythmias or cardiomyopathy, while others target the nervous system or reproductive organs. Chronic multi-organ exposure may result in cumulative systemic toxicity, increasing overall morbidity, mortality, and long-term health burdens. Mechanistically, organ-specific toxicity often involves oxidative stress, mitochondrial dysfunction, apoptosis, inflammation, and interference with enzymatic or hormonal pathways (Lavezzi and Ramos-Molina 2023).

1.5.2.7 Cumulative and Synergistic Effects

In everyday life, people rarely come into contact with just one chemical at a time. Instead, they are usually exposed to mixtures of different substances that interact in ways that change their effects. These interactions can be additive, meaning the total effect is simply the sum of each chemical's impact; synergistic, where the combined effect is greater than expected; or antagonistic, where one chemical reduces the effect of another. Due to these complexities, it is essential to assess the risks of chemical mixtures rather than examining single substances alone.

For example, combined exposure to Pb and As exacerbates neurotoxicity, while mixtures of organophosphate and pyrethroid pesticides may amplify endocrine-disrupting or carcinogenic effects (Wang et al. 2023). Airborne mixtures of particulate matter, PAHs, and VOCs have compounded respiratory and cardiovascular impacts. Even long-term exposure to low doses of several contaminants can lead to the production of subtle but significant "cocktail effects" that accumulate over time (Lavezzi and Ramos-Molina 2023). Therefore, understanding these cumulative and synergistic effects is critical for shaping public health policies, designing effective regulatory frameworks, and implementing risk mitigation strategies.

1.6 The Imperative for Effective Remediation

Crucial ecological disruptions and human health risks caused by xenobiotics require the urgent development and implementation of effective remediation strategies. Addressing these challenges is critical for responsible environmental management and the sustainable recovery of contaminated sites.

1.6.1 Limitations of Conventional Remediation Methods

Conventional remediation of contaminated sites primarily employs physical and chemical methods such as incineration, landfilling, soil washing, and pump-and-treat systems for groundwater remediation. Although these techniques reduce pollutant concentrations and sometimes restore local environmental conditions, they show several significant limitations. For example, the incineration process requires a large amount of energy, and it produces secondary pollutants, such as furans and dioxins, which pose further concerns for human health and the environment (Rath et al. 2024). Despite landfilling being a commonly used method, it only isolates contaminants rather than completely removing them. Over time, leaching from poorly managed landfills could lead to secondary contamination of soil and groundwater, perpetuating environmental hazards. Pump-and-treat systems, which involve the extraction of contaminated water followed by treatment, usually take a long time to achieve meaningful reductions in xenobiotic concentrations. Therefore, it is costly and logistically challenging for large-scale or diffuse contamination (Carroll et al. 2024).

Besides operational challenges, conventional remediation methods can harm ecosystems. Techniques such as soil excavation, chemical oxidation, and thermal treatments disrupt native microbial communities, alter soil structure, lower fertility, and decrease biodiversity.

Economic and resource limitations also reduce the effectiveness of conventional remediation methods. High cost and energy demand often limit large-scale implementation, especially in low- and middle-income regions, increasing global environmental inequities (Feng et al. 2023). These technical, ecological, and economic challenges underscore the urgent need for sustainable, cost-effective, and environmentally compatible remediation strategies.

1.6.2 The Search for Sustainable and Green Solutions

The ecological, technological, and financial limitations of traditional remediations have led to a growing interest in green and sustainable methods that utilize the abilities of living organisms. Using the metabolic processes of microorganisms, bioremediation provides an economical and environmentally beneficial alternative. Bacteria and archaea degrade a wide range of xenobiotics, such as hydrocarbons, solvents, pesticides, and dyes, under both aerobic and anaerobic conditions. Phytoremediation similarly utilizes plants to accumulate, seize, or metabolize contaminants. Species, such as *Brassica juncea* and *Populus*, have demonstrated efficacy in removing heavy metals and organic pollutants from soils and water (Ali et al. 2020).

Among biological strategies, mycoremediation has emerged as a promising approach employing fungi to remediate contaminated environments. Fungi have diverse enzymatic systems, such as ligninolytic enzymes, laccases, and peroxidases, that enable the breakdown of persistent xenobiotics (Khatua 2025). Mycoremediation offers several advantages, such as low cost, minimal energy input, and negligible secondary pollutants. In addition, it is applicable to both terrestrial and aquatic systems. Fungal hyphae penetrate soil matrices and substrates, which increases the contact with contaminants and facilitates degradation.

Recent case studies prove the applicability of mycoremediation globally. White-rot fungi, such as *Phanerochaete chrysosporium*, have been applied to petroleum-contaminated soils in North America and Europe, achieving significant hydrocarbon removal within a few months (Akpasi et al. 2023). Pleurotus species have been successfully used to remove textile dyes from wastewater systems in Asia, demonstrating both pollutant removal efficiency and an economic advantage (Sun et al. 2019). Combining mycoremediation with microbial consortia or plants further enhances remediation effectiveness, offering synergistic solutions that leverage multiple biological pathways.

Due to key characteristics of xenobiotics, sustainable approaches such as bioremediation and mycoremediation are essential. These methods reduce the environmental and health risks, restore ecosystem functions, enhance biodiversity, and support global sustainability objectives. Current researches contribute to developing these techniques and expanding their applicability, positioning green remediation strategies as a focal point to environmental management and policy.

1.7 Conclusion

Xenobiotics are among today's most serious environmental challenges due to their persistence, bioaccumulation potential, and wide-ranging impacts on ecosystems and human health. Although conventional remediation methods have achieved some local improvements, their limitations emphasize the need for safer and more sustainable alternatives. Bioremediation and mycoremediation have great potential for degrading or transforming persistent pollutants while minimizing environmental harm. Managing xenobiotic pollution requires an integrated strategy that combines scientific innovation, effective regulations, and community engagement to protect ecosystem and human health. Adopting sustainable remediation practices along with better global cooperation can significantly reduce the risks of xenobiotics, supporting a healthier balance between human activities and the well-being of ecosystems.

References

Aghapour M, Ubags ND, Bruder D, Hiemstra PS, Sidhaye V, Rezaee F, Heijink IH (2022) Role of air pollutants in airway epithelial barrier dysfunction in asthma and COPD. Eur Respir Rev 31(163):210112. https://doi.org/10.1183/16000617.0112-2021

Akpasi SO, Anekwe IMS, Tetteh EK, Amune UO, Shoyiga HO, Mahlangu TP, Kiambi SL (2023) Mycoremediation as a potentially promising technology: current status and prospects—a review. Appl Sci 13(8):4978. https://doi.org/10.3390/app13084978

Ali MM, Ali ML, Proshad R, Islam S, Rahman Z, Kormoker T (2020) Assessment of trace elements in the demersal fishes of a coastal river in Bangladesh: a public health concern. Thalassas 36(2):641–655. https://doi.org/10.1007/s41208-020-00227-7

Ali Z, Abdullah M, Hafeez F, Alasmari A, Sardar F, Ahmad J, Ahmad F, Alam M, & Yasin MT (2025) Advances in myco-remediation strategies for emerging contaminants: a comprehensive review. Biorem J 1–37. https://doi.org/10.1080/10889868.2025.2507431

Bajt O (2021) From plastics to microplastics and organisms. FEBS Open Bio 11(4):954–966. https://doi.org/10.1002/2211-5463.13120

Berríos-Rolón PJ, Cotto MC, Márquez F (2025) Polycyclic aromatic hydrocarbons (PAHs) in freshwater systems: a comprehensive review of sources, distribution, and ecotoxicological impacts. Toxics 13(4):321. https://doi.org/10.3390/toxics13040321

Bertram MG, Gore AC, Tyler CR, Brodin T (2022) Endocrine-disrupting chemicals. Curr Biol 32(13):R727–R730. https://doi.org/10.1016/j.cub.2022.05.063

Beyer J, Trannum HC, Bakke T, Hodson PV, Collier TK (2016) Environmental effects of the Deepwater Horizon oil spill: a review. Mar Pollut Bull 110(1):28–51. https://doi.org/10.1016/j.marpolbul.2016.06.027

Bhattacharya R, Flora SJS (2009) Cyanide toxicity and its treatment. In: Handbook of toxicology of chemical warfare agents. Elsevier, pp 255–270. https://doi.org/10.1016/B978-0-12-374484-5.00019-5

Biradar S, Kamble BM, Thakare R, Ingle S (2024) Overview of soil xenobiotics and their biological remediation strategies. Int J Environ Clim Chang 14(1):614–625. https://doi.org/10.9734/ijecc/2024/v14i13877

Bjørklund G, Tippairote T, Hangan T, Chirumbolo S, Peana M (2024) Early-life Lead exposure: risks and neurotoxic consequences. Curr Med Chem 31(13):1620–1633. https://doi.org/10.2174/0929867330666230409135310

Brunn H, Arnold G, Körner W, Rippen G, Steinhäuser KG, Valentin I (2023) PFAS: forever chemicals—persistent, bioaccumulative and mobile. Reviewing the status and the need for their phase out and remediation of contaminated sites. Environ Sci Eur 35(1):20. https://doi.org/10.1186/s12302-023-00721-8

Bukowska B, Mokra K, Michałowicz J (2022) Benzo[a]pyrene—environmental occurrence, human exposure, and mechanisms of toxicity. Int J Mol Sci 23(11):6348. https://doi.org/10.3390/ijms23116348

Capucciati A, Zucca FA, Monzani E, Zecca L, Casella L, Hofer T (2021) Interaction of neuromelanin with xenobiotics and consequences for neurodegeneration; promising experimental models. Antioxidants 10(6):824. https://doi.org/10.3390/antiox10060824

Carroll KC, Brusseau ML, Tick GR, Soltanian MR (2024) Rethinking pump-and-treat remediation as maximizing contaminated groundwater. Sci Total Environ 918:170600. https://doi.org/10.1016/j.scitotenv.2024.170600

Castañeda M, Avila BS, Gallego Ríos SE, Peñuela GA (2025) Analysis of organochlorine pesticides and polychlorinated biphenyls in tropical fruits and soils from Antioquia, Colombia and health risk assessment by consumption. J Food Compos Anal 142:107428. https://doi.org/10.1016/j.jfca.2025.107428

Chi L, Tu P, Ru H, Lu K (2021) Studies of xenobiotic-induced gut microbiota dysbiosis: from correlation to mechanisms. Gut Microbes 13(1):1921912. https://doi.org/10.1080/19490976.2021.1921912

Choukrallah M-A, Sewer A, Talikka M, Sierro N, Peitsch MC, Hoeng J, Ivanov NV (2018) Epigenomics in tobacco risk assessment: opportunities for integrated new approaches. Curr Opin Toxicol 11–12:67–83. https://doi.org/10.1016/j.cotox.2019.01.001

Degrendele C, Fiedler H, Kočan A, Kukučka P, Přibylová P, Prokeš R, Klánová J, Lammel G (2020) Multiyear levels of PCDD/Fs, dl-PCBs and PAHs in background air in Central Europe and implications for deposition. Chemosphere 240:124852. https://doi.org/10.1016/j.chemosphere.2019.124852

Dimitrov S, Hulteng E, Hong S (2017) Inflammation and exercise: inhibition of monocytic intracellular TNF production by acute exercise via β2-adrenergic activation. Brain Behav Immun 61:60–68. https://doi.org/10.1016/j.bbi.2016.12.017

Dinka DD (2018) Environmental xenobiotics and their adverse health impacts-a general review. J Environ Pollut Hum Health 6:77–88

Esther A, Schenke D, Heim W (2022) Noninvasively collected fecal samples as indicators of multiple pesticide exposure in wild birds. Environ Toxicol Chem 41(1):201–207. https://doi.org/10.1002/etc.5260

Feng Y, Hu J, Afshan S, Irfan M, Hu M, Abbas S (2023) Bridging resource disparities for sustainable development: a comparative analysis of resource-rich and resource-scarce countries. Resour Policy 85:103981. https://doi.org/10.1016/j.resourpol.2023.103981

Fromberg A, Granby K, Højgård A, Fagt S, Larsen JC (2011) Estimation of dietary intake of PCB and organochlorine pesticides for children and adults. Food Chem 125(4):1179–1187. https://doi.org/10.1016/j.foodchem.2010.10.025

Gore AC, Chappell VA, Fenton SE, Flaws JA, Nadal A, Prins GS, Toppari J, Zoeller RT (2015) EDC-2: the endocrine society's second scientific statement on endocrine-disrupting chemicals. Endocr Rev 36(6):E1–E150. https://doi.org/10.1210/er.2015-1010

Grandjean P, Landrigan PJ (2014) Neurobehavioural effects of developmental toxicity. Lancet Neurol 13(3):330–338. https://doi.org/10.1016/S1474-4422(13)70278-3

Guo H, Ahn S, Zhang L (2021) Benzene-associated immunosuppression and chronic inflammation in humans: a systematic review. Occup Environ Med 78(5):377–384. https://doi.org/10.1136/oemed-2020-106517

Hasan H, Parker A, Pollard SJT (2021) Whither regulation, risk and water safety plans? Case studies from Malaysia and from England and Wales. Sci Total Environ 755:142868. https://doi.org/10.1016/j.scitotenv.2020.142868

Iyevhobu KO, Okparaku SO, Obohwemu KO, Ken-Iyevhobu BA, Asibor E, Oseni DI, Momodu KO, Nnamudi P, Abayomi G, Chauhan R, Sharma S, Akabuokwu KC (2025) Xenobiotic distribution: a comprehensive review and toxicological implications. JORMA Int J Health Soc Sci 2(2):58–66. https://doi.org/10.64074/99wreq81

Jaurand M-C, Meiller C, Jean D (2021) Asbestos and mesothelioma: what is recent advance in research on asbestos-induced molecular carcinogenesis? In: Nakano T, Kijima T (eds) Malignant pleural mesothelioma. Springer Singapore, pp 17–31. https://doi.org/10.1007/978-981-15-9158-7_2

Jia Z, Shu Y, Huang R, Liu J, Liu L (2018) Enhanced reactivity of nZVI embedded into supermacroporous cryogels for highly efficient Cr(VI) and total Cr removal from aqueous solution. Chemosphere 199:232–242. https://doi.org/10.1016/j.chemosphere.2018.02.021

Jouquet P, Bultelle A, Djouraev I, Caquineau S, Hervé V, Vasseur-Cognet M (2022) Termite graveyards. Hidden geochemical patches? Soil Biol Biochem 170:108678. https://doi.org/10.1016/j.soilbio.2022.108678

Khalili R, Moridi A (2025) A comprehensive review of eutrophication in water resources: from identifying contributing factors to proposing management strategies. Interdiscip J Civ Eng 1(1):38–52. https://doi.org/10.48308/ijce.2025.240047.1003

Khatua S (2025) Myco-enzymes as a potent tool for the degradation of xenobiotics. In: Roy S, Mandal V (eds) Plant-microbe interaction under xenobiotic exposure. Springer Nature Singapore, pp 549–584. https://doi.org/10.1007/978-981-96-8260-7_21

Lamb KJ, Midway SR, Brasso RL, López-Duarte PC, Kimball ME, Polito MJ (2022) Mercury biomagnification in a coastal Louisiana food web following the 2010 Deepwater Horizon oil spill. Front Environ Sci 10:937124. https://doi.org/10.3389/fenvs.2022.937124

Lavezzi AM, Ramos-Molina B (2023) Environmental exposure science and human health. Int J Environ Res Public Health 20(10):5764. https://doi.org/10.3390/ijerph20105764

Lewis AJ, Yun X, Spooner DE, Kurz MJ, McKenzie ER, Sales CM (2022) Exposure pathways and bioaccumulation of per- and polyfluoroalkyl substances in freshwater aquatic ecosystems: key considerations. Sci Total Environ 822:153561. https://doi.org/10.1016/j.scitotenv.2022.153561

Li Y, Tao L, Wang Q, Wang F, Li G, Song M (2023) Potential health impact of microplastics: a review of environmental distribution, human exposure, and toxic effects. Environ Health 1(4):249–257. https://doi.org/10.1021/envhealth.3c00052

Lin T-A, Zhou C (2025) Epigenetic impact of endocrine-disrupting chemicals on atherosclerosis. Essays Biochem 69(03):211–228. https://doi.org/10.1042/EBC20253022

Macher GZ, Torma A, Beke D (2025) Examining the environmental ramifications of asbestos fiber movement through the water–soil continuum: a review. Int J Environ Res Public Health 22(4):505. https://doi.org/10.3390/ijerph22040505

Manisalidis I, Stavropoulou E, Stavropoulos A, Bezirtzoglou E (2020) Environmental and health impacts of air pollution: a review. Front Public Health 8:14. https://doi.org/10.3389/fpubh.2020.00014

Martyniuk CJ, Mehinto AC, Denslow ND (2020) Organochlorine pesticides: agrochemicals with potent endocrine-disrupting properties in fish. Mol Cell Endocrinol 507:110764. https://doi.org/10.1016/j.mce.2020.110764

Maxwell DL, Petriello MC, Pilsner JR (2024) PFAS exposure and male reproductive health: implications for sperm epigenetics. Semin Reprod Med 42(04):288–301. https://doi.org/10.1055/s-0044-1801363

McFarland MJ, Hauer ME, Reuben A (2022) Half of US population exposed to adverse lead levels in early childhood. Proc Natl Acad Sci U S A 119(11):e2118631119. https://doi.org/10.1073/pnas.2118631119

Mohapatra B, Dutta A, Phale PS, Kumar U (2025) Editorial: xenobiotics and emerging contaminants in ecosystems: innovative geo-microbial strategies for prevention, efficient clean-up and biosafety. Front Bioeng Biotechnol 13:1619769. https://doi.org/10.3389/fbioe.2025.1619769

Moreira MH, They NH, Rodrigues LR, Alvarenga-Lucius L, Pita-Barbosa A (2023) Salty freshwater macrophytes: the effects of salinization in freshwaters upon non-halophyte aquatic plants. Sci Total Environ 857:159608. https://doi.org/10.1016/j.scitotenv.2022.159608

Mukherjee S, Joshi SJ, Paul S, Kundu R (2024) Toxicity of aquatic system and remediation: the contemporary issues. CRC Press

Naidu R, Biswas B, Willett IR, Cribb J, Kumar Singh B, Paul Nathanail C, Coulon F, Semple KT, Jones KC, Barclay A, Aitken RJ (2021) Chemical pollution: a growing peril and potential catastrophic risk to humanity. Environ Int 156:106616. https://doi.org/10.1016/j.envint.2021.106616

Nath A, Bebarta V (2023) Redirecting intermediary metabolism to counteract cyanide poisoning. FASEB J 39(12):e70709. https://doi.org/10.31219/osf.io/n6bvs. Open Science Framework

Öberg C (2023) Disruption and the ecosystem: the changing roles of ecosystem stakeholders in the course of disruption. Technol Forecast Soc Chang 194:122679. https://doi.org/10.1016/j.techfore.2023.122679

Ojija F (2024) Emerging environmental contaminants: sources, effects on biodiversity and humans, remediation, and conservation implications. Sci Prog 107(2):00368504241253720. https://doi.org/10.1177/00368504241253720

Okorondu J, Umar NA, Ulor CO, Onwuagba CG, Diagi BE, Ajiere SI, Nwaogu C (2022) Anthropogenic activities as primary drivers of environmental pollution and loss of biodiversity a review. Int J Trend Sci Res Dev 6:621–643

Pan Z, Liu Q, Jiang R, Li W, Sun X, Lin H, Jiang S, Huang H (2021) Microplastic pollution and ecological risk assessment in an estuarine environment: the Dongshan Bay of China. Chemosphere 262:127876. https://doi.org/10.1016/j.chemosphere.2020.127876

Paolin E, Strlič M (2024) Volatile organic compounds (VOCs) in heritage environments and their analysis: a review. Appl Sci 14(11):4620. https://doi.org/10.3390/app14114620

Patel M, Kumar P (2025) Childhood exposure to heavy metals and long-term neurodevelopmental effects. Elsevier EBooks, pp 301–316. https://doi.org/10.1016/b978-0-443-36575-1.00016-9

Patterson WF, Robinson KL, Barnett BK, Campbell MD, Chagaris DC, Chanton JP, Daly KL, Hanisko DS, Hernandez FJ, Murawski SA, Pollack AG, Portnoy DS, Pulster EL (2023) Evidence of population-level impacts and resiliency for Gulf of Mexico shelf taxa following the Deepwater Horizon oil spill. Front Mar Sci 10:1198163. https://doi.org/10.3389/fmars.2023.1198163

Peller AJ, Hunt A, Holmes LB (2021) Physical features of newborns exposed during pregnancy to anticonvulsant medication and developmental monitoring. Birth Defects Research 113(12):995–1000. https://doi.org/10.1002/bdr2.1890

Persistent Organic Pollutants: A Global Issue, A Global Response | US EPA (2025). https://www.epa.gov/international-cooperation/persistent-organic-pollutants-global-issue-global-response

Rahman HH, Toohey W, Munson-McGee SH (2023) Exposure to arsenic, polycyclic aromatic hydrocarbons, metals, and association with skin cancers in the US adults. Environ Sci Pollut Res 30(45):101681–101708. https://doi.org/10.1007/s11356-023-29422-8

Rath S, Tripathy SK, Dubey D, Kumar Swain S, Panda S (2024) Microbial waste management in healthcare settings: a review. Int J Adv Life Sci Res 07(02):01–08. https://doi.org/10.31632/ijalsr.2024.v07i02.001

Rubin-Blum M, Sisma-Ventura G, Yudkovski Y, Belkin N, Kanari M, Herut B, Rahav E (2022) Diversity, activity, and abundance of benthic microbes in the Southeastern Mediterranean Sea. FEMS Microbiol Ecol 98(2):fiac009. https://doi.org/10.1093/femsec/fiac009

Singh M, Mishra RC, Shah I, Wadhwa V, Mor V (2023) Xenobiotics: sources, pathways, degradation, and risk associated with major emphasis on pharmaceutical compounds. Springer EBooks, pp 87–106. https://doi.org/10.1007/978-3-031-35775-6_5

Singh AG, Vedika S, Kumar P, Chaturvedi SK, Rai SN, Vamanu E, Kumar A (2024) Toxic heavy metal ions contamination in water and their sustainable reduction by eco-friendly methods: isotherms, thermodynamics and kinetics study. Sci Rep 14(1):7595. https://doi.org/10.1038/s41598-024-58061-3

Sobral AF, Cunha A, Costa I, Silva-Carvalho M, Silva R, Barbosa DJ (2025) Environmental xenobiotics and epigenetic modifications: implications for human health and disease. J Xenobiot 15(4):118. https://doi.org/10.3390/jox15040118

Sosnowska-Sienkiewicz P, Januszkiewicz-Lewandowska D, Mańkowski P (2021) The dramatic consequences of an accidental ligation of the celiac trunk during surgery performed on a child with neuroblastoma. Int J Environ Res Public Health 18(4):1841. https://doi.org/10.3390/ijerph18041841

Suleiman N, Ibrahim N (2024) Health and environmental impact of xenobiotics in water quality evaluation: a review. UMYU J Microbiol Res 9(3):299–307. https://doi.org/10.47430/ujmr.2493.036

Sun J, Dai X, Wang Q, Van Loosdrecht MCM, Ni B-J (2019) Microplastics in wastewater treatment plants: detection, occurrence and removal. Water Res 152:21–37. https://doi.org/10.1016/j.watres.2018.12.050

Tang FHM, Lenzen M, McBratney A, Maggi F (2021) Risk of pesticide pollution at the global scale. Nat Geosci 14(4):206–210. https://doi.org/10.1038/s41561-021-00712-5

Thiagarajan C, Devarajan Y (2025) The urgent challenge of ocean pollution: impacts on marine biodiversity and human health. Reg Stud Mar Sci 81:103995. https://doi.org/10.1016/j.rsma.2024.103995

Thilagam H, Gopalakrishnan S (2022) Environmental deterioration due to existing and emerging persistent organic pollutants: an overview. In: Vasanthy M, Sivasankar V, Sunitha TG (eds) Organic pollutants. Springer, pp 59–89. https://doi.org/10.1007/978-3-030-72441-2_3

Volatile Organic Compounds' Impact on Indoor Air Quality | US EPA (2025). https://www.epa.gov/indoor-air-quality-iaq/volatile-organic-compounds-impact-indoor-air-quality#intro

Wang S, Yuan C, Zafar FF, Wei M, Marrakchi F, Cao B, Fu Y (2023) Facile synthesis of chlorella-derived autogenous N-doped porous biochar for adsorption on tetracycline. Environ Pollut 330:121717. https://doi.org/10.1016/j.envpol.2023.121717

Wania F, Shunthirasingham C (2020) Passive air sampling for semi-volatile organic chemicals. Environ Sci: Processes Impacts 22(10):1925–2002. https://doi.org/10.1039/D0EM00194E

Xu Y, Elbakidze L, Yen H, Arnold JG, Gassman PW, Hubbart J, Strager MP (2022) Integrated assessment of nitrogen runoff to the Gulf of Mexico. Resour Energy Econ 67:101279. https://doi.org/10.1016/j.reseneeco.2021.101279

Yang Y, Hassan MF, Ali W, Zou H, Liu Z, Ma Y (2025) Effects of cadmium pollution on human health: a narrative review. Atmosphere 16(2):225. https://doi.org/10.3390/atmos16020225

Yorifuji T, Kadowaki T, Yasuda M, Kado Y (2023) Neurological and neurocognitive impairments in adults with a history of prenatal methylmercury poisoning: Minamata disease. Int J Environ Res Public Health 20(12):6173. https://doi.org/10.3390/ijerph20126173

Zhai J, Wang L, Liu Y, Wang C, Mao X (2023) Assessing the effects of China's three-north shelter forest program over 40 years. Sci Total Environ 857:159354. https://doi.org/10.1016/j.scitotenv.2022.159354

The Biology of Fungal Enzyme and the Foundation for Mycoremediation

2

Atreyi Pramanik and Pardeep Yadav

Abstract

Fungal enzyme systems are important in environmental sustainability as they potently degrade complex organic molecules, including cellulose, lignin, PAHs, and plastics. Enzymes like laccases, peroxidases, cellulases, and proteases are utilized in biofuel, food, textile, and pharmaceutical industries because of their stability and effectiveness. Mycoremediation is an environmentally friendly solution compared to conventional methods due to the sustainable utilization of mycelium to remediate pollutants. Specific enzymes, such as laccases and peroxidases, are particularly suitable in degrading dyes, hydrocarbons, and PCBs. These enzymes are generated by fungi: *P. chrysosporium, Pleurotus* spp., and *T. versicolor*. Also, lipases, esterases, and cutinases of such species as *Fusarium solani* or *Aspergillus niger* are promising in the degradation of plastics. *P. chrysosporium* specialized in degrading the persistent pollutants. This chapter emphasizes the prospect of fungal enzymes as a scalable, environmentally friendly formulation of biotechnological solutions but remark that fungal consortia present a relevant direction into the future of mycoremediation.

Keywords

Fungal enzymes · Mycoremediation · Biodegradation · Ligninolytic enzymes · Environmental biotechnology

A. Pramanik · P. Yadav (✉)
Department of Biotechnology, School of Applied and Life Sciences, Uttaranchal University, Dehradun, Uttarakhand, India

A. Roy et al. (eds.), *Mycoremediation of Xenobiotics*, Emerging Paradigms in Pharmaceutical Research, https://doi.org/10.1007/978-981-95-5743-1_2

2.1 Introduction

The environment is severely impacted by hazardous substances. Both human economic activity and natural processes release a wide variety of contaminants into the atmosphere. It encompasses a number of chemical classes, such as synthetic dyes, detergents, chlorinated compounds, plastics, dioxins, medicinal chemicals, polycyclic aromatic hydrocarbons (PAHs), and the mixtures they create, such as oil, creosote, etc. Contaminant leakage and unintentional release into the environment are major issues, particularly when the natural microbiome's biodegradation ability is inadequate to eliminate or neutralize the pollutants (Mishra et al. 2012). Compound process of microbial metal remediation is a multidimensional process that entails a number of factors, which include cell physiology, cell wall composition, and physicochemical parameters such as temperature, pH, metal concentration, ionic strength, and time. Since metals do not naturally break down as organic contaminants break down, their mean residence period in soil is thousands of years. They stay in the soil after undergoing intricate organic transformations or oxidation modifications (Gisbert et al. 2003). Microorganisms are able to extract metals from aqueous solutions. This action is referred to as biosorption. As recent studies suggest, mycosorption comprises of biosorption of fungi (Azin and Moghimi 2018). The outcome of the partial group of functional groups on the cell wall of the fungi and dyes to each other is the result of a physical or chemical process termed mycosorption that is dependent on the biomass of the fungi, depending on the amount of biomass available (Noman et al. 2019). Mycosorption is of great interest to scientists all over the world (Witkowska et al. 2021). Fungi can diminish the peril of the radionuclides, metalloids, and metals and also decompose organic molecules in the environment through chemical alteration or by altering chemical bioavailability, coupled with the fact that they do not need a pollutant to grow and, as such, are ideal in bioremediation. Since they are able to secrete extracellular enzymes to degrade complex carbohydrates without the need to hydrolyze them, most fungi can degrade a broad spectrum of contaminants (Akpasi et al. 2023).

In mycoremediation, a subclass of bioremediation, fungi are employed to decompose, rehabilitate, and heal destroyed environments (Asiriuwa et al. 2013; Mani and Kumar 2014). The hyphae fungi that are employed in mycoremediation bioremediation fix onto roots, rocks, and soil particles in the form of a filamentous body that is resistant to heavy metals and is sensitive to temperature, pH, and nutrients. The fungi are favored over bacteria due to their hyphae network, biomass, and long reproduction cycle, which makes them ideal in the clean-up of contaminated environments. Furthermore, the local bacteria in adverse environments have to compete with the metal-resistant fungus (Sun et al. 2012) The extensive metabolic diversity also enables fungi to be used in a wide variety of applications in the decontamination of various pollutants. The cell walls of fungi contain proteins and polysaccharides that contain amino, phosphate, hydroxyl, sulfate, and carboxyl groups that bind metal ions (Maheswari and Murugesan 2009). These functional groups provide the ligand atoms, which are needed during formation of the complexes with the metal ions that bind the metals and bind them in the biomass. The probability of metal removal depends on the selection of a metal-tolerant fungus in a polluted environment. An effective site-based bioremediation plan may be linked to bio-enhancement of fungi that are able to absorb metals. It has also been suggested by many researchers that studies of wild-type fungal strains are needed to be in a position to develop effective bioremediation methodologies that can retain high levels of metals and remove metals in the environment (Gentry et al. 2004; Mukherjee et al. 2010).

2.2 Types of Fungi in Remediation

Fungi are one of the organisms that are gaining growing popularity because of its potential applications in bioremediation. Fungi are a eukaryotic domain, a diverse kingdom of the tree of life. Nevertheless, the issue of taxonomy in fungi remains controversial. Fungi are widespread chemoheterotrophs in either terrestrial or aquatic environments. Nevertheless, fungi have been given less consideration in aquatic settings, particularly in the marine setting, even though they have been examined more in the terrestrial setting (Zeghal et al. 2021). Although anaerobic fungi have been reported in the digestive tract of animals and in oxygen-minimum layers of the ocean, even the known majority of fungi are aerobic and therefore are oxic in nature (Mura et al. 2019; Peng and Valentine 2021; Stief et al. 2014). As the fungi are the last to decompose multifarious organic matter, they are acknowledged to decompose wood, including lignin, cellulose, and other waste produced by plants that are usually utilized as agricultural waste. They also participate in decomposing. Because of their enzyme systems (both intracellular and extracellular) and the capacity to secrete acids, fungi can attack and metabolize a very large range of chemical classes, including inorganic and organic pollutants (Dashtban et al. 2010, 2016; Harms et al. 2011). The low substrate specificity is another typical feature of fungal enzymes (El-Gendi et al. 2021). This, however, is not contrary but rather beneficial because the nonspecific enzymes catalyze numerous reactions that enable the fungus to utilize a wide variety of materials as sources of carbon and energy (Harms et al. 2011). The number of oxidases and peroxidases facilitating oxidative enzymatic processes is very high, and almost all are always linked to the biodegradation possibilities of fungus. Subsequently, fungi generally require an oxic medium to grow (El-Gendi et al. 2021).

2.2.1 Wood Rot Fungi

Wood rot-causing fungi belong to the *Basidiomycota* (*Basidiomycetes*) (Li et al. 2022). These fungal species, besides their importance in the root of wood, are also important in the bioremediation of organic pollutants. Until the transformation of pollutants is done, there is no preconditioning required (Adenipekun and Lawal 2012). Consequently, they can disintegrate tissues, and they are most flexible. Depending on their capacities of colonizing novel environments, the wood-degrading fungal species can be divided into weak competitors, such as *Ganoderma applanatum* and *Dichomitus squalens*, and strong competitors, such as *T. versicolor*, *Phanerochaete* sp., *Pleurotus* ssp., etc. They can fall into two categories due to the influence they have on wood tissues: brown-rot fungus and white-rot fungi (Martínez et al. 2005). White-rot fungi (*Phanaerochaete* and *Polyporus* sp.) can absorb spilled explosives, organochlorine pesticides, polychlorinated biphenyls (PCBs), petroleum hydrocarbons, as well as PAHs in their entirety (Kristanti et al. 2011; Wu and Yu 2007). A study conducted by Adenipekun and Isikuhuemhen (2008) has found that *Lentinus squarrosula* can increase the amount of carbon, available phosphorus, and organic matter of motor oil contaminated soils (Adenipekun and Isikhuemhen 2008). The *Gloeophyllum trabeum, Fomitopsis pinicola*, and *Daedaleadic kinsii* were discovered to effectively degrade dichloro-diphenyl-trichloroethane (DDT) by the use of the Fenton reaction (Purnomo et al. 2011). Some of the numerous fungi that have been proved to degrade PAHs include *Agrocybe dura, Agrocybe praecox, Gymnopilus luteofolius, Irpex lacteus, Mycena galericulata, Trametes ochracea, Stropharia rugosoannulata, Phanerochaete velutina, and Physisporinus rivulosus* (Winquist et al. 2014). According to Zafra et al. (2016), Zafra et al. (2016), the degradation of pyrene and benzo [a] pyrene phenanthrene was vital with the help of the fungi *Aspergillus nomius, Rhizomucor variabilis, Aspergillus flavus*, and *Trichoderma asperellum*. Thus, polluted resources necessitate fungal colonies to be used in bioremediation. Hence, further investigation might be carried out to realise the particular processes and procedures applied to enhance the inborn talent of the performer.

2.2.1.1 Brown-Rot Fungi

Breaking down of the cellulose and hemicellulose of wood is carried out by cellulose-degrading fungi. The *Gloeophylales, Polyporales, Hymenochaetales,* and *Agaricales* all contain most of the brown-rot fungus (Anastasi et al. 2013). The decaying wood is brown in color due to the partially changed lignin (Zavarzina et al. 2010). Besides the generation of hydrogen peroxide (H_2O_2), a brown-rot fungus also generates hydroxyl (OH) free radicals, which lead to the mineralization and biodegradation of synthetic chemotherapeutic drugs (Arantes et al. 2012). It can be used to convert metals due to the synthesis of oxalic acid, and it is also resistant to antimicrobial drugs. Brown-rot fungi can be of great benefit in this bioremediation ability (Dunwell et al. 2000).

2.2.1.2 White-Rot Fungi (WRF)

Some varieties of WRF are unique because they can break down lignin along with cellulose and hemicellulose, which rots and bleaches wood (Martínez et al. 2005). These fungi that were first identified in the group in the course of studying mycoremediation generate the enzymes lignin peroxidase, laccase, Mn peroxidase, and H_2O_2 -generating enzymes (Akpasi et al. 2023). *Phanerochaete chrysosporium* has been widely used in the study of biodegradation of complex compounds with extracellular oxidative lignolytic enzymes. *P. chrysosporium* is superior to other fungi or microbes in breaking down dangerous or insoluble substances into H and CO_2. The recalcitrant materials are either oxidized or reduced in various ways in order to lower the toxicity of their existence in the environment. The ligninolytic enzymes are nonspecific and adaptive, and this uniqueness renders the contaminated soil easy to get rid of a range of aromatic complex and xenobiotic compounds. *Irpex lacteus, Lentinula edodes, Bjerkandera adusta, Trametes versicolor, and Pleurotus ostreatus (P. ostreatus)* are some of the WRF examples that have been known to degrade these materials. According to multiple studies, white-rot fungi participate in at least 30% of mycoremediation activities (Akpasi et al. 2023).

2.2.2 Leaf-Decaying Fungi

One of the most significant wooded area ecological elements is the fungus that destroys the leaves. These fungi actively mineralize, humify, and decay wood and soil organic compounds (Coleman 2008). As this population of fungi decomposes the litter in the forest, a rapid successional change occurs. *Ascomycota* fungi dominate in the initial phases of litter decay, but as the decay progresses, they slowly decline as *Basidiomycota* fungi increase in population (Das et al. 2007). Such fungi digest litter plants to generate a variety of lignocellulolytic chemicals. Basidiomycetous litter fungi require extracellular enzymes like peroxidase, laccase, and oxidase to break down lignocellulose, leading to the generation of H_2O_2 (hydrogen peroxide) (Purahong et al. 2016). The persistent organic contaminants in the soil that are biotransformed by the enzymes are pesticides and herbicides. This will therefore enhance the number of pollutants that can be found in the soil owing to the use of these fungi in bioremediation (Harms et al. 2011).

2.2.3 Endophytic Fungi

An example of endophytic microbial groups that may colonize plants without posing a threat to the host or the microorganism itself includes bacteria and fungi (Rout 2014). They are found on some plant tissues, including vascular bundles, apoplastic gaps, cortical roots, dead cortical cells, and immature buds, where they secrete various chemicals to carry out the process of methane assimilation and fixation of nitrogen. They can live in dead litter since they possess saprophytic qualities (Kubartová et al. 2009). In addition, they synthesized various groups of enzymes, including lipase and cellulase, peroxidase, and protease, to biotransform heavy metals in their reduced forms as well as to degrade environmental chemicals, including petrochemicals, polyaromatic hydrocarbons, herbicides, pesticides, polychlorobiphenyls, polyester polyurethane, and insecticides, among others (Purohit et al. 2018). They do this by a process of adaptation to increase the tolerance to toxins and other poisons, such as heavy metal. Such combinations are, therefore, good bioremediation sources (Hossain et al. 2012).

2.2.4 Mycorrhiza

The various types of interaction between fungi and plant roots include ectomycorrhiza, orchid mycorrhiza, ericoid mycorrhiza, monotropoid mycorrhiza, arbuscular mycorrhiza, ectendomycorrhiza, and so forth (Finlay 2008). Water is also connected to plant nutrient supply, and it also provides defense against environmental stress, like metal toxicity stress. The lowering of the metal translocation raised the toxicity of the heavy metals in the plant system (Sharma and Dietz 2006). This implies that translocation process allows plants to adapt and survive under the environment where heavy metal is highly contaminated. The fungus also relied on the host plant to sustain a contaminated environment around the fungus and also the enzyme degradation of several petroleum products, PAHs, chlorinated aromatic pesticides, atrazine, and 2, 4-dichlorophenoxy acetic acid (2, 4-D). Mycorrhizal is required in the destruction of parts of soil that may be hazardous (Lehmann et al. 2011).

2.2.5 Soil Fungi

This diverse group of soil fungi is classified into three: *Zygomycota, Chytridiomycota*, and *Ascomycota*. These groups are important in the soil ecology processes of carbon and nitrogen cycling and decomposing soil organic matter (Anastasi et al. 2013). They tend to be sapro-thalli, or cellulose-degrading, non-ligninolytic organisms. Some of the genera that belong to this classification include *Trichoderma, Stachybotrys, Rhizopus, Phlebia, Paecilium, Paecilomyces, Mortierella, Geomyces, Fusarium, Engyodontium, Cladosporium, Cunninghamella, Beauveria, Alternaria, Microsporum, Allescheriella*, and *Acremonium* (Purohit et al. 2018). Other fungi, such as non-ligninolytic, destroy the PAHs through hydroxylation by using an extracellular monooxygenase enzyme (Bamforth and Singleton 2005). At later phases of degradation, the fungus caused the degradation of the recalcitrant polymers. Thus, these fungi are termed xenobiotic-degrading fungi, and in laboratory studies, they have strikingly high levels of efficacy in soil bioremediation in multispecies consortia (Lehmann et al. 2011). Passarini et al. found 50% to 90% of pyrene and benzo[a]pyrene metabolized by *Aspergillus sclerotiorum* CBMAI 849 and *Mucor racemosus* CBMAI 847 (Passarini et al. 2011).

2.3 Principles of Mycoremediation

The worldwide demands for food and the necessity to avoid losses to pests of crops are the factors that render the usage of pesticides about 2 million tonnes each year. Pesticides are particularly used in an effort to eliminate unwanted plants, insects, rodents, and other living organisms that threaten cultivated crops (Lykogianni et al. 2021). Nonetheless, about 90% of agricultural pesticides used in the environment do not target organisms but are disseminated to the environment. The high population and huge application of modern farming techniques have resulted in pesticide presence in the atmosphere, the soil systems, groundwater, and the ocean. The carbamates, pyrethroids, the organochlorines, and the organophosphates are examples of the high-environmental-persistence, bioaccumulation, long-range transmission, and adverse effects on nontarget organisms (Kumari et al. 2014). Pesticide contact might cause various forms of human cancer, teratogenic and genotoxic malformations, endocrine and neurological impairment, and other long-term follow-up consequences on human health and the environment.

Many pesticides, including carbamates, pyrethroids, organochlorines, and organophosphorus, have been demonstrated to be degraded by fungi. Other fungi such as the fluxapyroxad, which is a broad-spectrum pyrazole-carboxamide commonly used to treat fungus in apple trees, are even suggested to be a possible fungicide degradester (Podbielska et al. 2020).

The fungi that break down pesticides have been known to generate the genera *Trametes, Ganoderma, Aspergillus, Fusarium, Pleurotus, Cladosporium, Rhizopus* and *Penicillium*, *Phlebia*, and *Mortiella*. Most of these genera belong to numerous families of fungi, such as filamentous fungi, yeasts, brown-rot fungi, and WRF. The capability of representative species of each group in degrading pesticides has been examined (Vaksmaa et al. 2023). Because of the capability of WRF nonspecific enzymes to degrade a large variety of pesticides and structurally highly dynamic organic contaminants, the number of studies on them has been very high (Ghosh and Thakur 2014). WRF are among the most studied bioremediation agents of pesticides due to their variety of oxidative and extracellular ligninolytic enzymes such as laccase (Lac), manganese peroxidase (MnP), lignin peroxidase (LiP), and versatile peroxidase (VP) (Zhuo and Fan 2021). These are synthetic pyrethroids, an organophosphorus, and an organochlorine, respectively. Their destruction efficacy at 25 °C stood at 95, 88, and 93% of chlorpyriphos, dicotyl, and cypermethrin, respectively, in 14 days. The herbicides that had been studied were all biodegraded following a short length of rapid adsorption to fungal pellets. It was proposed that hydrolyzation, dichlorination, and oxidation were important degradation mechanisms, depending on the identified metabolites (Hu et al. 2020). Also, it is shown that *T. versicolor* is capable of eliminating herbicides bentazon and diuron on agricultural wastewater (Beltrán-Flores et al. 2021). *Ganoderma lucidum* is another WRF that also exhibited similar diuron degradation with diuron increasing laccase activity (Coelho-Moreira et al. 2018).

2.3.1 Enzymes of Fungi in Mycoremediation

Applications of fungal enzymes include the production of organic compounds, hydrolysis of lignin-related compounds, decolorization of textile inks and dyes, wastewater treatment, laccase-based biosensors, bleaching of pulp, and pulp washing (Goala et al. 2021; Yesilada et al. 2018). WRFs has ability to degrade organic contaminants like trinitrotoluene (TNT), PAHs, PCBs, phenols, and dyes (Bharath and Gopalram 2023). The pathogen load and solids can be reduced by using enzymes produced by fungus, including laccases, cellulases, peroxidases, xylanases, and proteases. Moreover, enzymes are often useful in the more efficient deflocculation of the sludge. Hydrolases, oxidoreductases, and other contaminants are often mycoremediated by enzymes. Lignin peroxidase may be an extracellular enzyme secreted by fungi or be released into the aqueous phase as a consequence of an aerobic flooded fermentation process (Bhargava et al. 2003). Extracellular enzymes can be utilized with great success in an industrial environment to degrade organic and biodegradable waste as enzymes usually accelerate the decomposition of contaminants. Table 2.1 summarizes a list of enzymes involved in mycoremediation.

Table 2.1 List of enzymes from fungi in mycoremediation

Enzyme	Main function in mycoremediation	Representative PubMed references
Laccase	Multicopper oxidase; oxidizes phenolic & non-phenolic pollutants; dyes, PAHs, antibiotics, pesticides	Arregui et al. (2019), Dong et al. (2023)
Lignin peroxidase (LiP)	High-redox heme enzyme; oxidizes non-phenolic aromatics, PAHs, textile dyes	Gałązka et al. (2025)
Manganese peroxidase (MnP)	Oxidizes Mn(II) → Mn(III), diffusible oxidant; degrades phenols, PAHs, azo dyes	Mielgo et al. (2003)
Versatile peroxidase (VP)	Combines LiP + MnP activity; broad degradation spectrum (phenolic + non-phenolic)	Gałązka et al. (2025)
DyP-type peroxidase	Decolorizes synthetic dyes; oxidizes lignin-like aromatics	Sethupathy et al. (2023)
Unspecific peroxygenase (UPO)	Inserts oxygen into C–H bonds using H_2O_2; oxidizes PAHs, pesticides, pharmaceuticals	Karich et al. (2017)
Aryl-alcohol oxidase (AAO)	Oxidizes aromatic/aliphatic alcohols → aldehydes; generates H_2O_2 for peroxidases	Serrano et al. (2020)
Glyoxal oxidase (GLOX)	Produces H_2O_2 from aldehydes; supplies peroxidases in lignin/xenobiotic breakdown	Kadowaki et al. (2018)
Cellobiose dehydrogenase (CDH)	Electron donor to LPMOs; oxidizes cellobiose, supports biomass & pollutant degradation	Smuts et al. (2023)
Lytic polysaccharide monooxygenases (LPMOs)	Oxidatively cleave crystalline cellulose/chitin; enhance pollutant accessibility	Su et al. (2025)
Glutathione S-transferases (GSTs)	Conjugate GSH to electrophilic xenobiotics; detoxification and immobilization	Alnasser (2025)
Epoxide hydrolases	Hydrolyze toxic epoxides (from PAH oxidation), reducing mutagenicity and enabling metabolism	Gautheron and Jéru (2020)
Lipases/esterases/cutinases	Hydrolyze esters in oils, plastics, additives; degrade hydrocarbons, phthalates	Suresh et al. (2025)

2.4 Future Prospects

Further studies are required to enhance mycoremediation methods, which are cost-effective and ecologically friendly and which can be adopted individually or together with plants to treat mixed contaminations. Enhancements in these methods, both in aquatic and terrestrial settings, will maximize decontamination and the general success of the remediation process. To make bioremediation methods more competitive and applicable in practical applications, future studies must put more emphasis on discovering and screening new species of fungi and consortia of mixed microorganisms capable of degrading various pollutants and revealing high ecological flexibility. The application of mycoremediation to the real world is limited, as it remains largely applied in laboratory and greenhouse experiments only. As such, prior to the process of scaling up and commercializing green technology, comprehensive field tests on the remediation potential of each species are required. Complementary biological methods, such as natural attenuation and phytoremediation, and other mechanisms that help to decrease the stress of contaminants are also to be investigated and alternative ways of mycoremediation made viable using high-throughput methods, including enzyme engineering, next-generation sequencing (NGS), and microarray analysis. Finally, an extensive study of the phylogeny of microorganisms and genetic modification will help us to better understand the molecular processes that support these remediation processes.

References

Adenipekun CO, Isikhuemhen OS (2008) Bioremediation of engine oil polluted soil by the tropical white rot fungus, Lentinus squarrosulus Mont. (Singer). Pak J Biol Sci 11(12):1634–1637. https://doi.org/10.3923/PJBS.2008.1634.1637

Adenipekun CO, Lawal R (2012) Uses of mushrooms in bioremediation: a review. Biotechnol Mol Biol Rev 7(3):62–68. https://doi.org/10.5897/BMBR12.006

Akpasi SO, Anekwe IMS, Tetteh EK, Amune UO, Shoyiga HO, Mahlangu TP, Kiambi SL (2023) Mycoremediation as a potentially promising technology: current status and prospects—a review. Appl Sci 13(8):4978. https://doi.org/10.3390/APP13084978

Alnasser SM (2025) The role of glutathione S-transferases in human disease pathogenesis and their current inhibitors. Genes Dis 12(4):101482. https://doi.org/10.1016/J.GENDIS.2024.101482

Anastasi, A., Tigini, V., & Varese, G. C. (2013). The bioremediation potential of different ecophysiological groups of fungi, 29–49. https://doi.org/10.1007/978-3-642-33811-3_2

Arantes V, Jellison J, Goodell B (2012) Peculiarities of brown-rot fungi and biochemical Fenton reaction with regard to their potential as a model for bioprocessing biomass. Appl Microbiol Biotechnol 94(2):323–338. https://doi.org/10.1007/S00253-012-3954-Y/FIGURES/5

Arregui L, Ayala M, Gómez-Gil X, Gutiérrez-Soto G, Hernández-Luna CE, Herrera De Los Santos M et al (2019) Laccases: structure, function, and potential application in water bioremediation. Microb Cell Factories 18(1):1–33. https://doi.org/10.1186/S12934-019-1248-0/TABLES/1

Asiriuwa OD, Ikhuoria JU, Ilori EG (2013) Myco-remediation potential of heavy metals from contaminated soil. BEPLS 2(5)

Azin E, Moghimi H (2018) Efficient mycosorption of anionic azo dyes by Mucor circinelloides: surface functional groups and removal mechanism study. J Environ Chem Eng 6(4):4114–4123. https://doi.org/10.1016/J.JECE.2018.06.002

Bamforth SM, Singleton I (2005) Bioremediation of polycyclic aromatic hydrocarbons: current knowledge and future directions. J Chem Technol Biotechnol 80(7):723–736. https://doi.org/10.1002/JCTB.1276

Beltrán-Flores E, Sarrà M, Blánquez P (2021) Pesticide bioremediation by Trametes versicolor: application in a fixed-bed reactor, sorption contribution and bioregeneration. Sci Total Environ 794. https://doi.org/10.1016/J.SCITOTENV.2021.148386

Bharath Y, Gopalram K (2023) Mycoremediation of garbage contaminated municipal soil and garage contaminated oil soil using Pleurotus Ostreatus—a comparative study. AIP Conf Proc 2427(1). https://doi.org/10.1063/5.0101197

Bhargava S, Wenger KS, Marten MR (2003) Pulsed addition of limiting-carbon during Aspergillus oryzae fermentation leads to improved productivity of a recombinant enzyme. Biotechnol Bioeng 82(1):111–117. https://doi.org/10.1002/BIT.10548

Coleman DC (2008) From peds to paradoxes: linkages between soil biota and their influences on ecological processes. Soil Biol Biochem 40(2):271–289. https://doi.org/10.1016/J.SOILBIO.2007.08.005

da Coelho-Moreir JS, Brugnari T, Sá-Nakanishi AB, Castoldi R, de Souza CGM, Bracht A, Peralta RM (2018) Evaluation of diuron tolerance and biotransformation by the white-rot fungus Ganoderma lucidum. Fungal Biol 122(6):471–478. https://doi.org/10.1016/J.FUNBIO.2017.10.008

Das M, Royer TV, Leff LG (2007) Diversity of fungi, bacteria, and actinomycetes on leaves decomposing in a stream. Appl Environ Microbiol 73(3):756–767. https://doi.org/10.1128/AEM.01170-06

Dashtban M, Schraft H, Syed TA, Qin W (2010) Fungal biodegradation and enzymatic modification of lignin. Int J Biochem Mol Biol 1(1):36. Retrieved from https://pmc.ncbi.nlm.nih.gov/articles/PMC3180040/

Deshmukh R, Khardenavis AA, Purohit HJ (2016) Diverse metabolic capacities of fungi for bioremediation. Indian J Microbiol 56(3):247–264. https://doi.org/10.1007/S12088-016-0584-6

Dong C, Di, Tiwari A, Anisha GS, Chen CW, Singh A, Haldar D et al (2023) Laccase: A potential biocatalyst for pollutant degradation. Environ Pollut 319:120999. https://doi.org/10.1016/J.ENVPOL.2023.120999

Dunwell JM, Khuri S, Gane PJ (2000) Microbial relatives of the seed storage proteins of higher plants: conservation of structure and diversification of function during evolution of the cupin superfamily. Microbiol Mol Biol Rev 64(1):153–179. https://doi.org/10.1128/MMBR.64.1.153-179.2000

El-Gendi H, Saleh AK, Badierah R, Redwan EM, El-Maradny YA, El-Fakharany EM (2021) A comprehensive insight into fungal enzymes: structure, classification, and their role in mankind's challenges. J Fungi (Basel, Switzerland) 8(1). https://doi.org/10.3390/JOF8010023

Finlay RD (2008) Ecological aspects of mycorrhizal symbiosis: with special emphasis on the functional diversity of interactions involving the extraradical mycelium. J Exp Bot 59(5):1115–1126. https://doi.org/10.1093/JXB/ERN059

Gałązka A, Jankiewicz U, Orzechowski S (2025) The role of ligninolytic enzymes in sustainable agriculture: applications and challenges. Agronomy 15(2):451. https://doi.org/10.3390/AGRONOMY15020451

Gautheron J, Jéru I (2020) The multifaceted role of epoxide hydrolases in human health and disease. Int J Mol Sci 22(1):13. https://doi.org/10.3390/IJMS22010013

Gentry TJ, Rensing C, Pepper IL (2004) New approaches for bioaugmentation as a remediation technology. Crit Rev Environ Sci Technol 34(5):447–494. https://doi.org/10.1080/10643380490452362

Ghosh P, Swati, Thakur IS (2014) Enhanced removal of COD and color from landfill leachate in a sequential bioreactor. Bioresour Technol 170:10–19. https://doi.org/10.1016/J.BIORTECH.2014.07.079

Gisbert C, Ros R, De Haro A, Walker DJ, Bernal MP, Serrano R, Navarro-Aviñó J (2003) A plant genetically modified that accumulates Pb is especially promising for phytoremediation. Biochem Biophys Res Commun 303(2):440–445. https://doi.org/10.1016/S0006-291X(03)00349-8

Goala M, Yadav KK, Alam J, Adelodun B, Choi KS, Cabral-Pinto MMS et al (2021) Phytoremediation of dairy wastewater using Azolla pinnata: application of image processing technique for leaflet growth simulation. J Water Process Eng 42:102152. https://doi.org/10.1016/J.JWPE.2021.102152

Harms H, Schlosser D, Wick LY (2011) Untapped potential: exploiting fungi in bioremediation of hazardous chemicals. Nat Rev Microbiol 9(3):177–192. https://doi.org/10.1038/NRMICRO2519

Hossain MA, Piyatida P, da Silva JAT, Fujita M (2012) Molecular mechanism of heavy metal toxicity and tolerance in plants: central role of glutathione in detoxification of reactive oxygen species and methylglyoxal and in heavy metal chelation. J Bot 2012(1):872875. https://doi.org/10.1155/2012/872875

Hu K, Peris A, Torán J, Eljarrat E, Sarrà M, Blánquez P, Caminal G (2020) Exploring the degradation capability of Trametes versicolor on selected hydrophobic pesticides through setting sights simultaneously on culture broth and biological matrix. Chemosphere 250. https://doi.org/10.1016/J.CHEMOSPHERE.2020.126293

Kadowaki MAS, de Godoy MO, Kumagai PS, da Costa-Filho AJ, Mort A, Alexander Prade R, Polikarpov I (2018) Characterization of a new glyoxal oxidase from the thermophilic fungus Myceliophthora thermophila M77: hydrogen peroxide production retained in 5-Hydroxymethylfurfural oxidation. Catalysts 8(10):476. https://doi.org/10.3390/CATAL8100476

Karich A, Ullrich R, Scheibner K, Hofrichter M (2017) Fungal unspecific peroxygenases oxidize the majority of organic EPA priority pollutants. Front Microbiol 8(AUG):288117. https://doi.org/10.3389/FMICB.2017.01463/FULL

Kristanti RA, Hadibarata T, Toyama T, Tanaka Y, Mori K (2011) Bioremediation of crude oil by white rot fungi Polyporus sp. S133. J Microbiol Biotechnol 21(9):995–1000. https://doi.org/10.4014/JMB.1105.05047

Kubartová A, Ranger J, Berthelin J, Beguiristain T (2009) Diversity and decomposing ability of saprophytic fungi from temperate forest litter. Microb Ecol 58(1):98–107. https://doi.org/10.1007/S00248-008-9458-8/METRICS

Kumari M, Ghosh P, Swati, Thakur IS (2014) Microcosmic study of endosulfan degradation by Paenibacillus sp. ISTP10 and its toxicological evaluation using mammalian cell line. Int Biodeterior Biodegrad 96:33–40. https://doi.org/10.1016/J.IBIOD.2014.08.003

Lehmann J, Rillig MC, Thies J, Masiello CA, Hockaday WC, Crowley D (2011) Biochar effects on soil biota—a review. Soil Biol Biochem 43(9):1812–1836. https://doi.org/10.1016/J.SOILBIO.2011.04.022

Li T, Cui L, Song X, Cui X, Wei Y, Tang L et al (2022) Wood decay fungi: an analysis of worldwide research. J Soils Sediments 22(6):1688–1702. https://doi.org/10.1007/S11368-022-03225-9/TABLES/4

Lykogianni M, Bempelou E, Karamaouna F, Aliferis KA (2021) Do pesticides promote or hinder sustainability in agriculture? The challenge of sustainable use of pesticides in modern agriculture. Sci Total Environ 795. https://doi.org/10.1016/J.SCITOTENV.2021.148625

Maheswari S, Murugesan AG (2009) Remediation of arsenic in soil by Aspergillus nidulans isolated from an arsenic-contaminated site. Environ Technol 30(9):921–926. https://doi.org/10.1080/09593330902971279

Mani D, Kumar C (2014) Biotechnological advances in bioremediation of heavy metals contaminated ecosystems: an overview with special reference to phytoremediation. Int J Environ Sci Technol 11(3):843–872. https://doi.org/10.1007/S13762-013-0299-8/METRICS

Martínez ÁT, Speranza M, Ruiz-Dueñas FJ, Ferreira P, Camarero S, Guillén F et al (2005) Biodegradation of lignocellulosics: microbial, chemical, and enzymatic aspects of the fungal attack of lignin. Int Microbiol 8(3):195–204

Mielgo I, López C, Moreira MT, Feijoo G, Lema JM (2003) Oxidative degradation of azo dyes by manganese peroxidase under optimized conditions. Biotechnol Prog 19(2):325–331. https://doi.org/10.1021/BP020136W

Mishra V, Majumder CB, Agarwal VK (2012) Sorption of Zn(II) ion onto the surface of activated carbon derived from eucalyptus bark saw dust from industrial wastewater: isotherm, kinetics, mechanistic modeling, and thermodynamics. Desalin Water Treat 46(1–3):332–351. https://doi.org/10.1080/19443994.2012.677556

Mukherjee A, Das D, Kumar Mondal S, Biswas R, Kumar Das T, Boujedaini N, R. Khuda-Bukhsh A (2010) Tolerance of arsenate-induced stress in Aspergillus niger, a possible candidate for bioremediation. Ecotoxicol Environ Saf 73(2):172–182. https://doi.org/10.1016/J.ECOENV.2009.09.015

Mura E, Edwards J, Kittelmann S, Kaerger K, Voigt K, Mrázek J et al (2019) Anaerobic fungal communities differ along the horse digestive tract. Fungal Biol 123(3):240–246. https://doi.org/10.1016/J.FUNBIO.2018.12.004

Noman E, Al-Gheethi A, Mohamed RMSR, Talip BA (2019) Myco-remediation of xenobiotic organic compounds for a sustainable environment: a critical review. Top Curr Chem 377(3):1–41. https://doi.org/10.1007/S41061-019-0241-8/METRICS

Passarini MRZ, Rodrigues MVN, da Silva M, Sette LD (2011) Marine-derived filamentous fungi and their potential application for polycyclic aromatic hydrocarbon bioremediation. Mar Pollut Bull 62(2):364–370. https://doi.org/10.1016/J.MARPOLBUL.2010.10.003

Peng X, Valentine DL (2021) Diversity and N2O production potential of fungi in an oceanic oxygen minimum zone. J Fungi (Basel, Switzerland) *7*(3). https://doi.org/10.3390/JOF7030218

Podbielska M, Książek P, Szpyrka E (2020) Dissipation kinetics and biological degradation by yeast and dietary risk assessment of fluxapyroxad in apples. Sci Rep 10(1). https://doi.org/10.1038/S41598-020-78177-6

Purahong W, Wubet T, Lentendu G, Schloter M, Pecyna MJ, Kapturska D et al (2016) Life in leaf litter: novel insights into community dynamics of bacteria and fungi during litter decomposition. Mol Ecol 25(16):4059–4074. https://doi.org/10.1111/MEC.13739

Purnomo AS, Mori T, Takagi K, Kondo R (2011) Bioremediation of DDT contaminated soil using brown-rot fungi. Int Biodeterior Biodegradation 65(5):691–695. https://doi.org/10.1016/J.IBIOD.2011.04.004

Purohit, J., Chattopadhyay, A., Biswas, M. K., & Singh, N. K. (2018). Mycoremediation of agricultural soil: bioprospection for sustainable development, 91–120. https://doi.org/10.1007/978-3-319-77386-5_4

Rout ME (2014) The plant microbiome. Adv Bot Res 69:279–309. https://doi.org/10.1016/B978-0-12-417163-3.00011-1

Serrano A, Carro J, Martínez AT (2020) Reaction mechanisms and applications of aryl-alcohol oxidase. Enzyme 47:167–192. https://doi.org/10.1016/BS.ENZ.2020.05.005

Sethupathy S, Xie R, Liang N, Shafreen RMB, Ali MY, Zhuang Z et al (2023) Evaluation of a dye-decolorizing peroxidase from Comamonas serinivorans for lignin valorization potentials. Int J Biol Macromol 253:127117. https://doi.org/10.1016/J.IJBIOMAC.2023.127117

Sharma SS, Dietz KJ (2006) The significance of amino acids and amino acid-derived molecules in plant responses and adaptation to heavy metal stress. J Exp Bot 57(4):711–726. https://doi.org/10.1093/JXB/ERJ073

Smuts IE, Blakeway NJ, Rose SH, den Haan R, Viljoen-Bloom M, van Zyl WH (2023) Supplementation of recombinant cellulases with LPMOs and CDHs improves consolidated bioprocessing of cellulose. Enzym Microb Technol 164:110171. https://doi.org/10.1016/J.ENZMICTEC.2022.110171

Stief P, Fuchs-Ocklenburg S, Kamp A, Manohar CS, Houbraken J, Boekhout T et al (2014) Dissimilatory nitrate reduction by Aspergillus terreus isolated from the seasonal oxygen minimum zone in the Arabian Sea. BMC Microbiol 14(1). https://doi.org/10.1186/1471-2180-14-35

Su H, Guo C, Zhao H, Dong H, Sun J, Mao X (2025) Functional characterization of a lytic polysaccharide monooxygenase EbLPMO10A that contributes to β-chitin degradation by chitinase. Food Chem 474:143225. https://doi.org/10.1016/J.FOODCHEM.2025.143225

Sun J, Zou X, Ning Z, Sun M, Peng J, Xiao T (2012) Culturable microbial groups and thallium-tolerant fungi in soils with high thallium contamination. Sci Total Environ 441:258–264. https://doi.org/10.1016/J.SCITOTENV.2012.09.053

Suresh V, Shams R, Dash KK, Shaikh AM, Béla K (2025) Comprehensive review on enzymatic polymer degradation: A sustainable solution for plastics. J Agric Food Res 20:101788. https://doi.org/10.1016/J.JAFR.2025.101788

Vaksmaa A, Guerrero-Cruz S, Ghosh P, Zeghal E, Hernando-Morales V, Niemann H (2023) Role of fungi in bioremediation of emerging pollutants. Front Mar Sci 10:1070905. https://doi.org/10.3389/FMARS.2023.1070905/FULL

Winquist E, Björklöf K, Schultz E, Räsänen M, Salonen K, Anasonye F et al (2014) Bioremediation of PAH-contaminated soil with fungi—from laboratory to field scale. Int Biodeterior Biodegradation 86:238–247. https://doi.org/10.1016/J.IBIOD.2013.09.012

Witkowska D, Słowik J, Chilicka K (2021) Review heavy metals and human health: possible exposure pathways and the competition for protein binding sites. Molecules 26(19). https://doi.org/10.3390/MOLECULES26196060

Wu J, Yu HQ (2007) Biosorption of 2,4-dichlorophenol by immobilized white-rot fungus Phanerochaete chrysosporium from aqueous solutions. Bioresour Technol 98(2):253–259. https://doi.org/10.1016/J.BIORTECH.2006.01.018

Yesilada, O., Birhanli, E., & Geckil, H. (2018). Bioremediation and decolorization of textile dyes by white rot fungi and laccase enzymes, 121–153. https://doi.org/10.1007/978-3-319-77386-5_5

Zafra G, Taylor TD, Absalón AE, Cortés-Espinosa DV (2016) Comparative metagenomic analysis of PAH degradation in soil by a mixed microbial consortium. J Hazard Mater 318:702–710. https://doi.org/10.1016/J.JHAZMAT.2016.07.060

Zavarzina AG, Lisov AA, Zavarzin AA, Leontievsky AA (2010) Fungal oxidoreductases and Humification in forest soils:207–228. https://doi.org/10.1007/978-3-642-14225-3_11

Zeghal E, Vaksmaa A, Vielfaure H, Boekhout T, Niemann H (2021) The potential role of marine fungi in plastic degradation—a review. Front Mar Sci 8:738877. https://doi.org/10.3389/FMARS.2021.738877/FULL

Zhuo R, Fan F (2021) A comprehensive insight into the application of white rot fungi and their lignocellulolytic enzymes in the removal of organic pollutants. Sci Total Environ 778. https://doi.org/10.1016/J.SCITOTENV.2021.146132

Fungal Degradation Mechanisms of Xenobiotics

3

H. M. Kasuni Akalanka and Sagarika Ekanayake

Abstract

Fungal degradation of xenobiotics involves the enzymatic breakdown of persistent, human-made compounds resistant to natural biodegradation. These xenobiotics, including pesticides, pharmaceuticals, industrial dyes and synthetic polymers, are increasingly concerning due to their toxicological impact and long-term environmental persistence. Fungi, particularly through their rich enzymatic systems, offer a biologically sustainable solution to this challenge. This chapter explores the mechanisms and functional diversity of fungal enzymes involved in xenobiotic transformation, including laccases, lignin peroxidases, manganese peroxidases, versatile peroxidases, cytochrome P450 monooxygenases, and unspecific peroxygenases. These enzymes facilitate complex chemical reactions, such as aromatic ring cleavage, hydroxylation, and oxidative dichlorination, converting harmful compounds into less toxic or more bioavailable forms. The chapter further explores and highlights emerging tools, such as immobilization and genetic engineering, that improve efficiency and field application. It emphasizes the central role of fungal enzymes in developing innovative biotechnological solutions for pollution remediation.

H. M. K. Akalanka
Rural Health Research Institute, Charles Sturt University, Orange, New South Wales, Australia

S. Ekanayake (✉)
Department of Biochemistry, Faculty of Medical Sciences, University of Sri Jayewardenepura, Nugegoda, Sri Lanka
e-mail: sagarikae@sjp.ac.lk

A. Roy et al. (eds.), *Mycoremediation of Xenobiotics*, Emerging Paradigms in Pharmaceutical Research, https://doi.org/10.1007/978-981-95-5743-1_3

Keywords

Xenobiotic degradation · Fungal bioremediation · Laccase · Peroxidase · Cytochrome P450 monooxygenases · Enzyme immobilization

3.1 Introduction

3.1.1 Xenobiotics: Definition and Environmental Significance

Xenobiotics are synthetic chemical compounds that are foreign to biological systems and do not occur naturally in the environment. These compounds include industrial chemicals, pesticides, pharmaceuticals, synthetic dyes, plasticizers, and various organic pollutants that have been introduced into ecosystems through anthropogenic activities (Chugh et al. 2022) The term "xenobiotic" derives from the Greek words "xenos" (foreign) and "bios" (life), highlighting their alien nature to living organisms. The environmental significance of xenobiotics stems from several critical characteristics, including structural complexity that makes them resistant to conventional biodegradation pathways, persistence in environmental matrices leading to bioaccumulation, potential toxicity to nontarget organisms including humans, and widespread distribution through global transport mechanisms. Many xenobiotics exhibit recalcitrant properties due to the presence of aromatic rings, halogen substituents, nitro groups, or complex heterocyclic structures that are not so frequently found in natural compounds.

These foreign compounds are environmentally significant as many are chemically stable and resist biodegradation, persisting in soil and water. These also tend to bioaccumulate in living organisms and biomagnify through food webs, leading to higher concentrations in top predators (Mishra et al. 2021; Zhang et al. 2013). Such persistence and accumulation result in harmful ecological impacts, as xenobiotic pollutants exert toxic effects on wildlife and disrupt entire ecosystems while also posing long-term health risks to humans (Sharma et al. 2014). Furthermore, the complex and novel structures of many xenobiotics often defy breakdown by natural processes, creating serious challenges for their degradation and removal from the environment and necessitating specialized remediation strategies (Atashgahi et al. 2018).

3.1.2 Fungal Bioremediation Potential

Fungi have evolved sophisticated enzymatic machinery capable of degrading complex organic polymers, particularly lignin, which shares structural similarities with many xenobiotic compounds. This evolutionary adaptation has positioned fungi as exceptional candidates for bioremediation applications (Yu et al. 2023). Fungal species and their secreted enzymes act on many anthropogenic pollutants relevant to soil and water remediation, with white-rot fungi highlighted for broad substrate scope and extracellular action. Reviews and experimental studies document activity against synthetic dyes, polycyclic aromatic hydrocarbons (PAHs), chlorinated aromatics, pesticides, and a large set of pharmaceuticals and personal care compounds (Febbraio 2017; Hultberg and Golovko 2024).

Azo and phthalocyanine dyes are decolorized by fungal peroxidases and laccases, with fungal manganese peroxidases (MnP) and lignin peroxidases (LiP), transforming diverse dye classes in vitro and in culture (Heinfling et al. 1998). Moreover, polycyclic aromatic hydrocarbons (anthracene, phenanthrene, pyrene, fluoranthene) are substrates for extracellular LiP, MnP, and laccase; LiP converted ~59% anthracene and ~34% pyrene in vitro in one study, with MnP and laccase showing lower but measurable conversions enhanced by mediators (Wu et al. 2020). Highly chlorinated aromatics such as dioxins can be attacked by LiP/MnP, leading to dechlorination prior to ring cleavage in *Phanerochaete chrysosporium* cultures (Valli et al. 1992).

Pharmaceuticals (anti-inflammatories, β-blockers, antibiotics, and others) are transformed by whole fungi or purified enzymes, while removal efficiency depends on chemical structure and treatment mode (whole-cell *vs* enzyme) (Ezra et al. 2025).

White-rot basidiomycetes, in particular, produce an array of nonspecific oxidative enzymes that attack a broad range of organic pollutants through radical-mediated mechanisms. The advantages of fungal bioremediation include: broad substrate specificity of fungal enzymes, ability to operate under mild environmental conditions, complete mineralization potential for many compounds, cost effectiveness compared to physicochemical treatment methods, and environmental compatibility with minimal secondary pollution with the least hazardous by-products (Yu et al. 2023). Figure 3.1 illustrates the industrial and environmental applications of fungal enzymes.

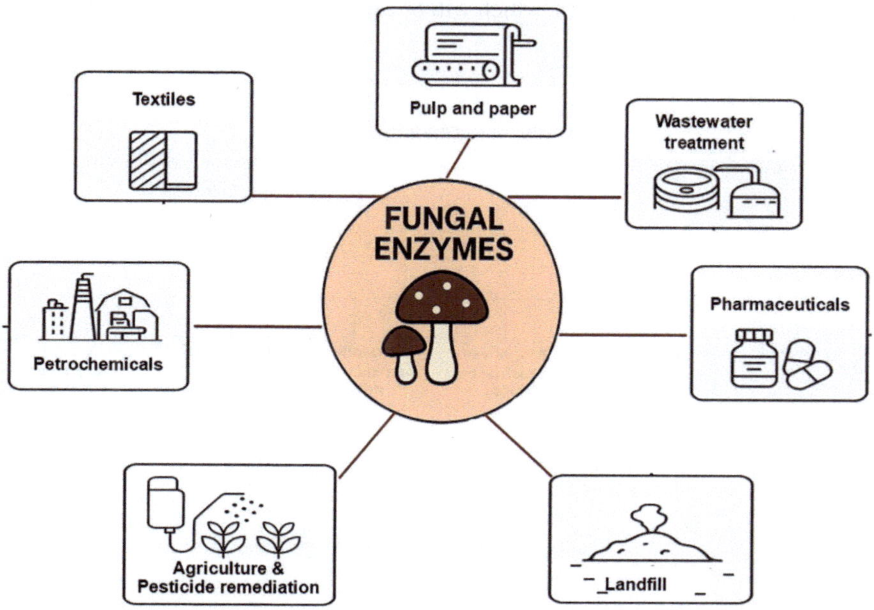

Fig. 3.1 Application domains of fungal enzymes in xenobiotic degradation

3.1.3 Chapter Objectives and Scope

This chapter presents a comprehensive review of fungal degradation mechanisms for xenobiotics, with specific focus on fungal enzymatic mechanisms. Further, the chapter examine the environmental factors that influence enzyme activity and explore genetic engineering and immobilization strategies to enhance bioremediation and outline emerging biotechnological solutions and future research directions.

3.2 Molecular Mechanisms of Fungal Enzymes

Fungi possess a diverse collection of extra- and intracellular enzymes enabling them to interact with, modify, and degrade a wide range of natural and synthetic compounds. Among these, oxidative enzymes, such as laccases, peroxidases (including lignin peroxidase, manganese peroxidase, and versatile peroxidase), and unspecific peroxygenases (UPOs) are of particular importance due to their broad substrate range and high redox potential. These enzymes, initially evolved for decomposition of lignin and other plant polymers, have been increasingly recognized for their ability to act on structurally complex xenobiotics, such as pesticides, dyes, pharmaceuticals, and industrial pollutants. Their catalytic versatility allows them to mediate reactions ranging from one-electron oxidations to oxygen insertion, resulting in transformation, detoxification, or complete mineralization of otherwise recalcitrant molecules. This enzymatic diversity underpins the ecological role of fungi in natural biodegradation processes and highlights their applied potential in environmental biotechnology and bioremediation. Figure 3.2 presents an overview of the impact of fungal enzymes on xenobiotics, which will be examined in greater detail throughout this chapter.

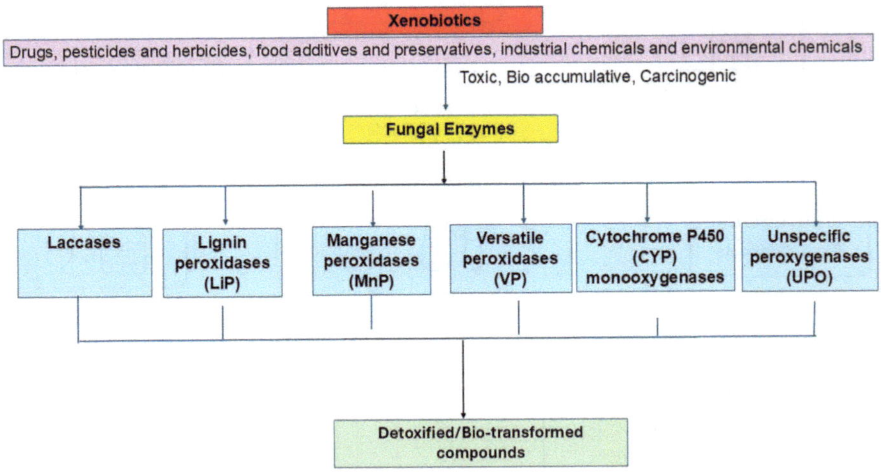

Fig. 3.2 Overview of the impact of fungal enzymes on xenobiotic degradation

3.2.1 Laccases: Multi-Copper Oxidases

Laccases (EC 1.10.3.2) are multi-copper oxidases capable of catalyzing the oxidation of a broad spectrum of phenolic and certain non-phenolic compounds through single-electron abstraction, coupled with the four-electron reduction of molecular oxygen to water. Structurally, laccases contain four copper atoms organized into three discrete sites: a type 1 (T1) copper that functions as the primary electron acceptor from substrates, and a trinuclear cluster composed of one type 2 (T2) and two type 3 (T3) copper atoms that together form the oxygen-reducing center. The catalytic cycle involves sequential steps: substrate oxidation at the T1 site, generating a substrate radical, internal electron transfer from T1 to the trinuclear cluster, oxygen binding and activation at the T2/T3 cluster, and the reduction of O_2 to two water molecules. The T1 copper site exhibits the characteristic intense blue coloration around 600 nm, arising from cysteine–copper charge transfer transitions (Reiss et al. 2013).

Laccases demonstrate broad substrate specificity toward phenolic compounds, including synthetic dyes, bisphenol A, nonylphenol, and various pharmaceuticals containing phenolic moieties. The enzyme's redox potential, typically ranging from 500 to 800 mV, largely determines its oxidative capacity toward different substrates. High-redox-potential laccases, such as those produced by *Pycnoporus sanguineus* and *Trametes villosa*, directly oxidize certain non-phenolic compounds. In addition, laccase–mediator systems significantly expand substrate versatility. By employing small redox mediators such as 1-hydroxybenzotriazole (HBT), 2,2′-azino-bis(3-ethylbenzothiazoline-6-sulfonic acid) (ABTS), or N-hydroxyphthalimide (NHPI), laccases penetrate substrate matrices and oxidize sterically hindered or otherwise inaccessible compounds. This mediator-assisted catalysis enables the transformation of a broad range of xenobiotics that would otherwise resist enzymatic attack.

Beyond their structural and mechanistic elegance, laccases exemplify the central enzymatic strategies employed by fungi for xenobiotic degradation. The radical-mediated reactions initiated at the T1 site destabilize recalcitrant groups, such as aromatic rings, halogen substituents, and azo bonds, thereby transforming persistent pollutants into less toxic or more bioavailable intermediates. Within fungal degradation systems, laccases frequently operate in concert with peroxidases and cytochrome P450 monooxygenases, providing complementary pathways that expand substrate specificity and enable near-complete mineralization. Thus, the laccase catalytic cycle not only illustrates the mechanistic basis of fungal oxidative metabolism but also underscores its pivotal role in the sustainable biotransformation and detoxification of diverse xenobiotic contaminants (Jones and Solomon 2015; Kjaergaard et al. 2015; Sakurai and Kataoka 2007).

3.2.1.1 Phenolic Pollutants and Endocrine Disruptors

A large number of xenobiotic pollutants have phenolic structures, making them amenable to direct oxidation by laccases. These include many endocrine disrupting compounds (EDCs) and industrial chemicals [(alkylphenols:nonylphenol and octylphenol found in surfactants), bisphenol A (a plasticizer monomer), chlorinated phenols, and phenolic antimicrobials such as triclosan] (Gałązka and Jankiewicz 2022). Fungal laccases readily oxidize such phenolic compounds to phenoxy radical intermediates, which then undergo further radical coupling, rearrangements leading to detoxification.

Bisphenol A (BPA) is efficiently transformed by white-rot fungal laccases into less harmful products. *Trametes* and *Pleurotus* laccases, for example, oxidize BPA to form bisphenyl-quinone derivative of BPA (2,2-bis(4-hydroxyphenyl)quinone-propane), wherein each aromatic ring is converted into a quinone (Gałązka et al. 2023). This dimeric quinone product is much less reactive biologically and no longer binds estrogen receptors, effectively eliminating BPA's endocrine activity. In some cases, laccase-mediated BPA degradation could progress even further. Studies have shown laccases cleaving BPA's central carbon bond or oxidizing its isopropylidene bridge, generating small aliphatic acids as final products. A laccase from *Coriolopsis gallica* was shown to oxidatively cleave BPA, producing β-hydroxybutyric acid without mediators. In the presence of HBT, additional products such as tartaric acid were formed, indicating aromatic ring fission of BPA (Mtibaà et al. 2018). These reactions demonstrate the ability of laccases not only to polymerize phenolics but also to break certain C–C bonds under appropriate conditions.

Laccases have been used to remove nonylphenol from water, typically by oxidizing the phenolic ring to quinone intermediates that polymerize or further degrade. Triclosan, an antimicrobial phenolic chlorinated diphenyl ether, is also susceptible to laccase oxidation: *Trametes versicolor* laccase converts triclosan into several oligomeric products (dimers, trimers, tetramers) *via* phenoxy radical coupling (Bankole et al. 2019). The proposed mechanism is that laccase first abstracts an electron from triclosan to form a phenoxy radical, which then undergoes C–C and C–O coupling with another radical, and propagates, forming oligomers. These oligomerization reactions effectively immobilize triclosan and drastically reduce its bioactivity (Wang et al. 2018). Importantly, the presence of Cu^{2+} was found to enhance triclosan degradation by stabilizing the laccase and possibly promoting radical coupling. Laccases have also shown the ability to degrade phenolic pesticides like 2,4-dichlorophenol and certain phenolic fungicides (Mtibaà et al. 2018). Overall, laccase-catalyzed oxidation of phenolic EDCs usually yields products like quinones, phenoxy dimers, or ring-opened acids that have lost the endocrine-disrupting properties of the parent compound.

3.2.1.2 Polycyclic Aromatic Hydrocarbons (PAHs)

PAHs are hydrophobic, fused-ring pollutants that are recalcitrant and carcinogenic, making their removal a key bioremediation priority. Laccases alone do not achieve complete mineralization but initiate critical oxidative steps that destabilize aromatic rings. By abstracting electrons or introducing oxygen, laccases convert PAHs into radical cations and quinones, which are more polar and biologically accessible. These intermediates undergo subsequent hydration, rearrangement, or enzymatic cleavage to yield diacids and carboxylates such as phthalic, protocatechuic, catechol, and benzoic acids. For example, *Pleurotus ostreatus* oxidizes naphthalene to benzoic acid, while *Ganoderma* spp. transform phenanthrene and fluorene into phthalic and catechol derivatives (Torres-Farradá et al. 2017). High-redox laccases from *Leucoagaricus gongylophorus* or *Ganoderma* spp. oxidize anthracene to anthraquinone and even partially degrade 5-ring PAHs, like benzo[a]pyrene. Synergy with manganese and lignin peroxidases facilitates further oxidation, ring fission, and eventual mineralization to CO_2 and H_2O. Redox mediators such as HBT or ABTS enhance degradation efficiency, particularly for large PAHs, though some fungi achieve high activity without mediators. In summary, laccase-driven oxidation is the crucial first step in fungal PAH degradation, converting stable aromatic rings into more biodegradable intermediates and enabling downstream pathways that reduce toxicity and persistence in contaminated environments.

3.2.1.3 Pesticides and Other Xenobiotics

Fungal laccases have shown the ability to transform a variety of pesticides, including herbicides, insecticides, and fungicides, especially those containing aromatic rings or other oxidizable functional groups. Many modern pesticides (e.g., certain phenylurea herbicides, organophosphate insecticides, and azole fungicides) have structures that make them susceptible to oxidation. Laccase action on pesticides often involves dealkylation, heteroatom oxidation, or cleavage of specific bonds.

Fungal laccases play a major role in the biotransformation of pesticides, particularly those with aromatic rings or oxidizable functional groups. They act on herbicides, insecticides, and fungicides through mechanisms, such as dealkylation, oxidative bond cleavage, and demethylation, leading to reduced toxicity and enhanced biodegradability (Baldrian 2006). A laccase from *Trametes versicolor* has been shown to degrade the phenylurea herbicide, isoproturon via demethylation and hydroxylation (Eichlerová et al. 2012) and completely remove atrazine and metolachlor in laboratory trials (Camarero et al. 2005). For example, the phenylurea herbicide isoproturon can be degraded by *T. versicolor* laccase in the presence of mediators, leading to demethylation of the urea moiety and ring hydroxylation (Margot et al. 2013). Moreover, atrazine and metolachlor, two herbicides, were completely degraded by *T. versicolor* laccase (Kupski et al. 2019).

Therefore, the mechanisms can be summarized by a few general routes including:

- Dealkylation of aromatic rings: by oxidizing alkylphenyl structures, effectively removing alkyl groups attached to aromatic rings as seen in certain herbicides (Amitai et al. 1998)
- Cleavage of thioether bonds: by breaking C–S bonds in thioether-containing pesticides (Vaithyanathan et al. 2021)
- Oxidative demethylation or deamination: by removing N-methyl or other amino substituents on aromatic rings by oxidizing the nitrogen or the adjacent carbon, which helps break down carbamate and phenylamide pesticides (Vaithyanathan et al. 2021)

3.2.2 Lignin Peroxidases: Heme-Dependent Oxidoreductases

Lignin peroxidases (LiP) (EC 1.11.1.14) are extracellular heme-containing enzymes that utilize hydrogen peroxide as an electron acceptor to oxidize high-redox-potential aromatic compounds. The enzyme contains a heme prosthetic group (iron protoporphyrin IX) coordinated by a proximal histidine and features a unique tryptophan residue (Trp171 in *Phanerochaete chrysosporium* LiP) that facilitates long-range electron transfer. Upon H_2O_2 binding, LiP undergoes a catalytic cycle involving formation of Compound I (Fe^{4+}=O with porphyrin π-cation radical) through two-electron oxidation, one-electron reduction of Compound I by substrate to form Compound II (Fe^{4+}=O), and reduction of Compound II back to the native enzyme through second substrate oxidation (Christian et al. 2005).

The enzymes are secreted by white-rot fungi. These enzymes function by accepting electrons from the substrate or *via* redox mediators to reduce H_2O_2 while producing oxidized products that become amenable to further fungal metabolism (Torres-Farradá et al. 2024).

Key catalytic features of LiP include:

- Heme dependence: LiP contains a heme prosthetic group that facilitates two-electron reduction of peroxide to form powerful oxidants capable of oxidizing non-phenolic aromatic rings.
- Oxidation mode: LiP produces extracellular oxidative equivalents that perform one-electron or two-electron oxidations on aromatic substrates, initiating ring-opening or demethylation reactions.
- Product fate: Oxidized xenobiotics are often converted into more polar intermediates that are further metabolized intracellularly by fungal enzyme catabolism or mineralized by other extracellular enzymes (Fig. 3.3).

The LiP system accepts a broad range of xenobiotics, including synthetic dyes, polycyclic aromatic hydrocarbons, chlorophenols, polychlorinated biphenyls (PCBs), pharmaceuticals, and emerging quinone pollutants, as documented across white-rot fungal studies (Torres-Farradá et al. 2024).

LiP activity relies on H_2O_2 as the terminal oxidant and is enhanced by small extracellular redox mediators and fungal physiological cofactors that extend substrate reach and turnover. Mediators and physicochemical cofactors increase access

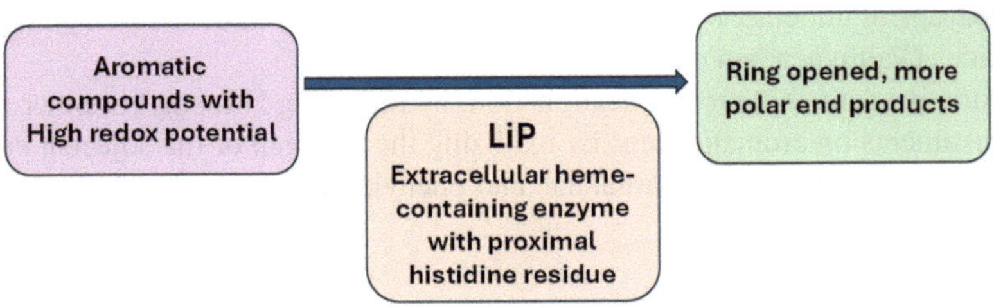

Fig. 3.3 Enzymatic reaction of lignin peroxidase

to buried or non-phenolic sites and promote electron transfer between the enzyme and the pollutant (Torres-Farradá et al. 2024; Yu et al. 2025).

- H_2O_2 requirement: initiate LiP catalysis by generating its oxidizing intermediates.
- Electron shuttles: secreted mediators (low-molecular aromatics produced by fungi) act as redox shuttles to bridge enzyme and poorly accessible substrates, expanding the effective substrate range.
- Adsorption effects: sorption to fungal biomass or particulates concentrate hydrophobic xenobiotics and, in combination with LiP catalysis, accelerate conversion as observed in 6PPD-Q [*N*-(1,3-dimethylbutyl)-*N*′-phenyl-*p*-phenylenediamine-quinone] removal, where adsorption plus enzymatic attack produced downstream metabolites.

Lignin peroxidases directly oxidize non-phenolic aromatic compounds with ionization potentials below 1.4 V, including polycyclic aromatic hydrocarbons, chlorinated phenols, and synthetic dyes. The enzyme generates highly reactive substrate radicals that undergo various reactions, including ring cleavage, demethylation, dehalogenation, and polymerization. The unique ability to oxidize non-phenolic substrates distinguishes LiP from other peroxidases and makes it particularly valuable for xenobiotic degradation (Falade et al. 2017; Olicón-Hernández et al. 2017; Yadav et al. 2006).

3.2.3 Manganese Peroxidases (MnP)

Manganese peroxidases (MnPs, EC 1.11.1.13) oxidize Mn^{2+} to diffusible Mn^{3+} (stabilized by fungal chelators), and the Mn^{3+}-chelate mediates one-electron oxidations that generate radicals on phenolic and related xenobiotics. MnP acts within a ligninolytic enzyme network (laccases, LiPs, intracellular enzymes, microbial partners) and is widely applied for dye, PAH, pesticide, mycotoxin, and other xenobiotic remediation under conditions that control H_2O_2, chelators, pH, and cofactors (Bilal et al. 2023; Scheel et al. 2000). The enzyme contains specific Mn^{2+} binding sites involving acidic amino acid residues (glutamate and aspartate) that position the metal ion for optimal electron transfer. The catalytic cycle involves: H_2O_2-mediated formation of Compound I, by oxidation of Mn^{2+} to Mn^{3+}. Subsequently resulting oxidized Mn^{3+} is released while Compound I is reduced to Compound II. This intermediate can then further oxidize another Mn^{2+} ion, completing the catalytic cycle. Compound II is regenerated back to its resting state by reacting with a reducing agent, such as another molecule of H_2O_2 or a substrate molecule, completing the cycle and allowing it to catalyze the oxidation of more Mn^{2+} ions (Glenn et al. 1986). The above-formed Mn^{3+} ions are stabilized by organic acid chelators, such as oxalate, malonate, or lactate, which prevent disproportionation and enhance oxidative capacity (Zhang et al. 2021).

MnP-generated Mn^{3+}-chelates oxidize phenolic compounds, initiate lipid peroxidation chains, and attack various organic pollutants through radical mechanisms. The enzyme's ability to generate diffusible, stable mobile oxidants allows MnP to act effectively within complex soil and aquatic matrices, extending its ecological and biotechnological significance in xenobiotic remediation (Qin et al. 2017).

MnP catalysis is central to fungal extracellular oxidation of recalcitrant organics and directly ties to the question of how fungi attack xenobiotics. MnPs are secreted heme peroxidases that use peroxide oxidants to drive catalytic turnover, oxidizing manganese and other substrates as part of a two-step redox cycle that yields diffusible oxidants for remote substrate attack. The enzyme oxidizes Mn^{2+} to Mn^{3+} using hydrogen peroxide as the electron acceptor, producing a Mn^{3+} species that leaves the enzyme and oxidizes environmental substrates *via* one-electron transfers, producing phenoxy and other radical intermediates that lead to bond cleavage, rearrangement, or polymerization (Chowdhary et al. 2019; Khajehzadeh et al. 2024) and the possible sequence is stated below.

- Oxidation step: $Mn^{2+} \rightarrow Mn^{3+}$ by MnP is the core redox event that converts a stable metal ion into a reactive diffusible oxidant.
- Chelation and stabilization: chelators (oxalate, dicarboxylates) bind Mn^{3+} preventing rapid disproportionation and enabling diffusion to substrates.
- Diffusible mediator function: Mn^{3+}-chelate performs remote one-electron oxidations, generating substrate radicals (phenoxy radicals, radical cations) that initiate depolymerization, dechlorination-like processes, or oxidative ring modifications depending on substrate chemistry.
- Indirect substrate oxidation: For non-phenolic or sterically shielded molecules, Mn^{3+}-mediated radical cascades or secondary mediators produced by partner enzymes enable oxidation that the enzyme active site cannot perform directly.

MnP participates in broad, nonspecific oxidative attack on many xenobiotic classes. White-rot fungal MnPs and associated ligninolytic systems have been shown to target phenolic and non-phenolic lignin fragments, synthetic dyes (azo, anthraquinone, and others), chlorinated phenols, PAHs such as anthracene, various pesticides including organochlorines (e.g., DDT derivatives), explosives, and certain mycotoxins, with evidence of activity or induction in environmental and laboratory studies (Egbewale et al. 2025; Giwa et al. 2025; Torres-Farradá et al. 2024).

MnP rarely acts alone in fungal xenobiotic degradation; it functions inside a ligninolytic enzyme network and in ecological consortia that enable multistep transformation and, for some pollutants, near-complete mineralization. Fungal systems coexpress laccases, lignin peroxidases, versatile peroxidases, and intracellular oxygenases/dehydrogenases that either produce mediators/H_2O_2, perform direct high-potential oxidations, or carry out intracellular metabolism of extracellularly formed intermediates (Egbewale et al. 2025; Giwa et al. 2025; Torres-Farradá et al. 2024).

3.2.4 Versatile Peroxidases (VP): Hybrid Functionality

Versatile peroxidases (VPs-EC 1.11.1.16) are secreted heme peroxidases with hybrid active sites that oxidize a very broad number of xenobiotics via H_2O_2 driven one and two electron routes, Mn^{2+}-mediated oxidation, and long-range electron transfer. They act synergistically with laccases and other fungal enzymes and are already applied and engineered for bioremediation deployments. This combination enables them to oxidize a remarkably wide range of substrates relevant to xenobiotic degradation.

VPs possess structural elements that combine features of manganese peroxidase and lignin peroxidase, giving VPs several substrate access routes and redox centers that underlie their ability to transform recalcitrant xenobiotics. The hybrid architecture explains how a single extracellular heme enzyme can oxidize both low- and high-redox substrates as well as Mn^{2+} (Torres-Farradá et al. 2024). Like MnP, VP could oxidize Mn^{2+} to Mn^{3+}, generating diffusible oxidants stabilized by organic acid chelators while also directly oxidizing high-redox-potential non-phenolic aromatic compounds in a manner similar to LiP. The enzyme contains multiple substrate-binding sites, including a Mn^{2+} binding pocket and exposed tryptophan residues that facilitate long-range electron transfer to bulky aromatic substrates. Its catalytic cycle involves sequential peroxide-dependent transitions through Compound I and Compound II intermediates, allowing oxidation of phenolic and non-phenolic xenobiotics, synthetic dyes, polycyclic aromatic hydrocarbons, chlorophenols, and endocrine-disrupting compounds. Importantly, VP's dual activity provides ecological and biotechnological advantages, enabling fungi such as *Pleurotus eryngii* to degrade structurally diverse pollutants under variable environmental conditions. This multifunctional capacity positions VP as a key player in fungal bioremediation strategies, bridging the complementary functions of LiP and MnP and expanding the enzymatic arsenal available for detoxification of recalcitrant xenobiotics (Ahsan et al. 2021; Ezra et al. 2025).

The key structural traits and their functional consequences are listed below.

- Heme prosthetic group: central iron porphyrin that performs H_2O_2 activation to generate reactive oxidizing intermediates
- Manganese-binding site: a surface-exposed acidic pocket coordinates Mn^{2+}, enabling formation of diffusible Mn^{3+} chelates that oxidize phenolic and aromatic xenobiotics at distance.
- Solvent-exposed aromatic residue: a surface tryptophan/tyrosine enables direct long-range electron transfer (LRET) to high redox-potential substrates that cannot access the heme pocket directly.
- Glycosylation and secretion signals: support extracellular stability and localization in fungal secretomes for environmental action.

VP catalysis follows classical heme-peroxidase chemistry (H_2O_2 activation) but integrates multiple reaction pathways (direct oxidation, Mn^{2+} oxidation, and LRET) and requires specific cofactors or mediators for some routes. The enzyme cycles through oxidized intermediates that accept electrons from substrate molecules, with alternative substrate access enabled by the manganese site and surface electron-transfer residues (Karich et al. 2017).

Mechanistic pathways and roles of electron donors/cofactors:

- H_2O_2-driven peroxidative cycle: H_2O_2 reacts with the resting Fe^{3+}heme to form Compound I and then Compound II; substrates are oxidized by sequential electron transfer from Compound I/II to regenerate Fe^{3+} (Karich et al. 2017).
- Mn^{2+} oxidation route: Mn^{2+} is bound/oxidized at the manganese site to produce Mn^{3+} chelates, which act as mobile oxidants that diffuse and oxidize phenolic and polymeric xenobiotics away from the enzyme surface.
- Surface tryptophan/tyrosine LRET route: solvent-exposed aromatic residues mediate long-range electron transfer to oxidize bulky or high-redox nonphenolic substrates that are unable to enter the heme pocket.
- Redox mediators and chelators: organic chelators stabilize Mn^{3+} (e.g., oxalate) and small-molecule redox mediators extend VP activity to substrates otherwise inaccessible; these cofactors modify product distribution and reaction reach.
- Electron donors: small phenolic compounds, lignin fragments, or exogenous mediators supply electrons; in many environmental contexts, co-secreted fungal metabolites and naturally present organic matter serve as mediators.

VPs oxidize a wide diversity of xenobiotic classes, acting on phenolics, non-phenolic aromatics, PAHs, dyes, chlorinated phenols, and persistent hydrophobic pollutants; several studies document VP or VP-rich secretomes, acting on PCBs, dyes, and endocrine disruptors.

3.2.5 Cytochrome P450 (CYPs) Monooxygenases

Fungal CYPs perform the initial oxidative activation of xenobiotics, enabling downstream detoxification and excretion; these reactions are integrated into multiphase cellular pathways that include conjugation and transport. CYP-catalyzed reactions are complemented by other oxidoreductases (laccases, peroxidases, dioxygenases) that together enable partial oxidation, detoxification, and mineralization of recalcitrant compounds (Khan et al. 2023; Park and Choi 2020).

- Phase I oxidation: CYP monooxygenases introduce oxygen (hydroxylation, epoxidation, N-/O-, or dealkylation) to increase polarity and create handles for conjugation (Khan et al. 2023; Park and Choi 2020).
- Phase II conjugation: oxidized products are frequently conjugated (glutathione, glucuronide-like, or sulfate equivalents in fungi) to increase solubility for export (Sang et al. 2018).
- Phase III secretion: conjugates and unmetabolized xenobiotics are removed by efflux transporters; transcriptional programs often coordinate CYPs with ABC transporters (Sang et al. 2018).

- Enzyme interplay: ligninolytic enzymes (laccases, manganese/lignin peroxidases) often act alongside CYPs in aromatic xenobiotic breakdown, producing different intermediate types that feed CYP or conjugation steps (Torres-Farradá et al. 2024; Zhang et al. 2021).
- Alternative CYP roles: some fungal P450s catalyze noncanonical reactions (e.g., P450nor-mediated NO reduction) demonstrating catalytic diversity beyond monooxygenation (Torres-Farradá et al. 2024).

Cytochrome P450 enzymes (CYPs) constitute a superfamily of heme-containing monooxygenases characterized by axial coordination of the heme iron by a cysteine thiolate ligand (Manikandan and Nagini 2018). This unique coordination environment enables the formation of highly reactive iron-oxo intermediates capable of inserting oxygen atoms into C-H bonds under mild conditions. The catalytic cycle involves the following: substrate binding and low-spin to high-spin iron transition, NADPH-mediated reduction of Fe^{3+} to Fe^{2+}, oxygen binding to form Fe^{2+}-O_2 complex, second electron transfer and protonation to generate Compound I (Fe^{4+}=O), and oxygen insertion into substrate C–H bonds (Li et al. 2020).

Fungal P450s exhibit remarkable substrate diversity, catalyzing hydroxylation, epoxidation, dealkylation, dehalogenation, and oxidative cleavage reactions on various xenobiotics (Wang et al. 2018). These enzymes play crucial roles in Phase I metabolism, often introducing polar functional groups that facilitate subsequent conjugation reactions or enhance compound bioavailability for further degradation. Notable P450-mediated xenobiotic transformations include the following: hydroxylation of PAHs at specific positions, O-dealkylation of organophosphate pesticides, oxidative dehalogenation of chlorinated compounds, and N oxidation of nitrogen-containing heterocycles (Behrendorff et al. 2015; Sang et al. 2018; Shin et al. 2017). Various fungal cytochrome P450 systems involved in xenobiotic degradations are given in Table 3.1.

3.2.6 Unspecific Peroxygenases (UPO): Emerging Oxidative Enzymes

Unspecific peroxygenases (EC 1.11.2.1) are secreted heme-thiolate enzymes from fungi with structural features that enable peroxide-driven oxygen transfer for extracellular catalysis. These extracellular heme-containing enzymes utilize hydrogen peroxide as both an electron acceptor and an oxygen donor for substrate oxygenation. Their active site architecture resembles other heme enzymes but is adapted for direct use of H_2O_2 and for accommodating diverse substrates. Unlike P450s, UPOs do not require NAD(P)H cofactors or electron transport chains, making them attractive for biotechnological applications. UPOs catalyze various oxygenation reactions, including hydroxylation, epoxidation, and heteroatom oxidation with high regio- and stereoselectivity. The enzyme's ability to perform P450-like chemistry without cofactor requirement represents a significant advantage for practical applications. Comparative analyses combining experimental activity profiles with

Table 3.1 Fungal cytochrome P450 systems involved in xenobiotic degradation

Species	Xenobiotic classes targeted	P450 family/function
Fusarium graminearum (filamentous pathogen)	Diverse xenobiotics; fungicides; secondary-metabolite biosynthesis	Large CYPome (≈100 genes); specific CYPs affect xenobiotic response, virulence and ergosterol biosynthesis
Phanerochaete chrysosporium (white-rot)	PAHs, various aromatics and pollutants	Multiple CYP63/CYPome members implicated in aromatic oxidation and coaction with ligninolytic enzymes
Sclerotinia homoeocarpa	Chlorothalonil (multi-site fungicide)	CYPs (CYP561, CYP65) biotransform chlorothalonil to hydroxylated intermediates
Tolypocladium sp., marine ascomycetes	Pyrene and other PAHs	Transcriptomic and metabolite evidence points to CYP monooxygenases initiating pyrene degradation
Thielavia sp.	Phenolic xenobiotics (NP, 4-tert-OP, 2,4-DCP)	Degradation linked to laccase activity and P450 participation for chlorinated phenols
Scedosporium sp. (environmental)	Tetrahydrofuran (THF)	Metagenomics identifies a P450 candidate and cognate reductase associated with THF degradation

structural data indicate that a more compact and constrained active-site architecture enhances peroxygenase activity. Within the *Ascomycota* phylum, UPOs belonging to a high-activity clade exhibit a distinctive aliphatic pocket in their catalytic center, which is thought to facilitate substrate positioning and improve the efficiency of oxygen transfer reactions (Barros-Reguera et al. 2025).

UPO mechanistic and structural highlights include:

- Heme-thiolate active site UPOs use a heme cofactor with a thiolate proximal ligand that supports formation of high-valent oxo-ferryl intermediates (Compound I) supplied by peroxide (Khan et al. 2023)
- Peroxide dependence: accept oxygen from hydrogen peroxide (or organic peroxides) rather than molecular O_2, obviating the need for auxiliary redox partners and intracellular electron transfer chains (Hofrichter and Ullrich 2014; Khan et al. 2023).
- Secreted glycoproteins: UPOs are extracellular enzymes secreted by fungi, which positions them to oxidize substrates outside the cell and to function in complex environmental matrices (Costa and Liang 2023).
- Active site access and selectivity with structural and simulation studies show substrate cavities and flexible access channels that control selectivity and reactivity across different UPO isoforms (Costa and Liang 2023; Hofrichter and Ullrich 2014; Khan et al. 2023).

Each reaction follows peroxide activation to Compound I with subsequent substrate oxidation; substrate positioning in the active site governs whether

Table 3.2 Reaction types and mechanistic summary of unspecific peroxygenases

Reaction type	Mechanistic summary
Benzylic and aliphatic hydroxylation	Direct H-atom abstraction and rebound to give alcohols (peroxygenation)
Aromatic hydroxylation and phenol formation	Electrophilic oxygenation of aromatic rings *via* ferryl species
Epoxidation of double bonds	Oxygen insertion across C=C yielding epoxides
Sulfoxidation and S-oxidation	Two-electron oxygen transfer to sulfur centers
Dealkylation and oxidative cleavage	Sequential oxygenation leading to scission or removal of alkyl groups

hydrogen-atom abstraction, electrophilic attack, or two-electron insertion predominates (Khan et al. 2023).

UPOs catalyze peroxide-dependent oxygen transfer via high-valent heme intermediates, enabling both one- and two-electron oxidations. Their reaction scope stems from a common mechanism: H_2O_2 activates the heme to an oxo-ferryl species that inserts oxygen into substrate bonds. Multiple reaction types and mechanistic pathways of unspecific peroxygenases are proposed (Costa and Liang 2023; Hofrichter and Ullrich 2014; Santacruz-Juárez et al. 2021) and are summarized in Table 3.2.

Furthermore, UPO activity toward various xenobiotics, including pharmaceuticals, pesticides, and industrial chemicals, is reported (Gao et al. 2025). The enzyme's broad substrate acceptance and operational simplicity make it a promising candidate for bioremediation applications, though commercial availability remains limited compared to other fungal enzymes.

3.3 Field Applications: Through Enzyme Immobilization, Genetic Engineering

The remarkable diversity of fungal enzymes, including laccases, peroxidases, and cytochrome P450 monooxygenases, provides a robust foundation for biotechnological applications. Industrial implementations have demonstrated the feasibility and effectiveness of fungal treatment systems for various waste streams, with removal efficiencies often exceeding 90% for target contaminants. Recent advances in genetic engineering, particularly CRISPR/Cas9 technologies, are accelerating the development of enhanced fungal strains with improved degradation capabilities. Comparative studies across fungal species reveal distinct advantages for different applications, with white-rot basidiomycetes excelling in broad-spectrum xenobiotic degradation and ascomycetes offering advantages for specific applications requiring acid tolerance or rapid growth. The integration of emerging technologies, including enzyme immobilization, advanced bioreactor designs, and process monitoring systems, is enhancing the practical viability of fungal bioremediation. However, many challenges remain in extending laboratory successes to industrial applications, achieving economic competitiveness, and navigating regulatory requirements.

Future development should focus on addressing fundamental knowledge gaps in enzyme mechanisms, developing robust scale-up strategies, and integrating fungal technologies with circular economy principles. The convergence of biotechnology, nanotechnology, and artificial intelligence offers exciting opportunities for next-generation bioremediation systems that are more efficient, economical, and environmentally sustainable. The field is poised for significant growth as environmental regulations become more stringent, public awareness of pollution impacts increases, and the limitations of conventional treatment technologies become more apparent. Continued research and development investment, coupled with supportive regulatory frameworks, will be essential for realizing the full potential of fungal bioremediation technologies.

3.3.1 Field Application of Fungal Enzymes in Xenobiotic Degradation

As explained above, fungal oxidoreductases, including laccases, manganese peroxidases, and other lignin-degrading enzymes, exhibit exceptional potential for bioremediation by targeting some of the most recalcitrant xenobiotic pollutants. These enzymes can effectively oxidize and decompose a wide spectrum of hazardous organic compounds, such as dyes, phenols, pesticides, and polyaromatic hydrocarbons, frequently converting them into environmentally benign or significantly less toxic derivatives (Yamaguchi and Miyazaki 2024). However, deploying fungal enzymes in field matrices, such as soil and wastewater, poses challenges due to the enzymes' sensitivity and the complexity of the environment. The two main strategies for enabling effective field applications are enzyme immobilization and genetic engineering, encompassing recombinant expression and enzyme modification. Below, each strategy is examined in terms of its ability to enhance catalytic performance and stability for in situ xenobiotic degradation, with examples drawn from bioreactors, soil and wastewater treatments, and real-world case studies. Current challenges in scaling these approaches are also discussed, along with emerging trends (e.g., co-substrate delivery systems and advanced materials) aimed at overcoming these limitations.

3.3.2 Enzyme Immobilization: Enhancing Stability and Reusability

Enzyme immobilization involves attaching or encapsulating fungal enzymes on solid supports, forming a heterogeneous catalyst that could be deployed in contaminated environments. Immobilization is a well-established technique to improve enzyme stability and reusability, addressing the inherent instability and one-time use limitations of free enzymes (Sheldon and Woodley 2018). By fixing enzymes onto durable carriers (such as polymers, clays, or nanomaterials), the enzyme's conformation is better preserved under field conditions, and it is protected from proteolytic degradation or washout. The result is an enzyme system that remains active over longer periods and multiple reaction cycles, even under varying pH, temperature, or the presence of interferents in soil or wastewater. Immobilized enzymes often show enhanced operational stability and reused many times, which also minimizes costs in large-scale applications (Bianchini et al. 2015).

There are several methods of immobilization—including carrier binding (physical adsorption, ionic or covalent binding to support surfaces), entrapment within gels or fibers, and carrier-free approaches like cross-linked enzyme aggregates (CLEAs). Selecting a suitable support and method is critical and depends on the enzyme and the field application matrix (Nájera-Martínez et al. 2022). Common supports range from natural materials (e.g., cellulose, alginate, or clays) to synthetic beads, porous glass, and emerging nanostructured materials. For instance, inexpensive minerals, such as kaolinite clay, have been used to immobilize fungal laccase *via* adsorption; the negatively charged clay surface binds the enzyme and provides structurally stable, reusable support. Laccase immobilized on kaolinite retained ~50% of its original activity after multiple reuse cycles, demonstrating good operational stability (Yamaguchi and Miyazaki 2024). Kaolinite-immobilized laccase, when applied with a mediator, achieved nearly 80% degradation of malachite green across five cycles, demonstrating both sustained activity in dye-contaminated water and the advantage of recyclability (Wen et al. 2019). This illustrates how immobilization maintain enzyme effectiveness in a complex aqueous matrix with the added benefit of recyclability.

Laccase immobilized on magnetic nanoparticles has achieved 60–94% decolorization of textile dyes across multiple cycles, with the added advantage of easy recovery by magnetic separation (Peñaranda et al. 2022). Enzyme immobilization not only improves stability and reusability but also enables integrated co-substrate supply and multistep catalysis. Co-immobilized systems, such as peroxidase with glucose oxidase, create self-sufficient cascades where glucose oxidase generates H_2O_2 to drive peroxidase activity, achieving up to 99.5% dye degradation in wastewater within hours and maintaining activity over multiple reuse cycles. Likewise, laccase systems co-entrapped with redox mediators (e.g., ABTS or natural phenolics) broaden the spectrum of degradable xenobiotics by providing in situ electron shuttles. These approaches demonstrate how immobilization transforms enzymes into more robust, efficient, and versatile biocatalysts, suitable for reactors or soil applications under real environmental conditions (Yang et al. 2017).

3.3.3 Genetic Engineering and Recombinant Expression of Fungal Enzymes

Genetic engineering has become a cornerstone in advancing fungal enzyme applications for xenobiotic degradation. Native fungal strains often secrete low enzyme yields and lack the robustness needed for field conditions; recombinant DNA technology addresses these issues by enabling high-yield, cost-effective production and tailoring of enzymes (Piscitelli et al. 2010). Heterologous expression in hosts such as *E.coli*, *Pichia pastoris*, *Saccharomyces cerevisiae*, *Aspergillus*, and *Trichoderma* provides several advantages: bacterial systems offer speed and scalability (with codon optimization, chaperone co-expression, and cofactor supplementation overcoming folding and activity barriers), while fungal and yeast hosts confer posttranslational modifications like glycosylation that enhance stability and secretion. Case studies demonstrate these benefits engineered *Trichoderma atroviride* strains overproducing laccase achieved efficient degradation of phenolics, endocrine disruptors (e.g., bisphenol A), synthetic dyes, and PAHs, outperforming wild-type fungi. Similarly, recombinant *E. coli* systems have been optimized to express active multicopper oxidases capable of rapidly removing textile dyes from wastewater, highlighting the feasibility of industrial-scale enzyme generation (Balcázar-López et al. 2016).

Beyond yield, genetic engineering enables the creation of enzymes tailored to real-world conditions. Protein engineering strategies directed evolution, rational design, and computational modeling—have produced laccase and peroxidase variants with higher redox potential, broader pH tolerance, greater thermal stability, and improved mediator affinity, allowing effective action in diverse effluents and soils. Specific examples include evolved fungal laccases capable of oxidizing tetracyclines under alkaline conditions and mutants designed for enhanced degradation of high-redox-potential pollutants. These engineered enzymes not only survive but perform optimally in harsh and variable environments (Balcázar-López et al. 2016).

Genetic engineering also extends beyond purified enzyme production to whole-cell biocatalysts. Engineered bacteria, yeasts, or fungi expressing fungal enzyme genes can continuously produce and secrete enzymes in situ, sometimes combined with transporter genes or cofactor regeneration systems to form self-sustaining catalysts. Recombinant microbial consortia allow division of labor, with different strains handling sequential degradation steps, while transgenic plants engineered to secrete fungal laccases from their roots demonstrate a phytoremediation-based delivery platform. Such systems integrate stability, continuous activity, and adaptability, offering a living alternative to enzyme dosing.

Together, recombinant expression, protein engineering, and whole-cell biocatalyst design vastly expand the practicality of fungal enzymes. These strategies address production scalability, catalytic efficiency, environmental stability, and pollutant specificity, making genetic engineering a central driver in transforming fungal biocatalysts from laboratory tools into field-ready solutions for large-scale xenobiotic remediation (Guan et al. 2025; Xu et al. 2024).

3.4 Real-World Applications and Successes

Field applications of fungal enzyme strategies are showing promising outcomes in pollution mitigation. Laccase-based treatments have been widely investigated for industrial wastewater detoxification, particularly in textile and pulp/paper sectors, where they effectively decolorize and detoxify effluents by breaking down aromatic xenobiotics with oxygen as the sole electron acceptor, producing only water. Immobilized laccase bioreactors have demonstrated efficient removal of reactive dyes under continuous flow, including systems using metal–organic frameworks that achieved 60–70% decolorization without mediators, even at elevated temperatures. Pilot-scale demonstrations further validate the potential of these approaches, with crude fungal laccase successfully reducing pesticide residues in agricultural wastewater, underscoring the feasibility of integrating enzymatic treatments into conventional wastewater management (Birhanlı et al. 2022; Vaithyanathan et al. 2021).

Fungal enzymes have been applied in soil remediation through approaches such as biopiles and soil bioreactors, where white-rot fungi (e.g., *Phanerochaete chrysosporium*, *Trametes versicolor*) and their ligninolytic enzymes facilitate the breakdown of polycyclic aromatic hydrocarbons (PAHs) and other persistent pollutants. Full-scale biopile treatments have demonstrated enhanced PAH degradation with fungal inocula or enzyme extracts. Enzyme immobilization further improves soil applications: carriers like functionalized biochar or silica provide stability, prolong activity, and concentrate pollutants near active sites, accelerating pesticide degradation. Innovative strategies, such as engineering insect larvae to express fungal laccase for in situ detoxification, highlight the potential of combining biological and engineered systems for soil xenobiotic remediation (Wang et al. 2024; Yamaguchi and Miyazaki 2024).

Genetically engineered microbes are being tested in real-world bioremediation. A laccase-overexpressing *Trichoderma* strain accelerated phenol removal in contaminated soils, while *Pichia pastoris* engineered to secrete versatile peroxidase showed promise for pharmaceutical pollutant degradation in bioreactors. Pilot-scale case studies demonstrate that enzyme-based treatments can lower pollutant loads and toxicity in dye effluents, pesticide-contaminated soils, and wastewater, often producing fewer harmful byproducts than chemical methods. Integrated systems, such as immobilized laccase reactors coupled with conventional processes, have already achieved regulatory-compliant water discharge, highlighting the potential of enzymatic approaches within existing treatment frameworks (Vaithyanathan et al. 2021).

3.5 Challenges and Limitations in Field Deployment

Despite their promise, fungal enzyme strategies face major challenges in large-scale deployment. High production and purification costs, limited enzyme stability in uncontrolled environments, and activity loss over time hinder economic feasibility. Immobilized enzymes suffer from diffusion barriers, fouling, and the high cost or environmental risks of engineered support materials. Field application also raises mass transfer, mixing, and co-substrate supply issues, with some enzymes inactivated by their own cofactors. Genetically engineered organisms encounter regulatory, ecological, and survival constraints, and many fungal enzymes remain difficult to express at scale due to folding and host toxicity problems (Sodhi et al. 2024).

Finally, economic and integration barriers persist, as enzymatic methods must match the efficiency, cost, and practicality of established physicochemical. Addressing these technical, regulatory, and economic hurdles is essential for routine field application (Mayer and Staples 2002; Rybarczyk et al. 2023).

3.6 Emerging Trends and Future Directions

Ongoing research is developing innovative immobilization strategies to overcome scalability and stability challenges. Low-cost nanostructured biochars, derived from waste biomass, offer high surface area, functional group tailoring, and dual roles as enzyme supports and pollutant sorbents, thereby enhancing degradation efficiency. In parallel, 3D-printed supports enable custom porous structures that optimize enzyme–substrate contact, while patented designs, such as magnetic 3D-printed carriers, further facilitate enzyme recovery and reuse. These advances promise to reduce costs and improve the feasibility of large-scale bioremediation systems (Armanu et al. 2025; Rybarczyk et al. 2023).

Carrier-free immobilization using cross-linked enzyme aggregates (CLEAs) offers a cost-effective alternative to traditional supports. By precipitating and cross-linking enzymes into insoluble granules, CLEAs achieve high enzyme loading, retain strong activity, and can be applied in stirred reactors or directly onto contaminated sites. Eliminating costly carriers enhances scalability, while ongoing optimization, such as adding stabilizers during formation, improves the robustness of ligninolytic enzyme CLEAs for field applications (Yamaguchi and Miyazaki 2024). Synthetic biology enables new genetic approaches to bioremediation. Engineered microbial consortia divide pollutant degradation steps among different strains, creating synergistic pathways that can be deployed in encapsulated carriers as "living immobilized systems." Parallel efforts in phytoremediation include transgenic plants engineered to secrete fungal enzymes, such as laccases, from their roots, providing continuous in situ pollutant breakdown within the rhizosphere.

Protein engineering is advancing fungal enzyme performance by enhancing stability and functionality under harsh environmental conditions. Directed evolution has produced laccase variants tolerant to solvents and salts, while computational design accelerates the identification of stabilizing mutations. Novel approaches include engineering self-assembling or surface-binding peptides, enabling enzymes to spontaneously immobilize on target surfaces and simplifying field deployment (Yang et al. 2017). Innovative systems are being developed to address cofactor and mediator limitations. Encapsulated mini-reactors can provide a steady in situ supply of H_2O_2 for peroxidase-driven reactions, while mediator recycling approaches using tethered supports or electrochemical regeneration reduce reliance on costly continuous mediator addition.

Fungal enzymes are advancing toward field-ready solutions for xenobiotic degradation through hybrid strategies that integrate immobilized enzymes, engineered microbes, and complementary physicochemical treatments. Progress in materials science, molecular biology, and environmental engineering is enhancing enzyme stability, efficiency, and delivery, bringing biocatalytic systems closer to large-scale deployment. The ultimate aim is to establish robust, recyclable enzymatic tools that can sustainably replace or complement chemical methods, enabling cleaner soils and waters through fungal biocatalysis.

References

Ahsan Z, Kalsoom U, Bhatti HN, Aftab K, Khalid N, Bilal M (2021) Enzyme-assisted bioremediation approach for synthetic dyes and polycyclic aromatic hydrocarbons degradation. J Basic Microbiol 61(11):960–981. https://doi.org/10.1002/jobm.202100218

Amitai G, Adani R, Sod-Moriah G, Rabinovitz I, Vincze A, Leader H, Chefetz B, Leibovitz-Persky L, Friesem D, Hadar Y (1998) Oxidative biodegradation of phosphorothiolates by fungal laccase. FEBS Lett 438(3):195–200. https://doi.org/10.1016/s0014-5793(98)01300-3

Armanu EG, Bertoldi S, Chrzanowski Ł, Volf I, Heipieper HJ, Eberlein C (2025) Benefits of immobilized bacteria in bioremediation of sites contaminated with toxic organic compounds. Microorganisms 13(1). https://doi.org/10.3390/microorganisms13010155

Atashgahi S, Shetty SA, Smidt H, de Vos WM (2018) Flux, impact, and fate of halogenated xenobiotic compounds in the gut. Front Physiol 9:888. https://doi.org/10.3389/fphys.2018.00888

Balcázar-López E, Méndez-Lorenzo LH, Batista-García RA, Esquivel-Naranjo U, Ayala M, Kumar VV, Savary O, Cabana H, Herrera-Estrella A, Folch-Mallol JL (2016) Xenobiotic compounds degradation by heterologous expression of a Trametes sanguineus Laccase in Trichoderma atroviride. PLoS One 11(2):e0147997. https://doi.org/10.1371/journal.pone.0147997

Baldrian P (2006) Fungal laccases - occurrence and properties. FEMS Microbiol Rev 30(2):215–242. https://doi.org/10.1111/j.1574-4976.2005.00010.x

Bankole PO, Adekunle AA, Govindwar SP (2019) Demethylation and desulfonation of textile industry dye, Thiazole Yellow G by Aspergillus niger LAG. Biotechnol Rep (Amst) 23:e00327. https://doi.org/10.1016/j.btre.2019.e00327

Barros-Reguera M, Lopez-Tavera E, Schröder GC, Nardini G, Kristoffersen KA, Ayuso-Fernández I, Eijsink VGH, Sørlie M (2025) Structure-function relationships in unspecific peroxygenases revealed by a comparative study of their action on the phenolic lignin monomer 4-propylguaiacol. Biotechnol Biofuels Bioprod 18(1):83. https://doi.org/10.1186/s13068-025-02675-w

Behrendorff JB, Huang W, Gillam EM (2015) Directed evolution of cytochrome P450 enzymes for biocatalysis: exploiting the catalytic versatility of enzymes with relaxed substrate specificity. Biochem J 467(1):1–15. https://doi.org/10.1042/bj20141493

Bianchini LF, Arruda MF, Vieira SR, Campelo PM, Grégio AM, Rosa EA (2015) Microbial biotransformation to obtain new antifungals. Front Microbiol 6:1433. https://doi.org/10.3389/fmicb.2015.01433

Bilal M, Zdarta J, Jesionowski T, Iqbal HMN (2023) Manganese peroxidases as robust biocatalytic tool - An overview of sources, immobilization, and biotechnological applications. Int J Biol Macromol 234:123531. https://doi.org/10.1016/j.ijbiomac.2023.123531

Birhanlı E, Noma SAA, Boran F, Ulu A, Yeşilada Ö, Ateş B (2022) Design of laccase-metal-organic framework hybrid constructs for biocatalytic removal of textile dyes. Chemosphere 292:133382. https://doi.org/10.1016/j.chemosphere.2021.133382

Camarero S, Ibarra D, Martínez MJ, Martínez AT (2005) Lignin-derived compounds as efficient laccase mediators for decolorization of different types of recalcitrant dyes. Appl Environ Microbiol 71(4):1775–1784. https://doi.org/10.1128/aem.71.4.1775-1784.2005

Chowdhary P, Shukla G, Raj G (2019) Microbial manganese peroxidase: a ligninolytic enzyme and its ample opportunities in research. SN Appl Sci 1:45. https://doi.org/10.1007/s42452-018-0046-3

Christian V, Shrivastava R, Shukla D, Modi HA, Vyas BR (2005) Degradation of xenobiotic compounds by lignin-degrading white-rot fungi: enzymology and mechanisms involved. Indian J Exp Biol 43(4):301–312

Chugh M, Kumar L, Bharadvaja N (2022) Fungal diversity in the bioremediation of toxic effluents. In: Environmental contamination remediation and management, pp 185–203. https://doi.org/10.1016/b978-0-323-85839-7.00009-8

Costa GJ, Liang R (2023) Understanding the multifaceted mechanism of compound I formation in unspecific peroxygenases through multiscale simulations. J Phys Chem B 127(41):8809–8824. https://doi.org/10.1021/acs.jpcb.3c04589

Egbewale SO, Kumar A, Mokoena MP, Olaniran AO (2025) Mycotransformation of anthracene by indigenous Trichoderma lixii and Talaromyces pinophilus isolates: insights into the metabolic pathways, enzyme profiles and acute toxicity. Biodegradation 36(4):51. https://doi.org/10.1007/s10532-025-10147-z

Eichlerová I, Šnajdr J, Baldrian P (2012) Laccase activity in soils: considerations for the measurement of enzyme activity. Chemosphere 88(10):1154–1160. https://doi.org/10.1016/j.chemosphere.2012.03.019

Ezra R, Vanti G, Masaphy S (2025) Sustainable, targeted, and cost-effective laccase-based bioremediation technologies for antibiotic residues in the ecosystem: a comprehensive review. Biomolecules 15(8). https://doi.org/10.3390/biom15081138

Falade AO, Nwodo UU, Iweriebor BC, Green E, Mabinya LV, Okoh AI (2017) Lignin peroxidase functionalities and prospective applications. Microbiol Open 6(1). https://doi.org/10.1002/mbo3.394

Febbraio F (2017) Biochemical strategies for the detection and detoxification of toxic chemicals in the environment. World J Biol Chem 8(1):13–20. https://doi.org/10.4331/wjbc.v8.i1.13

Gałązka A, Jankiewicz U (2022) Endocrine disrupting compounds (Nonylphenol and Bisphenol A)-sources, harmfulness and laccase-assisted degradation in the aquatic environment. Microorganisms 10(11). https://doi.org/10.3390/microorganisms10112236

Gałązka AJ, Jankiewicz U, Szczepkowski A (2023) Biochemical characteristics of laccases and their practical application in the removal of xenobiotics from water. Appl Sci 13:4394. https://doi.org/10.3390/app13074394

Gao J, Wang W, Yin Z, Ai J, Lu F, Qin HM, Mao S (2025) Unspecific peroxygenases and P450-derived peroxygenases as green biocatalysts driven by H(2)O(2) for sustainable applications: a review. Int J Biol Macromol 322(Pt 1):146548. https://doi.org/10.1016/j.ijbiomac.2025.146548

Giwa AS, Waheed M, Khalid HV, Shafique E, Rahman SU, Ali N (2025) Mechanistic insights into fungal-bacterial synergy for DDT biotransformation. Antonie Van Leeuwenhoek 118(9):137. https://doi.org/10.1007/s10482-025-02149-x

Glenn JK, Akileswaran L, Gold MH (1986) Mn(II) oxidation is the principal function of the extracellular Mn-peroxidase from Phanerochaete chrysosporium. Arch Biochem Biophys 251(2):688–696. https://doi.org/10.1016/0003-9861(86)90378-4

Guan G, Li B, Xu L, Qian J, Zou B, Huo S, Ding Z, Cui K, Wang F (2025) Improving the properties of laccase through heterologous expression and protein engineering. Microorganisms 13(6). https://doi.org/10.3390/microorganisms13061422

Heinfling A, Martínez MJ, Martínez AT, Bergbauer M, Szewzyk U (1998) Transformation of industrial dyes by manganese peroxidases from Bjerkandera adusta and Pleurotus eryngii in a manganese-independent reaction. Appl Environ Microbiol 64(8):2788–2793. https://doi.org/10.1128/aem.64.8.2788-2793.1998

Hofrichter M, Ullrich R (2014) Oxidations catalyzed by fungal peroxygenases. Curr Opin Chem Biol 19:116–125. https://doi.org/10.1016/j.cbpa.2014.01.015

Hultberg M, Golovko O (2024) Use of sawdust for production of ligninolytic enzymes by white-rot fungi and pharmaceutical removal. Bioprocess Biosyst Eng 47(4):475–482. https://doi.org/10.1007/s00449-024-02976-8

Jones SM, Solomon EI (2015) Electron transfer and reaction mechanism of laccases. Cell Mol Life Sci 72(5):869–883. https://doi.org/10.1007/s00018-014-1826-6

Karich A, Ullrich R, Scheibner K, Hofrichter M (2017) Fungal unspecific peroxygenases oxidize the majority of organic EPA priority pollutants. Front Microbiol 8:1463. https://doi.org/10.3389/fmicb.2017.01463
Khajehzadeh M, Ghobad-Nejhad M, Moghimi H, Abolhasani Soorki A, Dai YC, Si J (2024) Degradation of high concentrations of anthracene using white-rot wood-inhabiting fungi and investigation of enzyme activities. Mycobiology 52(5):298–305. https://doi.org/10.1080/12298093.2024.2409485
Khan MF, Hof C, Niemcová P, Murphy CD (2023) Recent advances in fungal xenobiotic metabolism: enzymes and applications. World J Microbiol Biotechnol 39(11):296. https://doi.org/10.1007/s11274-023-03737-7
Kjaergaard CH, Jones SM, Gounel S, Mano N, Solomon EI (2015) Two-electron reduction versus one-electron oxidation of the type 3 pair in the multicopper oxidases. J Am Chem Soc 137(27):8783–8794. https://doi.org/10.1021/jacs.5b04136
Kupski L, Salcedo GM, Caldas SS, de Souza TD, Furlong EB, Primel EG (2019) Optimization of a laccase-mediator system with natural redox-mediating compounds for pesticide removal. Environ Sci Pollut Res Int 26(5):5131–5139. https://doi.org/10.1007/s11356-018-4010-y
Li Z, Jiang Y, Guengerich FP, Ma L, Li S, Zhang W (2020) Engineering cytochrome P450 enzyme systems for biomedical and biotechnological applications. J Biol Chem 295(3):833–849. https://doi.org/10.1074/jbc.REV119.008758
Manikandan P, Nagini S (2018) Cytochrome P450 structure, function and clinical significance: a review. Curr Drug Targets 19(1):38–54. https://doi.org/10.2174/1389450118666170125144557
Margot J, Bennati-Granier C, Maillard J, Blánquez P, Barry DA, Holliger C (2013) Bacterial versus fungal laccase: potential for micropollutant degradation. AMB Express 3(1):63. https://doi.org/10.1186/2191-0855-3-63
Mayer AM, Staples RC (2002) Laccase: new functions for an old enzyme. Phytochem 60(6):551–565. https://doi.org/10.1016/s0031-9422(02)00171-1
Mishra S, Lin Z, Pang S, Zhang W, Bhatt P, Chen S (2021) Recent advanced technologies for the characterization of xenobiotic-degrading microorganisms and microbial communities. Front Bioeng Biotechnol 9:632059. https://doi.org/10.3389/fbioe.2021.632059
Mtibaà R, Olicón-Hernández DR, Pozo C, Nasri M, Mechichi T, González J, Aranda E (2018) Degradation of bisphenol A and acute toxicity reduction by different thermo-tolerant ascomycete strains isolated from arid soils. Ecotoxicol Environ Saf 156:87–96. https://doi.org/10.1016/j.ecoenv.2018.02.077
Nájera-Martínez EF, Melchor-Martínez EM, Sosa-Hernández JE, Levin LN, Parra-Saldívar R, Iqbal HMN (2022) Lignocellulosic residues as supports for enzyme immobilization, and biocatalysts with potential applications. Int J Biol Macromol 208:748–759. https://doi.org/10.1016/j.ijbiomac.2022.03.180
Olicón-Hernández DR, González-López J, Aranda E (2017) Overview on the biochemical potential of filamentous fungi to degrade pharmaceutical compounds. Front Microbiol 8:1792. https://doi.org/10.3389/fmicb.2017.01792
Park H, Choi IG (2020) Genomic and transcriptomic perspectives on mycoremediation of polycyclic aromatic hydrocarbons. Appl Microbiol Biotechnol 104(16):6919–6928. https://doi.org/10.1007/s00253-020-10746-1
Peñaranda PA, Noguera MJ, Florez SL, Husserl J, Ornelas-Soto N, Cruz JC, Osma JF (2022) Treatment of wastewater, phenols and dyes using novel magnetic torus microreactors and laccase immobilized on magnetite nanoparticles. Nanomaterials (Basel) 12(10). https://doi.org/10.3390/nano12101688
Piscitelli A, Pezzella C, Giardina P, Faraco V, Giovanni S (2010) Heterologous laccase production and its role in industrial applications. Bioeng Bugs 1(4):252–262. https://doi.org/10.4161/bbug.1.4.11438
Qin X, Sun X, Huang H, Bai Y, Wang Y, Luo H, Yao B, Zhang X, Su X (2017) Oxidation of a nonphenolic lignin model compound by two Irpex lacteus manganese peroxidases: evidence for implication of carboxylate and radicals. Biotechnol Biofuels 10:103. https://doi.org/10.1186/s13068-017-0787-z

Reiss R, Ihssen J, Richter M, Eichhorn E, Schilling B, Thöny-Meyer L (2013) Laccase versus laccase-like multi-copper oxidase: a comparative study of similar enzymes with diverse substrate spectra. PLoS One 8(6):e65633. https://doi.org/10.1371/journal.pone.0065633

Rybarczyk A, Smułek W, Grzywaczyk A, Kaczorek E, Jesionowski T, Nghiem LD, Zdarta J (2023) 3D printed polylactide scaffolding for laccase immobilization to improve enzyme stability and estrogen removal from wastewater. Bioresour Technol 381:129144. https://doi.org/10.1016/j.biortech.2023.129144

Sakurai T, Kataoka K (2007) Basic and applied features of multicopper oxidases, CueO, bilirubin oxidase, and laccase. Chem Rec 7(4):220–229. https://doi.org/10.1002/tcr.20125

Sang H, Hulvey JP, Green R, Xu H, Im J, Chang T, Jung G (2018) A xenobiotic detoxification pathway through transcriptional regulation in filamentous fungi. mBio 9(4). https://doi.org/10.1128/mBio.00457-18

Santacruz-Juárez E, Buendia-Corona RE, Ramírez RE, Sánchez C (2021) Fungal enzymes for the degradation of polyethylene: molecular docking simulation and biodegradation pathway proposal. J Hazard Mater 411:125118. https://doi.org/10.1016/j.jhazmat.2021.125118

Scheel T, Höfer M, Ludwig S, Hölker U (2000) Differential expression of manganese peroxidase and laccase in white-rot fungi in the presence of manganese or aromatic compounds. Appl Microbiol Biotechnol 54(5):686–691. https://doi.org/10.1007/s002530000427

Sharma BM, Bharat GK, Tayal S, Nizzetto L, Cupr P, Larssen T (2014) Environment and human exposure to persistent organic pollutants (POPs) in India: a systematic review of recent and historical data. Environ Int 66:48–64. https://doi.org/10.1016/j.envint.2014.01.022

Sheldon RA, Woodley JM (2018) Role of biocatalysis in sustainable chemistry. Chem Rev 118(2):801–838. https://doi.org/10.1021/acs.chemrev.7b00203

Shin JY, Bui DC, Lee Y, Nam H, Jung S, Fang M, Kim JC, Lee T, Kim H, Choi GJ, Son H, Lee YW (2017) Functional characterization of cytochrome P450 monooxygenases in the cereal head blight fungus Fusarium graminearum. Environ Microbiol 19(5):2053–2067. https://doi.org/10.1111/1462-2920.13730

Sodhi AS, Bhatia S, Batra N (2024) Laccase: sustainable production strategies, heterologous expression and potential biotechnological applications. Int J Biol Macromol 280(Pt 1):135745. https://doi.org/10.1016/j.ijbiomac.2024.135745

Torres-Farradá G, Manzano León AM, Rineau F, Ledo Alonso LL, Sánchez-López MI, Thijs S, Colpaert J, Ramos-Leal M, Guerra G, Vangronsveld J (2017) Diversity of ligninolytic enzymes and their genes in strains of the genus ganoderma: applicable for biodegradation of xenobiotic compounds? Front Microbiol 8:898. https://doi.org/10.3389/fmicb.2017.00898

Torres-Farradá G, Thijs S, Rineau F, Guerra G, Vangronsveld J (2024) White rot fungi as tools for the bioremediation of xenobiotics: a review. J Fungi (Basel) 10(3). https://doi.org/10.3390/jof10030167

Vaithyanathan VK, Vaidyanathan VK, Cabana H (2021) Laccase-driven transformation of high priority pesticides without redox mediators: towards bioremediation of contaminated wastewaters. Front Bioeng Biotechnol 9:770435. https://doi.org/10.3389/fbioe.2021.770435

Valli K, Wariishi H, Gold MH (1992) Degradation of 2,7-dichlorodibenzo-p-dioxin by the lignin-degrading basidiomycete Phanerochaete chrysosporium. J Bacteriol 174(7):2131–2137. https://doi.org/10.1128/jb.174.7.2131-2137.1992

Wang K, Huang K, Jiang G (2018) Enhanced removal of aqueous acetaminophen by a laccase-catalyzed oxidative coupling reaction under a dual-pH optimization strategy. Sci Total Environ 616–617:1270–1278. https://doi.org/10.1016/j.scitotenv.2017.10.191

Wang L, Chen G, Du X, Li M, Zhang Z, Liang H, Gao D (2024) Bioremediation of PAHs-contaminated site in a full-scale biopiling system with immobilized enzymes: removal efficiency and microbial communities. Environ Res 262(Pt 1):119763. https://doi.org/10.1016/j.envres.2024.119763

Wen X, Du C, Wan J, Zeng G, Huang D, Yin L, Deng R, Tan S, Zhang J (2019) Immobilizing laccase on kaolinite and its application in treatment of malachite green effluent with the coexistence of Cd (II). Chemosphere 217:843–850. https://doi.org/10.1016/j.chemosphere.2018.11.073

Wu S, Gong G, Yan K, Sun Y, Zhang L (2020) Polycyclic aromatic hydrocarbons in edible oils and fatty foods: occurrence, formation, analysis, change and control. Adv Food Nutr Res 93:59–112. https://doi.org/10.1016/bs.afnr.2020.02.001

Xu J, Zhang Y, Zhu X, Shen C, Liu S, Xiao Y, Fang Z (2024) Direct evolution of an alkaline fungal laccase to degrade tetracyclines. Int J Biol Macromol 277(Pt 4):134534. https://doi.org/10.1016/j.ijbiomac.2024.134534

Yadav JS, Doddapaneni H, Subramanian V (2006) P450ome of the white rot fungus Phanerochaete chrysosporium: structure, evolution and regulation of expression of genomic P450 clusters. Biochem Soc Trans 34(Pt 6):1165–1169. https://doi.org/10.1042/bst0341165

Yamaguchi H, Miyazaki M (2024) Bioremediation of hazardous pollutants using enzyme-immobilized reactors. Molecules 29(9). https://doi.org/10.3390/molecules29092021

Yang J, Li W, Ng TB, Deng X, Lin J, Ye X (2017) Laccases: production, expression regulation, and applications in pharmaceutical biodegradation. Front Microbiol 8:832. https://doi.org/10.3389/fmicb.2017.00832

Yu J, Lai J, Neal BM, White BJ, Banik MT, Dai SY (2023) Genomic diversity and phenotypic variation in fungal decomposers involved in bioremediation of persistent organic pollutants. J Fungi (Basel) 9(4). https://doi.org/10.3390/jof9040418

Yu H, Luo L, Wu B, He J, Wang H, Chen R, Ji M, Yang Q, Zeng G, Wu W, Sun D (2025) Efficient catalytic degradation and detoxification of 6PPD-quinone by the multifunctional enzyme system of phanerochaete chrysosporium. J Hazard Mater 494:138634. https://doi.org/10.1016/j.jhazmat.2025.138634

Zhang K, Wei YL, Zeng EY (2013) A review of environmental and human exposure to persistent organic pollutants in the Pearl River Delta, South China. Sci Total Environ 463–464:1093–1110. https://doi.org/10.1016/j.scitotenv.2012.10.104

Zhang Y, Dong Z, Luo Y, Yang E, Xu H, Chagan I, Yan J (2021) The Manganese Peroxidase Gene Family of Trametes trogii: gene identification and expression patterns using various metal ions under different culture conditions. Microorganisms 9(12). https://doi.org/10.3390/microorganisms9122595

Mycoremediation of Recalcitrant Pollutants

4

Shoaib Khan, Arpita Roy, MadhuKriti, Soumya Pandit, and Ram Chander Goud Merugu

Abstract

Pollution is a severe issue which affects human health, environment, and sustainable development. Pollutants are the causing agents which are responsible for the environmental pollution. Conventional processes used to remove pollutants are expensive, less efficient, and tend to form new toxic byproducts. Mycoremediation uses fungi to eliminate or degrade pollutants, is a low-cost and environmentally friendly alternative. Fungi possess unique characteristics, including fast growth, extensive mycelial networks, and various enzyme productions, like laccase, lignin peroxidase, and manganese peroxidase. These enable them to degrade most hazardous pollutants, such as hydrocarbons, pesticides, herbicides, dyes, heavy metals, medicines, and other toxic chemicals. This chapter provides a summary of mechanism of action of fungi and their enzymes in the remediation of different pollutants, their utilization in water treatment, and its limitations.

S. Khan · MadhuKriti
Department of Biotechnology, Sharda School of Bioscience & Technology, Sharda University, Greater Noida, UP, India

A. Roy (✉)
Department of Biotechnology, Sharda School of Bioscience & Technology, Sharda University, Greater Noida, UP, India

Centre for Phytochemical Research, Sharda University, Greater Noida, India

S. Pandit
Centre for Phytochemical Research, Sharda University, Greater Noida, India

Department of Life Science, Sharda School of Bioscience and Technology, Sharda University, Greater Noida, UP, India

R. C. G. Merugu
Department of Biochemistry, Mahatma Gandhi University, Nalgonda, Telangana, India

A. Roy et al. (eds.), *Mycoremediation of Xenobiotics*, Emerging Paradigms in Pharmaceutical Research, https://doi.org/10.1007/978-981-95-5743-1_4

Keywords
Mycoremediation · Enzymes · Heavy metals · Pesticides · Dyes · Pharmaceuticals

4.1 Introduction

Pollution has become one of the most urgent global challenges of the twenty-first century, posing severe risks to human health, biodiversity, natural ecosystems, and economic stability. Uncontrolled generation and poor management of wastes and effluents from households, industries, and agricultural practices have accelerated environmental deterioration, pushing already fragile ecosystems toward irreversible damage. According to the World Health Organization (WHO), nearly 2.2 billion people worldwide still lack access to safe drinking water services, and approximately 144 million individuals are forced to rely on water contaminated with hazardous pollutants. Beyond its direct effects on health and economic productivity, it also interferes with impact food security, threatens biodiversity, and reduces the availability of safe drinking water, thereby destabilizing ecological and social systems. Addressing these challenges requires the development of innovative, sustainable, and cost-effective remediation strategies. Conventional physical and chemical treatment methods, although widely applied but often inadequate (Khan et al. 2019a). Therefore an alternative approaches are required to overcome these problems while providing effective and environmentally friendly solutions.

One such approach is mycoremediation where fungi is used for detoxification and degradation of pollutants. It's an eco-friendly and economically viable method which takes advantage of the biological and ecological characteristics of fungi that are effective in environmental cleanup. Fungi are known for their rapid growth, extensive mycelial networks that interact with large volumes of soil and water, and secretion of versatile extracellular ligninolytic enzymes capable of degrading a wide range of recalcitrant and structurally complex pollutants (Bhattacharya et al. 2011; Kapahi and Sachdeva 2017; Singh et al. 2015). Their high surface-area-to-volume ratio enhances pollutant adsorption and transformation, while their ability to tolerate heavy metals, acclimate to variable pH and thermal environments, and produce metal-ion binding proteins further strengthens their remediation potential. Together, these traits make fungi ideal candidates for the treatment of pollutants, such as synthetic dyes, pesticides, herbicides, pharmaceuticals, and other industrial contaminants.

This chapter disscuss the role of fungi in degrading various hazardous pollutants, persistent, and recalcitrant which includes polycyclic aromatic hydrocarbons (PAHs), insecticides, antifungal, pesticides, herbicides agents, detergents, cyanotoxins, synthetic dyes, pharmaceuticals, antibiotics, heavy metals, and phthalates. Biochemical and physiological mechanisms which enable fungi to detoxify and mineralize these contaminants has been disscused. In addition, it discusses the current challenges and limitations that hinder large-scale application of mycoremediation and outlines strategies to overcome these barriers.

4.2 Mycoremediation of Polycyclic Aromatic Hydrocarbon (PAH)

Polycyclic aromatic hydrocarbons (PAHs) are primarily produced by incomplete combustion of organic materials like coal, wood, and petroleum. PAHs are harmful substances that are released into the air, water, and soil by both natural and anthropogenic activities (Pozdnyakova 2012; Samanta et al. 2002). Compounds of PAHs are composed of several benzene rings that are fused together, with typical examples being pyrene, anthracene, chrysene, phenanthrene, 2-methyl naphthalene, and acenaphthene. Depending on their source, they are generally divided into biological, petrogenic, and pyrogenic sources. The toxicological fate of PAHs is of particular concern because they are persistent and can bioaccumulate. They have been linked to immunosuppression as well as the formation of several different types of cancer, such as skin, lung, and stomach cancers. Other health effects include jaundice, inflammation, allergy, cataracts, hemolysis, respiratory impairment, and kidney damage. Due to their chemical stability, they can remain in the environment for extended periods of time, thus causing long-term ecological stress. During the last decades, several PAH remediation processes have been explored biodegradation using microorganisms to metabolize and mineralize recalcitrant pollutants has proved to be a very promising and environmentally friendly (Bhattacharya et al. 2014). However, microbial degradation of high-molecular-weight PAHs poses still a challenge, since most bacterial strains present low metabolic versatility, with consequent slow degradation rates (Czaplicki et al. 2018).

Ligninolytic fungi have emerged as a viable, environmentally friendly, and cost-effective method for PAHs degradation. These fungi able to degrade a vast range of PAHs, from low molecular mass to high-molecular-weight PAHs with up to six fused aromatic rings (Bhattacharya et al. 2014). This biodegradation capacity is attributed to their production of various lignin-degrading enzymes such as laccase, lignin peroxidase, and manganese peroxidase. These enzymes due to their wide substrate specificity, are able to act on PAHs because of the structural similarity of these contaminants to lignin, the complex aromatic polymer naturally attacked by the fungi (Pozdnyakova 2012; Baldrian et al. 2000). Besides ligninolytic enzymes, intracellular enzyme systems, like cytochrome P450 monooxygenases, also play a crucial role, especially in the initial steps of oxidation of PAH metabolism, allowing for subsequent degradation into intermediates (Bhattacharya et al. 2013). A study has also broadened the scope to incorporate non-ligninolytic enzymes in PAH biodegradation, where *Dentipellis* sp. KUC8613 uses a multienzyme degradation pathway, starting with PAH oxidation through cytochrome P450, followed by dioxygenases' catalytic activities, dehydrogenases, FAD-monooxygenases, GSTs, and epoxide hydrolases (Park et al. 2019). The common route for ligninolytic fungi includes the oxidation of PAHs to quinone derivatives, which are channeled through ring-cleavage routes for subsequent degradation (Pozdnyakova 2012). Certain species have the ability to completely mineralize PAHs into carbon dioxide, while others have only partial degradation, giving rise to intermediates like anthraquinone, diphenic acid, and phthalic acid. Bhattacharya et al. (2013) developed a biphasic treatment protocol based on the white-rot fungus *Phanerochaete chrysosporium*. In the initial phase, nutrient supply was employed to induce the upregulation of cytochrome P450 monooxygenase genes (pc2 and pah4), which increases early-stage oxidation of PAH. During the following nutrient-starving stage, the fungus altered its metabolism to favor ligninolytic enzyme production, thus allowing the degradation of more recalcitrant PAH structures. This successive enzymatic process, P450 monooxygenases initiating initial degradation followed by ligninolytic enzymes performing secondary breakdown, led to significantly enhanced PAH removal efficiency.

One of the major problems in the mycoremediation of PAHs is their naturally low bioavailability, that is because of their hydrophobic character and slow diffusion of the molecules through the membranes of fungal cells (Czaplicki et al. 2018). This restricted accessibility usually hinders the effectiveness of biodegradation processes. To overcome this impediment, certain fungi developed cooperative mechanisms, for example, by transferring PAHs to neighboring soil bacteria more suited for degradation (Czaplicki et al. 2018; Schamfuß et al. 2013), while others use specialized surface-active proteins called hydrophobins to bind and accumulate hydrophobic molecules (Czaplicki et al. 2018). Young et al. (2015) reported the overexpression of a small hydrophobin in *Punctularia strigosozonata* during fuel oil hydrocarbon degradation. Hydrophobins were also implicated in bioremediation of heavy metals by *Trichoderma harzianum* (Puglisi et al. 2012). Surprisingly, the qid74 hydrophobin gene was not notably upregulated during degradation of PAH by *T. harzianum* (Czaplicki et al. 2018), indicating that although hydrophobins were shown to increase the bioavailability of PAH through solubilization (Czaplicki et al. 2018), their role directly in all fungi remains uncertain. Apart from hydrophobins, bioavailability may be increased by surfactants, both synthetic and biological ways, e.g., Tween-20, Tween-80, and saponin, to improve PAH solubility and thus access to enzymes (Bhattacharya et al. 2013). Some fungi have the ability to produce emulsifying agents naturally in response to PAH exposure to develop microenvironments that enhance substrate access and degradative efficiency (Nikiforova et al. 2009). Moreover, the addition of stimulants such as cellulose and chitin can enhance fungal degradation activity, but high supplementation could possibly have a reverse effect, since fungi tend to channel metabolic energy to these easily metabolizable carbon sources rather than targeting PAHs (Czaplicki et al. 2018). Ligninolytic enzymes are primary initiators of PAH degradation, typically produced during the secondary metabolism phase, their initial induction can be achieved through nutrient restriction, especially of carbon and nitrogen (Singh and Chen 2008). Adding fungal cultures with low-molecular-weight molecules, like the compounds glycerol, veratryl alcohol, ferulic acid, polyethylene glycol, humic acid, p-coumaric acid, copper sulfate, vanillin, and citric acid, has been demonstrated to promote enzyme activity and speed up PAH degradation (Pozdnyakova 2012; Bhattacharya et al. 2014; Singh and Chen 2008). Similarly, supplementation with manganese and glutathione can directly induce manganese peroxidase activity and enhance ligninolytic capability (Pozdnyakova 2012).

Environmental stressors such as high salinity, heavy metal contamination, and low temperature a limit efficiency of fungal remediation. Low temperature can inhibit enzymatic activity, decrease nutrient availability, and enhance PAH viscosity, makes the pollutants further inaccessible (Robichaud et al. 2019a, b). In a study, *Trametes versicolor* did not grow at a low-temperature PAH-contaminated location in Canada, while some *Psathyrella* species survived, highlighting their potential utility in bioremediation within polar and alpine environments (Robichaud et al. 2019b). Heavy metals interfere with fungal membranes, suppress metabolism, and disrupt ligninolytic enzyme activity; yet, organisms like *Pleurotus ostreatus* and *P. eryngii* have shown tolerance under co-contamination conditions by triggering antioxidant defense pathways (Baldrian et al. 2000; Wu et al. 2016).

Marine fungi also have potential in PAH cleanup, due to their salt tolerance, nutrient limitation, and temperature tolerances. *Cochliobolus lunatus* reported to degrade chrysene, a heavy polycyclic aromatic hydrocarbon, in salinity conditions (Puglisi et al. 2012). Additionally, plant-fungus associations possess synergistic benefits, such as between *P. ostreatus* and *Zea mays*, which increases manganese peroxidase activity, fungal biomass, and the overall pollutant removal performance (Košnář et al. 2019).

4.3 Mycoremediation of Heavy Metals

Heavy metal pollution is still one of the most serious hazards to environmental and human health because of the toxic, persistent, and nonbiodegradable nature. Arsenic (As), copper (Cu), lead (Pb), chromium (Cr), mercury (Hg), cadmium (Cd), silver (Ag), and nickel (Ni) (Morais et al. 2012) are usually present in soils, sediments, and aquatic environments. Contamination of the environment with heavy metals is a major concern due to the serious threat it poses to biodiversity, ecosystem processes, and human health. Arsenic, copper, lead, chromium, mercury, cadmium, silver, and nickel are some of the most common pollutants found in contaminated environments. These poisonous elements get into soil and water systems as a result of a mix of natural processes and anthropogenic activity. Effluents from industries like paint, textiles, metal components manufacturing, and fertilizer plants are the most significant sources (Morais et al. 2012). After heavy metals leach into the environment, they inevitably enter the body by eating or drinking contaminated food and water (Morais et al. 2012; Jaishankar et al. 2014). The majority of these metals possess mutagenic and carcinogenic properties (Khan et al. 2019b), and exposure to these heavy metals causes damage to essential organs like the kidney, liver, spleen, heart, and reproductive system (Masindi and Muedi 2018).

The production of reactive oxygen species (ROS) upon exposure to heavy metals results in oxidative damage to vital biomolecules, like nucleic acids, proteins, and lipids (Gupta et al. 2012; Tchounwou et al. 2012). In microbes, heavy metals are known to inhibit the function of critical enzymes like alkaline phosphatase, catalase, glucosidase, and urease (Khan 2000). In plants, the accumulation of heavy metals can impede critical physiological activities, like water and nutrient uptake, alongside the process of photosynthesis (Chibuike and Obiora 2014). Growth and metabolism can be inhibited by metals such as mercury, lead, and arsenic; they deform flowers and fruits, retard leaf, stem, and root growth, decrease crop yield, and impair nitrogen-fixing capacity (Chibuike and Obiora 2014). In aquatic organisms, heavy metal-induced oxidative stress leads to cell-damaging lethal injury (Gheorghe et al. 2017). Various chemical and physical processes exist for the removal of heavy metals from polluted environments, including strategies involving reverse osmosis and redox transformations and solutions merging reverse osmosis filtration and redox mechanisms. These traditional processes possess some limitations, like high energy consumption, high running costs, lower efficiency at low metal concentrations, and unwanted changes in soil chemistry (Khan et al. 2019a). Mycoremediation can overcome these restrictions. Fungi are particularly deemed effective because of their increased tolerance toward heavy metals, greater surface area-to-volume ratio, and overall strength when compared to bacteria and algae (Kapahi and Sachdeva 2017; Khan et al. 2019a; Singh et al. 2015).

Heavy metals transformation in fungi is mainly done by biosorption, bioaccumulation, and biovolatilization (Baldrian et al. 2000; Boriová et al. 2014).

Aspergillus, Rhizomucor, Fusarium, and *Emericella* genera were isolated from arsenic-contaminated agricultural soils. These fungi exhibited excellent arsenic tolerance and even promoted plant growth and yield when irrigated with arsenic-laden sterile water. These fungi also enhanced soil enzyme activity and physicochemical properties (Singh et al. 2015).

The mycorrhizal fungus *Funneliformis geosporum* enhanced soil health and wheat crop yield in zinc-contaminated soil by suppressing zinc accumulation in crops (Abu-Elsaoud et al. 2017). The edible fungus *Pleurotus ostreatus* partitioned out manganese from polluted water containing surfactants-these surface-active agents enhanced surface area and the number of metal-binding sites on fungal hyphae, thus enhancing bioaccumulation (Wu et al. 2016). *Pleurotus ostreatus* has further been demonstrated to eliminate several metals (lead, zinc, chromium, cobalt, copper, and nickel) from coal washery effluent, with high antioxidant enzyme activity being likely to help the fungus withstand metal toxicity. Other fungal species such as *Fomitopsis meliae, Absidia cylindroslora, Rhizopus microsporus,* and *Trichoderma ghanense* have shown resilience to copper, cadmium, arsenic, lead, and iron (Albert et al. 2018; Oladipo et al. 2018). *Trichoderma harzianum*, which was isolated from soil adjacent to an old mining site, was able to tolerate and accumulate silver (Cecchi et al. 2017), and indigenous fungi from gold, silver, and gemstone mines have also been found to accumulate noble metals, providing potential economic opportunities (Cecchi et al. 2017). In industrially contaminated soils of Pakistan, *Aspergillus* isolates eliminated lead and mercury (Khan et al. 2019b), and *Aspergillus* eliminated cadmium and chromium (Khan et al. 2019a). Collectively, these investigations show that native fungi are important tools for bioremediation of a variety of toxic heavy metals, with the advantage of enhancing the quality of soil and promoting plant growth in contaminated agricultural fields. However, one limitation remains: fungi tend to retain the metals they remove, meaning that the fungal biomass itself must be safely recovered after remediation to prevent recontamination. Developing efficient strategies to address this challenge is critical for the practical application of mycoremediation.

4.4 Mycoremediation of Pesticides and Herbicides

The extensive and general application of herbicides and pesticides has become the characteristic of contemporary farming. Agrochemicals are very efficient in regulating plant pathogens, inhibiting weeds, and regulating insect pests, which have played a prominent role in raising yields and world food production. Pesticides and fertilizers that are used tend to leach into groundwater and rivers. They initiate process of eutrophication, which removes dissolved oxygen from aquatic life and endangers aquatic organisms' lives. Additionally, pesticide and herbicide residues often mix in food crops and vegetables and mixed with human food chain and possessess possible health hazards.

Numerous synthetic herbicides and pesticides, such as glyphosate, endosulfan, paraquat, fipronil, and aldrin, have been extensively utilized in agriculture to manage weed growth, insect pests, and plant diseases. Increasing evidence has, however, associated these substances with catastrophic health hazards and ecosystem degradation. Studies point toward the fact that some of these compounds hold carcinogenic potential, interfere with the endocrine system, possess neurotoxic potential, and create irreversible harm to reproductive abilities. They have also been found to damage essential organs such as the liver and the kidneys in humans and animals. Because of these widely documented risks, some pesticides and herbicides have been prohibited or are under strict regulatory measures in most states (Maqbool et al. 2016; Purnomo et al. 2011). In spite of such controls, application of these agrochemicals is continued on a large scale, especially in those areas where agricultural output is highly dependent upon chemical control of pests. In addition, very persistent pesticides like DDT, aldrin, and dieldrin are still a major environmental concern due to their remarkable chemical stability. These chemicals are resistant to natural degradation processes and therefore remain active for long periods in soil, sediment, and water for decades. After release into the environment, they often leach into surface waters and groundwater or become soil matrix-embedded. Subsequently, they bioaccumulate in humans, animals, and plants via the food chain, eventually disrupting microbial communities, affecting biodiversity, and posing a threat to ecosystem services (Maqbool et al. 2016).

Conventional remediation strategies, such as chemical reduction or oxidation, high-temperature incineration, and volatilization, have been used to degrade or eliminate these poisonous substances from polluted areas. These are usually costly and involve extensive energy input which make them inapplicable to large-scale use within agricultural environments. Mycoremediation has been demonstrated to be a cost-effective and environmentally friendly option for the remediation of pesticide- and herbicide-contaminated environments. Mycoremediation has the benefit of in situ use, without the necessity of excavation of the soil and decreased generation of secondary pollutants. This strategy utilizes the inherent metabolic potential of fungi, which includes its secretion of extracellular enzymes, including laccases, peroxidases, and hydrolases, to degrade complex and recalcitrant chemical structures.

Increasing evidence has proven the potential of indigenous fungal strains, which have been derived from polluted environments, to degrade and withstand certain pesticides and herbicides. For instance, *Aspergillus tamarii* and *Botryosphaeria laricina*, derived from endosulfan-contaminated agricultural soils, not only exhibited survival in the presence of endosulfan but also decomposed it as well as its toxic metabolites (Silambarasan and Abraham 2013). *Aspergillus glaucus* can degrade fipronil and its sulfone metabolite via ligninolytic enzyme pathways (Gajendiran and Abraham 2017). In a study, *Trametes maxima* enzyme extracts, and their coculture with *Paecilomyces carneus*, fully decomposed the herbicide atrazine (Chan-Cupul et al. 2016). White-rot fungi such as *Trametes versicolor, Hypholoma dispersum,* and *T. pavonia* were found to involve ligninolytic enzymes in paraquat degradation (Camacho-Morales et al. 2017). Similarly, *P. verruculosum, Penicillium spiculisporus*, and *Aspergillus flavus* were highly capable of degrading glyphosate. In addition, white-rot fungus *Pleurotus ostreatus* was found to convert aldrin and dieldrin via epoxidation and hydroxylation reactions, thus lowering their toxicity and persistence in the environment.

4.5 Mycoremediation of Cyanotoxins

Eutrophication, supported mainly by a range of anthropogenic activities, has become one of the major causes of deleterious cyanobacterial blooms in water bodies (Zanchett and Oliveira-Filho 2013). The cause of this nutrient enrichment is usually farming runoff laden with nitrogen and phosphorus, the release of untreated or partially treated sewage, and industrial effluent. As a result of these nutrient-surplus conditions, cyanobacteria grow extremely fast and form massive blooms that are toxic to the ecosystem. In addition to their ability to shade out underwater vegetation and rob the water of dissolved oxygen, most of these blooms have the ability to generate a wide range of toxic secondary compounds, such as cylindrospermopsin, microcystin-LR, nodularin, and β-N-methylamino-L-alanine (BMAA) (Balsano et al. 2017). They are all known as cyanotoxins, and they pose grave threats to the health of people, animals, and the integrity of aquatic communities in general. In aquatic food webs, cyanotoxins have been found to exert a variety of adverse effects. For example, they can induce acute mortality in small crustaceans like *Daphnia* spp., interfere with photosynthetic activities in other species of cyanobacteria, and disturb reproductive activities in mollusks by inhibiting egg production (Zanchett and Oliveira-Filho 2013).

There are some aquatic fungal species which showed significant cyanotoxin mitigation potential. *Mucor hiemalis* not only resistant to the oxidative stress induced by toxins, like microcystin and BMAA, but also uptake and degrade these toxins actively (Balsano et al. 2017; Jia et al. 2012). This oxidative stress tolerance is mediated through the induction of antioxidant defence mechanism such as production of superoxide dismutase, catalase, and glutathione peroxidase. Apart from the direct degradation of cyanotoxins, some fungi also possess strong algicidal activities that can suppress directly toxic cyanobacterial blooms. *Trametes versicolor* and *Trichoderma citrinoviride* were reported to inhibit algal cells by secreting extracellular enzymes (Dai et al. 2018; Mohamed et al. 2014). Transcriptomic and proteomic research on these organisms has identified a wide range of decomposition enzymes, such as laccases, hydrolases, proteases, cellulases, and manganese peroxidases, that allow them to break down cell walls, proteins, and other key structural and metabolic elements of algae (Dai et al. 2018; Du et al. 2015; Gao et al. 2017).

4.6 Mycoremediation of Pharmaceutical Wastes

Pharmaceutical wastes are discharged into the environment through various effluents coming from the pharmaceutical industries. These effluents usually comprise a mixture of a wide range of active pharmaceutical ingredients (APIs) and their metabolites, ranging from antibiotics to analgesics, anti-inflammatory substances, and hormones. Release of these effluents without proper treatment can lead to surface waters, groundwater, and soils contamination for a long time and possesses serious public and animal health risks (Larsson 2014; Sathya et al. 2022). Certain chemicals, especially endocrine-disrupting chemicals, like synthetic hormones, can disrupt reproductive cycles, induce developmental malformations, and change population dynamics in aquatic organisms. The effects often propagate across the food web with possible destabilization of entire ecosystems.

Adding to the problem is the ineffectiveness of traditional water treatment technologies in removing these recalcitrant xenobiotic chemicals (Esterhuizen-Londt et al. 2016). Traditional treatment processes, including coagulation-flocculation, sedimentation, filtration, and chlorination, are mainly capable of removing particulate matter and pathogens, not intricate, chemically stable pharmaceuticals. Thus, much of these pollutants evade treatment plants and persist in natural water bodies, sediments, and agricultural soils irrigated with reclaimed water. One of the greatest related issues is the contribution of environmental pharmaceutical residues, especially antibiotics, to fueling the transmission of antimicrobial resistance (AMR). AMR has been recognized as one of the greatest twenty-first-century global public health threats. It is estimated that resistant microbial strains cause half a million deaths globally in a year. In the USA alone, the financial cost of treating drug-resistant microbe infections is mind-boggling, with antimicrobial therapy expenditures of around USD 35 billion per year (Kraemer et al. 2019). The issue goes beyond the clinic: environmental exposure to antibiotics creates selective pressure on microbial populations, allowing resistant strains to spread within natural communities. These selective pressures have environmental implications beyond human disease. Antibiotics and other drugs can disrupt the functioning of ecologically important microorganisms upon which vital biogeochemical processes, such as carbon cycling, soil respiration, iron reduction, denitrification, and nitrification (Cycoń et al. 2019) ed. These disturbances to the microbial-mediated processes can decrease soil fertility, change nutrient dynamics, and reduce ecosystem resilience to environmental stressors.

Antibiotics are one of the most commonly used drug classes globally; however, large quantities are discharged into the environment by various sources, effluents from pharmaceutical manufacturers, accidental dumping of expired drugs, human excretion, hospital waste, and unauthorized use in livestock and aquaculture (Gothwal and Shashidhar 2015; Kraemer et al. 2019). Their discharge causes severe economic, environmental, and human health impacts (Gothwal and Shashidhar 2015; Kraemer et al. 2019). In aquatic systems, antibiotic buildup leads to chronic toxicity toward fish, amphibians, and other aquatic organisms (Liu et al. 2018). Antibiotics in plants affect electron transport, block mitochondrial and chloroplast protein synthesis, damage plastid replication, lower chlorophyll and carotenoid levels, induce chromosomal abnormalities, and eventually inhibit growth and photosynthesis (Gothwal and Shashidhar 2015). A number of physical and chemical methods have been used for the treatment of antibiotic-polluted water, Methods such as ozonation, potassium permanganate oxidation, Fenton reactions, sonolysis, titanium dioxide photocatalysis, membrane bioreactors, and activated carbon adsorption. Though useful in certain scenarios, they are generally expensive, energy-requiring, and produce toxic by-products (Gothwal and Shashidhar 2015). Additionally, some antibiotics' structural stability causes them to be resistant to these traditional methods (Copete-Pertuz et al. 2018).

Various species of fungi have been found to degrade antibiotics, like bifonazole, clotrimazole, sulfonamides, oxacillin, oxytetracycline, and fluoroquinolones (Copete-Pertuz et al. 2018; Gothwal and Shashidhar 2015; Kryczyk-Poprawa et al. 2019). For example, the ligninolytic fungus *Pleurotus ostreatus* adsorbed oxytetracycline and removed it entirely from liquid culture within 2 weeks, though this was independent of the enzyme laccase. Conversely, a Colombian isolate of *Leptosphaerulina* sp. degraded isoxazolyl-penicillins, like oxacillin and dicloxacillin, using versatile peroxidase and laccase to eliminate both the compounds and the antibacterial activity without forming toxic byproducts (Copete-Pertuz et al. 2018). *Trametes versicolor* biodegraded fluoroquinolone antibiotics, ciprofloxacin and ofloxacin, through ligninolytic enzymes and the cytochrome P450 system (Prieto et al. 2011) and reduced sulfamethazine to zero levels (García-Galán et al. 2011). The ligninolytic fungus *Irpex lacteus* also showed quick and efficient degradation of fluoroquinolones, like flumequine, ciprofloxacin, and ofloxacin, with manganese peroxidase being an important enzyme (Čvančarová et al. 2015, 2013). Also, *I. lacteus* removed the antibacterial effect of norfloxacin and ofloxacin (Čvančarová et al. 2015). The edible mushroom *Lentinula edodes* was found to uptake and excrete antifungal drugs, such as bifonazole and clotrimazole, using its mycelia (Kryczyk-Poprawa et al. 2019).

4.7 Mycoremediation of Phthalates, Dyes, Surfactants, and XOCs

4.7.1 Phthalates

Phthalates are man-made chemicals widely employed in the manufacture of products, like vinyl tiles, capacitors, films, toys, and medical equipment, where they exist mainly as plasticizers to enhance flexibility (Benjamin et al. 2015). Their unrestricted release into the environment, however, allows their entry into the food chain with serious health implications. Phthalate exposure has been correlated with lower testosterone levels, thyroid dysfunction, lowered lung function, male infertility, and carcinogenicity (Moore et al. 2001; Nagao et al. 2000). Traditional degradation methods, such as hydrolysis, UV irradiation, and titanium dioxide-assisted photocatalysis, can break down phthalates but tend to be slow, inefficient, and of limited efficacy (Benjamin et al. 2015). In contrast, mycoremediation presents a viable and environmentally friendly alternative for their rapid degradation. Cutinase and esterase enzymes from *Fusarium oxysporum* have been investigated for their capacity to hydrolyze phthalates such as di-hexyl phthalate, dipropyl phthalate, di-2-ethylhexyl phthalate (DEHP), butyl benzyl phthalate, and dipentyl phthalate. Cutinase worked rapidly and completely without the formation of toxic byproducts, while enzyme-mediated reactions involving esterase were slower and formed intermediates that caused oxidative stress and inhibited enzyme function (Ahn et al. 2006; Kim et al. 2007, 2005, 2003, 2002). In addition, complete degradation of DEHP leached from blood storage bags made from polyvinyl chloride was shown by fungal isolates, like *Purpureocillium lilacinum*, *Fusarium subglutinans, Aspergillus japonicus, Aspergillus parasiticus, Penicillium brocae*, and *Penicillium funiculosum*, all derived from plastic-polluted soils (Pradeep et al. 2013). These results highlight the high potential of mycoremediation, using whole fungal biomass as well as purified enzymes, for phthalate-contaminated area treatment on a large scale. Although these advances have been made, the molecular processes and gene regulation involved in phthalate degradation are still not well understood. Research should thus focus on clarifying entire biochemical pathways, maximizing fungal biomass-based bioreactor systems, and optimizing operating conditions to realize efficient, fast, and scalable phthalate degradation.

4.7.2 Surfactants

The release of detergents into the environment poses a major ecological hazard. Most detergents are highly biodegradable and cause little direct health hazard to humans; however, they can cause huge damage to aquatic life, compromising the health and richness of aquatic plants and animals. Regular wastewater treatment processes can be used to deal with detergent pollution; however, mycoremediation is a cheaper option that is more environmentally friendly. Some fungal species, such as *Penicillium verrucosum, Cladosporium cladosporioides,* and *Geotrichum candidum,* have shown the capacity to degrade some of the commercially used detergents (Jakovljević and Vrvić 2018, 2017). While promising, the mechanisms of detergent degradation by fungi are yet to be fully understood. To realize the full potential of mycoremediation for the removal of detergents, more research will be needed to understand the molecular mechanisms of degradation. Detailed transcriptomic, proteomic, and biochemical studies are needed to define the major enzymes, metabolic pathways, and regulatory networks involved. These findings would open the door to the rational optimization of fungal strains and treatment conditions, allowing the production of effective, scalable, and environmentally friendly methods for deterring detergent pollution.

4.7.3 Dye

Dye pollution is a serious environmental issue, with millions of liters of dye wastewater being released annually from manufacturing, textile, and printing facilities. Because of their chemical stability, dyes are highly recalcitrant and stay in the environment for a long time. For most industries, direct discharge is more economical compared to the installation of proper treatment systems (Bhattacharya et al. 2011). Most dye-containing effluents are toxic, mutagenic, and carcinogenic, posing serious risks to ecosystems and human health (Ngieng et al. 2013). Fungi that secrete lignin-degrading enzymes like laccase, manganese peroxidase, and lignin peroxidase have shown immense capability for dye degradation because of the wide substrate specificity of these enzymes (Bhattacharya et al. 2011). *Aspergillus flavus,* which was isolated from soil surrounding a paper processing plant, was capable of accumulating and biodegrading Congo red dye (Bhattacharya et al. 2011). Likewise, Melastoma malabathricum-derived endophytic fungus *Marasmius cladophyllus* broke down several classes of synthetic dyes, although the particular enzymic pathways involved were not explained (Ngieng et al. 2013). *Phlebia acerina* was found to eliminate dye-polluted wastewater efficiently, decreasing its toxicity to a remarkable extent (Kumar et al. 2018), whereas *Trametes versicolor* broke down various azo and anthraquinone dyes (Yang et al. 2017). In bioreactor experiments, *Bjerkandera adusta* demonstrated fast decolorization of some acid and reactive dyes (Sodaneath et al. 2017). Enzyme profiling showed increased laccase and manganese peroxidase activity in *T. versicolor* and increased lignin peroxidase activity in *B. adusta*, supporting that ligninolytic enzyme production was increased to enable efficient breakdown of the dyes (Sodaneath et al. 2017; Yang et al. 2017).

4.7.4 Xenobiotic Organic Compounds (XOC)

Mycoremediation of XOCs is carried out by biosorption, biodegradation, and enzymatic mineralization, which can occur individually or in combination (Wesenberg 2003). Laccase, MnP, and LiP primarily involve enzymatic degradation, as in

P. chrysosporium, which deploys peroxidases or laccase to break down XOCs (Evans et al. 1994). Laccase, although initially evolved for the degradation of lignin, is low in substrate specificity and can oxidize a wide variety of pollutants, including aromatic/aliphatic amines, aminophenols, polyphenols, chlorophenolics, and pesticides (Pozdnyakova et al. 2004). Mechanism includes cleavage of Cα oxidation, Cα–Cβ, and aryl–alkyl bonds of lignin, producing water by removal of proton and electron (Thurston 1994). Laccase in methoxy-substituted monophenols and aromatic amines forms quinone radicals that also become further oxidized, polymerized, or precipitated (Archibald et al. 1997; Bourbonnais et al. 1995; Niku-Paavola and Viikari 2000). It also oxidizes non-phenolic aromatics, taking O_2 to form dihydroxy and ring-opened products, which are used by fungi as carbon and energy sources (Anastasi et al. 2009).

Mycoremediation of xenobiotic organic compounds (XOCs) is done by myco-degradation and myco-sorption processes (Brugnari et al. 2018; Lonappan et al. 2018). Myco-degradation is based on the action of fungal enzymes or pellets degrading pollutants, whereas myco-sorption is based on fungal biomass adsorbing dyes through cell wall functional groups. For differentiation, the fungal pellets are pretreated with DMSO, followed by enzyme activity and spectrophotometric measurements (Chenthamarakshan et al. 2017). Degradation efficiency determined in terms of COD reduction or GC–MS may be as high as 97.8%. Some fungi exhibit high degradation efficiencies: *T. versicolor* removed diclofenac, indomethacin, and naproxen through laccase in sodium acetate buffer (Tran et al. 2010); *P. sanguineus* broke down 92–94% of nonylphenol and triclosan in 4–6 h (Ramírez-Cavazos et al. 2014); *Trametes* sp. removed 93–99% of PAHs through MnP and laccase (Lu et al. 2017). Fungal pellets occasionally perform better than purified enzymes, as demonstrated by *A. fumigatus* in azo dye degradation (58–100% vs. 64–96%) (Chenthamarakshan et al. 2017). Industrial wastewater research also supports fungal effectiveness: *P. chrysosporium* broke down indigo dye 95%, *P. incarnata* eliminated 90% of PAHs (Lee et al. 2014), and *Pleurotus ostreatus* detoxified phenols in olive by-product without external nutrient supplementation (Martirani et al. 1996). It is essential to have optimal conditions, particularly pH 4.5–5.5, for enzyme-catalyzed breakdown (Wang et al. 2018). A study reported that *P. sajor-caju* degraded 84% of RB-5 dye in 36 h (Murugesan et al. 2007), while *C. militaris* degrade 73% in 48 h (Bhamare et al. 2018). For mycosorption, *Trichoderma, Alternaria, Aspergillus, Rhizopus*, and *Penicillium* are commonly used (Bowman and Free 2006). In another study A. alternata was able to remove 99.99% Congo red (Chakraborty et al. 2013), whereas *A. niger* was able to remove 30% of Procion red (Almeida and Corso 2014). pH also significantly affects biosorption: 98.5% Congo red was degraded at pH 2–5 wherease only 60% at pH 9 in wastewater (Lu et al. 2017; Mahmoud et al. 2017). In a study *P. janthinellum* (344.83 mg g^{-1}) showed better results than *A. niger* (263.2 mg g^{-1}), although *A. niger* showed overall better removal efficiency (Lu et al. 2017; Wang et al. 2015).

4.8 Conclusion

Mycoremediation is a promising, environmentally friendly, and sustainable method. It facilitates degradation and detoxification of various pollutants, such as polycyclic aromatic hydrocarbons (PAHs), herbicides, pesticides, dyes, pharmaceuticals, surfactants, heavy metals, and cyanotoxins. Fungi show extreme resistance to diverse environmental conditions. Their multi-specific mechanisms, such as biosorption, bioaccumulation, and enzymatic mineralization help them in water and soil remediation. Latest developments in mycoremediation includes bioreactor-based technology, enzyme optimization, and agricultural residues as low-cost substrates to improved the viability of scaling up processes. Despite these advantages, various challenges such as limited bioavailability of hydrophobic contaminants, environmental stresses, and risk of secondary contamination limit the widespread application of mycoremediation. These challenges must be overcome by combining molecular biology, omics tools, genetic engineering, and new bioprocess designs to enhance efficiency and reliability.

References

Abu-Elsaoud AM, Nafady NA, Abdel-Azeem AM (2017) Arbuscular mycorrhizal strategy for zinc mycoremediation and diminished translocation to shoots and grains in wheat. PLoS One 12:e0188220. https://doi.org/10.1371/journal.pone.0188220

Ahn J-Y, Kim Y-H, Min J, Lee J (2006) Accelerated degradation of dipentyl phthalate by Fusarium oxysporum f. sp. pisi Cutinase and toxicity evaluation of its degradation products using bioluminescent bacteria. Curr Microbiol 52:340–344. https://doi.org/10.1007/s00284-005-0124-9

Albert Q, Leleyter L, Lemoine M, Heutte N, Rioult J-P, Sage L, Baraud F, Garon D (2018) Comparison of tolerance and biosorption of three trace metals (Cd, Cu, Pb) by the soil fungus Absidia cylindrospora. Chemosphere 196:386–392. https://doi.org/10.1016/j.chemosphere.2017.12.156

Almeida EJR, Corso CR (2014) Comparative study of toxicity of azo dye Procion Red MX-5B following biosorption and biodegradation treatments with the fungi Aspergillus Niger and Aspergillus terreus. Chemosphere 112:317–322. https://doi.org/10.1016/j.chemosphere.2014.04.060

Anastasi A, Prigione V, Casieri L, Varese GC (2009) Decolourisation of model and industrial dyes by mitosporic fungi in different culture conditions. World J Microbiol Biotechnol 25:1363–1374. https://doi.org/10.1007/s11274-009-0023-5

Archibald FS, Bourbonnais R, Jurasek L, Paice MG, Reid ID (1997) Kraft pulp bleaching and delignification by Trametes versicolor. J Biotechnol 53:215–236. https://doi.org/10.1016/S0168-1656(97)01675-1

Baldrian P, In Der Wiesche C, Gabriel J, Nerud F, Zadražil F (2000) Influence of cadmium and mercury on activities of Ligninolytic enzymes and degradation of polycyclic aromatic hydrocarbons by *Pleurotus ostreatus* in soil. Appl Environ Microbiol 66:2471–2478. https://doi.org/10.1128/AEM.66.6.2471-2478.2000

Balsano E, Esterhuizen-Londt M, Hoque E, Lima SP (2017) Responses of the antioxidative and biotransformation enzymes in the aquatic fungus Mucor hiemalis exposed to cyanotoxins. Biotechnol Lett 39:1201–1209. https://doi.org/10.1007/s10529-017-2348-8

Benjamin S, Pradeep S, Sarath Josh M, Kumar S, Masai E (2015) A monograph on the remediation of hazardous phthalates. J Hazard Mater 298:58–72. https://doi.org/10.1016/j.jhazmat.2015.05.004

Bhamare HM, Jadhav HP, Sayyed RZ (2018) Statistical optimization for enhanced production of extracellular laccase from Aspergillus sp. HB_RZ4 isolated from bark scrapping. Environ Sustain 1:159–166. https://doi.org/10.1007/s42398-018-0015-1

Bhattacharya SS, Garlapati VK, Banerjee R (2011) Optimization of laccase production using response surface methodology coupled with differential evolution. New Biotechnol 28:31–39. https://doi.org/10.1016/j.nbt.2010.06.001

Bhattacharya SS, Syed K, Shann J, Yadav JS (2013) A novel P450-initiated biphasic process for sustainable biodegradation of benzo[a]pyrene in soil under nutrient-sufficient conditions by the white rot fungus Phanerochaete chrysosporium. J Hazard Mater 261:675–683. https://doi.org/10.1016/j.jhazmat.2013.07.055

Bhattacharya S, Das A, Prashanthi K, Palaniswamy M, Angayarkanni J (2014) Mycoremediation of Benzo[a]pyrene by Pleurotus ostreatus in the presence of heavy metals and mediators. 3 Biotech 4:205–211. https://doi.org/10.1007/s13205-013-0148-y

Boriová K, Čerňanský S, Matúš P, Bujdoš M, Šimonovičová A (2014) Bioaccumulation and biovolatilization of various elements using filamentous fungus *Scopulariopsis brevicaulis*. Lett Appl Microbiol 59:217–223. https://doi.org/10.1111/lam.12266

Bourbonnais R, Paice MG, Reid ID, Lanthier P, Yaguchi M (1995) Lignin oxidation by laccase isozymes from Trametes versicolor and role of the mediator 2,2′-azinobis(3-ethylbenzthiazoline-6-sulfonate) in kraft lignin depolymerization. Appl Environ Microbiol 61:1876–1880. https://doi.org/10.1128/aem.61.5.1876-1880.1995

Bowman SM, Free SJ (2006) The structure and synthesis of the fungal cell wall. BioEssays 28:799–808. https://doi.org/10.1002/bies.20441

Brugnari T, Pereira MG, Bubna GA, De Freitas EN, Contato AG, Corrêa RCG, Castoldi R, De Souza CGM, Polizeli MDLTDM, Bracht A, Peralta RM (2018) A highly reusable MANAE-agarose-immobilized Pleurotus ostreatus laccase for degradation of bisphenol A. Sci Total Environ 634:1346–1351. https://doi.org/10.1016/j.scitotenv.2018.04.051

Camacho-Morales RL, Guillén-Navarro K, Sánchez JE (2017) Degradation of the herbicide paraquat by macromycetes isolated from southeastern Mexico. 3 Biotech 7:324. https://doi.org/10.1007/s13205-017-0967-3

Cecchi G, Marescotti P, Di Piazza S, Zotti M (2017) Native fungi as metal remediators: silver myco-accumulation from metal contaminated waste-rock dumps (Libiola Mine, Italy). J Environ Sci Health Part B 52:191–195. https://doi.org/10.1080/03601234.2017.1261549

Chakraborty S, Basak B, Dutta S, Bhunia B, Dey A (2013) Decolorization and biodegradation of Congo red dye by a novel white rot fungus Alternaria alternata CMERI F6. Bioresour Technol 147:662–666. https://doi.org/10.1016/j.biortech.2013.08.117

Chan-Cupul W, Heredia-Abarca G, Rodríguez-Vázquez R (2016) Atrazine degradation by fungal co-culture enzyme extracts under different soil conditions. J Environ Sci Health Part B 51:298–308. https://doi.org/10.1080/03601234.2015.1128742

Chenthamarakshan A, Parambayil N, Miziriya N, Soumya PS, Lakshmi MSK, Ramgopal A, Dileep A, Nambisan P (2017) Optimization of laccase production from Marasmiellus palmivorus LA1 by Taguchi method of design of experiments. BMC Biotechnol 17:12. https://doi.org/10.1186/s12896-017-0333-x

Chibuike GU, Obiora SC (2014) Heavy metal polluted soils: effect on plants and bioremediation methods. Appl Environ Soil Sci 2014:1–12. https://doi.org/10.1155/2014/752708

Copete-Pertuz LS, Plácido J, Serna-Galvis EA, Torres-Palma RA, Mora A (2018) Elimination of Isoxazolyl-Penicillins antibiotics in waters by the ligninolytic native Colombian strain Leptosphaerulina sp. considerations on biodegradation process and antimicrobial activity removal. Sci Total Environ 630:1195–1204. https://doi.org/10.1016/j.scitotenv.2018.02.244

Čvančarová M, Moeder M, Filipová A, Reemtsma T, Cajthaml T (2013) Biotransformation of the antibiotic agent Flumequine by Ligninolytic fungi and residual antibacterial activity of the transformation mixtures. Environ Sci Technol 47:14128–14136. https://doi.org/10.1021/es403470s

Čvančarová M, Moeder M, Filipová A, Cajthaml T (2015) Biotransformation of fluoroquinolone antibiotics by ligninolytic fungi—metabolites, enzymes and residual antibacterial activity. Chemosphere 136:311–320. https://doi.org/10.1016/j.chemosphere.2014.12.012

Cycoń M, Mrozik A, Piotrowska-Seget Z (2019) Antibiotics in the soil environment—degradation and their impact on microbial activity and diversity. Front Microbiol 10:338. https://doi.org/10.3389/fmicb.2019.00338

Czaplicki LM, Dharia M, Cooper EM, Ferguson PL, Gunsch CK (2018) Evaluating the mycostimulation potential of select carbon amendments for the degradation of a model PAH by an ascomycete strain enriched from a superfund site. Biodegradation 29:463–471. https://doi.org/10.1007/s10532-018-9843-z

Dai W, Chen X, Wang X, Xu Z, Gao X, Jiang C, Deng R, Han G (2018) The algicidal fungus Trametes versicolor F21a eliminating blue algae via genes encoding degradation enzymes and metabolic pathways revealed by transcriptomic analysis. Front Microbiol 9:826. https://doi.org/10.3389/fmicb.2018.00826

Du J, Pu G, Shao C, Cheng S, Cai J, Zhou L, Jia Y, Tian X (2015) Potential of extracellular enzymes from Trametes versicolor F21a in Microcystis spp. degradation. Mater Sci Eng C 48:138–144. https://doi.org/10.1016/j.msec.2014.11.004

Esterhuizen-Londt M, Schwartz K, Pflugmacher S (2016) Using aquatic fungi for pharmaceutical bioremediation: uptake of acetaminophen by Mucor hiemalis does not result in an enzymatic oxidative stress response. Fungal Biol 120:1249–1257. https://doi.org/10.1016/j.funbio.2016.07.009

Evans CS, Dutton MV, GuillÃ©n F, Veness RG (1994) Enzymes and small molecular mass agents involved with lignocellulose degradation. FEMS Microbiol Rev 13:235–239. https://doi.org/10.1111/j.1574-6976.1994.tb00044.x

Gajendiran A, Abraham J (2017) Biomineralisation of fipronil and its major metabolite, fipronil sulfone, by Aspergillus glaucus strain AJAG1 with enzymes studies and bioformulation. 3 Biotech 7:212. https://doi.org/10.1007/s13205-017-0820-8

Gao X, Wang C, Dai W, Ren S, Tao F, He X, Han G, Wang W (2017) Proteomic analysis reveals large amounts of decomposition enzymes and major metabolic pathways involved in algicidal process of Trametes versicolor F21a. Sci Rep 7:3907. https://doi.org/10.1038/s41598-017-04251-1

García-Galán MJ, Rodríguez-Rodríguez CE, Vicent T, Caminal G, Díaz-Cruz MS, Barceló D (2011) Biodegradation of sulfamethazine by Trametes versicolor: removal from sewage sludge and identification of intermediate products by UPLC–QqTOF-MS. Sci Total Environ 409:5505–5512. https://doi.org/10.1016/j.scitotenv.2011.08.022

Gheorghe S, Stoica C, Vasile GG, Nita-Lazar M, Stanescu E, Lucaciu IE (2017) Metals toxic effects in aquatic ecosystems: modulators of water quality. In: Tutu H (ed) Water quality. InTech. https://doi.org/10.5772/65744

Gothwal R, Shashidhar T (2015) Antibiotic pollution in the environment: a review. CLEAN Soil Air Water 43:479–489. https://doi.org/10.1002/clen.201300989

Gupta SC, Hevia D, Patchva S, Park B, Koh W, Aggarwal BB (2012) Upsides and downsides of reactive oxygen species for cancer: the roles of reactive oxygen species in tumorigenesis, prevention, and therapy. Antioxid Redox Signal 16:1295–1322. https://doi.org/10.1089/ars.2011.4414

Jaishankar M, Tseten T, Anbalagan N, Mathew BB, Beeregowda KN (2014) Toxicity, mechanism and health effects of some heavy metals. Interdiscip Toxicol 7:60–72. https://doi.org/10.2478/intox-2014-0009

Jakovljević VD, Vrvić MM (2017) Penicillium verrucosum as promising candidate for bioremediation of environment contaminated with synthetic detergent at high concentration. Appl Biochem Microbiol 53:368–373. https://doi.org/10.1134/S0003683817030164

Jakovljević VD, Vrvić MM (2018) Potential of pure and mixed cultures of Cladosporium cladosporioides and Geotrichum candidum for application in bioremediation and detergent industry. Saudi J Biol Sci 25:529–536. https://doi.org/10.1016/j.sjbs.2016.01.020

Jia Y, Du J, Song F, Zhao G, Tian X (2012) A fungus capable of degrading microcystin-LR in the algal culture of microcystis aeruginosa PCC7806. Appl Biochem Biotechnol 166:987–996. https://doi.org/10.1007/s12010-011-9486-6

Kapahi M, Sachdeva S (2017) Mycoremediation potential of Pleurotus species for heavy metals: a review. Bioresour Bioprocess 4:32. https://doi.org/10.1186/s40643-017-0162-8

Khan M (2000) Effect of metals contamination on soil microbial diversity, enzymatic activity, organic matter decomposition and nitrogen mineralisation (a review). Pak J Biol Sci 3:1950–1956. https://doi.org/10.3923/pjbs.2000.1950.1956

Khan I, Aftab M, Shakir S, Ali M, Qayyum S, Rehman MU, Haleem KS, Touseef I (2019a) Mycoremediation of heavy metal (Cd and Cr)–polluted soil through indigenous metallotolerant fungal isolates. Environ Monit Assess 191:585. https://doi.org/10.1007/s10661-019-7769-5

Khan I, Ali M, Aftab M, Shakir S, Qayyum S, Haleem KS, Tauseef I (2019b) Mycoremediation: a treatment for heavy metal-polluted soil using indigenous metallotolerant fungi. Environ Monit Assess 191:622. https://doi.org/10.1007/s10661-019-7781-9

Kim Y-H, Lee J, Ahn J-Y, Gu MB, Moon S-H (2002) Enhanced degradation of an endocrine-disrupting chemical, butyl benzyl phthalate, by *Fusarium oxysporum* f. sp. *pisi* Cutinase. Appl Environ Microbiol 68:4684–4688. https://doi.org/10.1128/AEM.68.9.4684-4688.2002

Kim Y-H, Lee J, Moon S-H (2003) Degradation of an endocrine disrupting chemical, DEHP [di-(2-ethylhexyl)-phthalate], by Fusarium oxysporum f. sp. pisi cutinase. Appl Microbiol Biotechnol 63:75–80. https://doi.org/10.1007/s00253-003-1332-5

Kim Y-H, Min J, Bae K-D, Gu MB, Lee J (2005) Biodegradation of dipropyl phthalate and toxicity of its degradation products: a comparison of Fusarium oxysporum f. sp. pisi cutinase and Candida cylindracea esterase. Arch Microbiol 184:25–31. https://doi.org/10.1007/s00203-005-0026-z

Kim Y-H, Seo H-S, Min J, Kim Y-C, Ban Y-H, Han KY, Park J-S, Bae K-D, Gu MB, Lee J (2007) Enhanced degradation and toxicity reduction of dihexyl phthalate by Fusarium oxysporum f. sp. pisi cutinase. J Appl Microbiol 102:221–228. https://doi.org/10.1111/j.1365-2672.2006.03095.x

Košnář Z, Částková T, Wiesnerová L, Praus L, Jablonský I, Koudela M, Tlustoš P (2019) Comparing the removal of polycyclic aromatic hydrocarbons in soil after different bioremediation approaches in relationto the extracellular enzyme activities. J Environ Sci 76:249–258. https://doi.org/10.1016/j.jes.2018.05.007

Kraemer SA, Ramachandran A, Perron GG (2019) Antibiotic pollution in the environment: from microbial ecology to public policy. Microorganisms 7:180. https://doi.org/10.3390/microorganisms7060180

Kryczyk-Poprawa A, Żmudzki P, Maślanka A, Piotrowska J, Opoka W, Muszyńska B (2019) Mycoremediation of azole antifungal agents using in vitro cultures of Lentinula edodes. 3 Biotech 9:207. https://doi.org/10.1007/s13205-019-1733-5

Kumar R, Negi S, Sharma P, Prasher IB, Chaudhary S, Dhau JS, Umar A (2018) Wastewater cleanup using Phlebia acerina fungi: an insight into mycoremediation. J Environ Manag 228:130–139. https://doi.org/10.1016/j.jenvman.2018.07.091

Larsson DGJ (2014) Pollution from drug manufacturing: review and perspectives. Philos Trans R Soc B Biol Sci 369:20130571. https://doi.org/10.1098/rstb.2013.0571

Lee H, Jang Y, Choi Y-S, Kim M-J, Lee J, Lee H, Hong J-H, Lee YM, Kim G-H, Kim J-J (2014) Biotechnological procedures to select white rot fungi for the degradation of PAHs. J Microbiol Methods 97:56–62. https://doi.org/10.1016/j.mimet.2013.12.007

Liu L, Wu W, Zhang J, Lv P, Xu L, Yan Y (2018) Progress of research on the toxicology of antibiotic pollution in aquatic organisms. Acta Ecol Sin 38:36–41. https://doi.org/10.1016/j.chnaes.2018.01.006

Lonappan L, Liu Y, Rouissi T, Brar SK, Verma M, Surampalli RY (2018) Adsorptive immobilization of agro-industrially produced crude laccase on various micro-biochars and degradation of diclofenac. Sci Total Environ 640–641:1251–1258. https://doi.org/10.1016/j.scitotenv.2018.06.005

Lu T, Zhang Q, Yao S (2017) Efficient decolorization of dye-containing wastewater using mycelial pellets formed of marine-derived Aspergillus Niger. Chin J Chem Eng 25:330–337. https://doi.org/10.1016/j.cjche.2016.08.010

Mahmoud MS, Mostafa MK, Mohamed SA, Sobhy NA, Nasr M (2017) Bioremediation of red azo dye from aqueous solutions by A spergillus niger strain isolated from textile wastewater. J Environ Chem Eng 5:547–554. https://doi.org/10.1016/j.jece.2016.12.030

Maqbool Z, Hussain S, Imran M, Mahmood F, Shahzad T, Ahmed Z, Azeem F, Muzammil S (2016) Perspectives of using fungi as bioresource for bioremediation of pesticides in the environment: a critical review. Environ Sci Pollut Res 23:16904–16925. https://doi.org/10.1007/s11356-016-7003-8

Martirani L, Giardina P, Marzullo L, Sannia G (1996) Reduction of phenol content and toxicity in olive oil mill waste waters with the ligninolytic fungus Pleurotus ostreatus. Water Res 30:1914–1918. https://doi.org/10.1016/0043-1354(95)00330-4

Masindi V, Muedi KL (2018) Environmental contamination by heavy metals. In: Saleh HE-DM, Aglan RF (eds) Heavy metals. InTech. https://doi.org/10.5772/intechopen.76082

Mohamed ZA, Hashem M, Alamri SA (2014) Growth inhibition of the cyanobacterium Microcystis aeruginosa and degradation of its microcystin toxins by the fungus Trichoderma citrinoviride. Toxicon 86:51–58. https://doi.org/10.1016/j.toxicon.2014.05.008

Moore RW, Rudy TA, Lin TM, Ko K, Peterson RE (2001) Abnormalities of sexual development in male rats with in utero and lactational exposure to the antiandrogenic plasticizer Di(2-ethylhexyl) phthalate. Environ Health Perspect 109:229–237. https://doi.org/10.1289/ehp.01109229

Morais S, Costa e FG, Lourdes Pereir MD (2012) Heavy metals and human health. In: Oosthuizen J (ed) Environmental health - emerging issues and practice. InTech. https://doi.org/10.5772/29869

Murugesan K, Dhamija A, Nam I-H, Kim Y-M, Chang Y-S (2007) Decolourization of reactive black 5 by laccase: optimization by response surface methodology. Dyes Pigments 75:176–184. https://doi.org/10.1016/j.dyepig.2006.04.020

Nagao T, Ohta R, Marumo H, Shindo T, Yoshimura S, Ono H (2000) Effect of butyl benzyl phthalate in Sprague-Dawley rats after gavage administration: a two-generation reproductive study. Reprod Toxicol 14:513–532. https://doi.org/10.1016/S0890-6238(00)00105-2

Ngieng NS, Zulkharnain A, Roslan HA, Husaini A (2013) Decolourisation of synthetic dyes by endophytic fungal flora isolated from Senduduk plant (*Melastoma malabathricum*). ISRN Biotechnol 2013:1–7. https://doi.org/10.5402/2013/260730

Nikiforova SV, Pozdnyakova NN, Turkovskaya OV (2009) Emulsifying agent production during PAHs degradation by the white rot fungus Pleurotus Ostreatus D1. Curr Microbiol 58:554–558. https://doi.org/10.1007/s00284-009-9367-1

Niku-Paavola M-L, Viikari L (2000) Enzymatic oxidation of alkenes. J Mol Catal B Enzym 10:435–444. https://doi.org/10.1016/S1381-1177(99)00117-4

Oladipo OG, Awotoye OO, Olayinka A, Bezuidenhout CC, Maboeta MS (2018) Heavy metal tolerance traits of filamentous fungi isolated from gold and gemstone mining sites. Braz J Microbiol 49:29–37. https://doi.org/10.1016/j.bjm.2017.06.003

Park H, Min B, Jang Y, Kim J, Lipzen A, Sharma A, Andreopoulos B, Johnson J, Riley R, Spatafora JW, Henrissat B, Kim KH, Grigoriev IV, Kim J-J, Choi I-G (2019) Comprehensive genomic and transcriptomic analysis of polycyclic aromatic hydrocarbon degradation by a mycoremediation fungus, Dentipellis sp. KUC8613. Appl Microbiol Biotechnol 103:8145–8155. https://doi.org/10.1007/s00253-019-10089-6

Pozdnyakova NN (2012) Involvement of the Ligninolytic system of white-rot and litter-decomposing fungi in the degradation of polycyclic aromatic hydrocarbons. Biotechnol Res Int 2012:1–20. https://doi.org/10.1155/2012/243217

Pozdnyakova NN, Rodakiewicz-Nowak J, Turkovskaya OV (2004) Catalytic properties of yellow laccase from Pleurotus ostreatus D1. J Mol Catal B Enzym 30:19–24. https://doi.org/10.1016/j.molcatb.2004.03.005

Pradeep S, Faseela P, Josh MKS, Balachandran S, Devi RS, Benjamin S (2013) Fungal biodegradation of phthalate plasticizer in situ. Biodegradation 24:257–267. https://doi.org/10.1007/s10532-012-9584-3

Prieto A, Möder M, Rodil R, Adrian L, Marco-Urrea E (2011) Degradation of the antibiotics norfloxacin and ciprofloxacin by a white-rot fungus and identification of degradation products. Bioresour Technol 102:10987–10995. https://doi.org/10.1016/j.biortech.2011.08.055

Puglisi I, Faedda R, Sanzaro V, Lo Piero AR, Petrone G, Cacciola SO (2012) Identification of differentially expressed genes in response to mercury I and II stress in Trichoderma harzianum. Gene 506:325–330. https://doi.org/10.1016/j.gene.2012.06.091

Purnomo AS, Mori T, Takagi K, Kondo R (2011) Bioremediation of DDT contaminated soil using brown-rot fungi. Int Biodeterior Biodegrad 65:691–695. https://doi.org/10.1016/j.ibiod.2011.04.004

Ramírez-Cavazos LI, Junghanns C, Ornelas-Soto N, Cárdenas-Chávez DL, Hernández-Luna C, Demarche P, Enaud E, García-Morales R, Agathos SN, Parra R (2014) Purification and characterization of two thermostable laccases from Pycnoporus sanguineus and potential role in degradation of endocrine disrupting chemicals. J Mol Catal B Enzym 108:32–42. https://doi.org/10.1016/j.molcatb.2014.06.006

Robichaud K, Girard C, Dagher D, Stewart K, Labrecque M, Hijri M, Amyot M (2019a) Local fungi, willow and municipal compost effectively remediate petroleum-contaminated soil in the Canadian North. Chemosphere 220:47–55. https://doi.org/10.1016/j.chemosphere.2018.12.108

Robichaud K, Stewart K, Labrecque M, Hijri M, Cherewyk J, Amyot M (2019b) An ecological microsystem to treat waste oil contaminated soil: using phytoremediation assisted by fungi and local compost, on a mixed-contaminant site, in a cold climate. Sci Total Environ 672:732–742. https://doi.org/10.1016/j.scitotenv.2019.03.447

Samanta SK, Singh OV, Jain RK (2002) Polycyclic aromatic hydrocarbons: environmental pollution and bioremediation. Trends Biotechnol 20:243–248. https://doi.org/10.1016/S0167-7799(02)01943-1

Sathya K, Nagarajan K, Carlin Geor Malar G, Rajalakshmi S, Raja Lakshmi P (2022) A comprehensive review on comparison among effluent treatment methods and modern methods of treatment of industrial wastewater effluent from different sources. Appl Water Sci 12:70. https://doi.org/10.1007/s13201-022-01594-7

Schamfuß S, Neu TR, Van Der Meer JR, Tecon R, Harms H, Wick LY (2013) Impact of mycelia on the accessibility of Fluorene to PAH-degrading bacteria. Environ Sci Technol 47:6908–6915. https://doi.org/10.1021/es304378d

Silambarasan S, Abraham J (2013) Mycoremediation of Endosulfan and its metabolites in aqueous medium and soil by Botryosphaeria laricina JAS6 and Aspergillus tamarii JAS9. PLoS One 8:e77170. https://doi.org/10.1371/journal.pone.0077170

Singh D, Chen S (2008) The white-rot fungus Phanerochaete chrysosporium: conditions for the production of lignin-degrading enzymes. Appl Microbiol Biotechnol 81:399–417. https://doi.org/10.1007/s00253-008-1706-9

Singh M, Srivastava PK, Verma PC, Kharwar RN, Singh N, Tripathi RD (2015) Soil fungi for mycoremediation of arsenic pollution in agriculture soils. J Appl Microbiol 119:1278–1290. https://doi.org/10.1111/jam.12948

Sodaneath H, Lee J-I, Yang S-O, Jung H, Ryu HW, Cho K-S (2017) Decolorization of textile dyes in an air-lift bioreactor inoculated with *Bjerkandera adusta* OBR105. J Environ Sci Health Part A 52:1099–1111. https://doi.org/10.1080/10934529.2017.1340753

Tchounwou PB, Yedjou CG, Patlolla AK, Sutton DJ (2012) Heavy metal toxicity and the environment. In: Luch A (ed) Molecular, clinical and environmental toxicology, Experientia Supplementum. Springer Basel, Basel, pp 133–164. https://doi.org/10.1007/978-3-7643-8340-4_6

Thurston CF (1994) The structure and function of fungal laccases. Microbiology 140:19–26. https://doi.org/10.1099/13500872-140-1-19

Tran NH, Urase T, Kusakabe O (2010) Biodegradation characteristics of pharmaceutical substances by whole fungal culture Trametes versicolor and its laccase. J Water Environ Technol 8:125–140. https://doi.org/10.2965/jwet.2010.125

Wang M-X, Zhang Q-L, Yao S-J (2015) A novel biosorbent formed of marine-derived Penicillium janthinellum mycelial pellets for removing dyes from dye-containing wastewater. Chem Eng J 259:837–844. https://doi.org/10.1016/j.cej.2014.08.003

Wang K, Huang K, Jiang G (2018) Enhanced removal of aqueous acetaminophen by a laccase-catalyzed oxidative coupling reaction under a dual-pH optimization strategy. Sci Total Environ 616–617:1270–1278. https://doi.org/10.1016/j.scitotenv.2017.10.191

Wesenberg D (2003) White-rot fungi and their enzymes for the treatment of industrial dye effluents. Biotechnol Adv 22:161–187. https://doi.org/10.1016/j.biotechadv.2003.08.011

Wu M, Xu Y, Ding W, Li Y, Xu H (2016) Mycoremediation of manganese and phenanthrene by Pleurotus eryngii mycelium enhanced by tween 80 and saponin. Appl Microbiol Biotechnol 100:7249–7261. https://doi.org/10.1007/s00253-016-7551-3

Yang S-O, Sodaneath H, Lee J-I, Jung H, Choi J-H, Ryu HW, Cho K-S (2017) Decolorization of acid, disperse and reactive dyes by *Trametes versicolor* CBR43. J Environ Sci Health Part A 52:862–872. https://doi.org/10.1080/10934529.2017.1316164

Young D, Rice J, Martin R, Lindquist E, Lipzen A, Grigoriev I, Hibbett D (2015) Degradation of bunker C fuel oil by white-rot fungi in sawdust cultures suggests potential applications in bioremediation. PLoS One 10:e0130381. https://doi.org/10.1371/journal.pone.0130381

Zanchett G, Oliveira-Filho E (2013) Cyanobacteria and cyanotoxins: from impacts on aquatic ecosystems and human health to anticarcinogenic effects. Toxins 5:1896–1917. https://doi.org/10.3390/toxins5101896

Uses of Mycoremediation in Large-Scale Applications

5

Dharmasoth Rama Devi, Kanikella Sivaji, Keloth Basavaiah, A. Anjali, and P. S. S. Prasanna Kumar

Abstract

Mycoremediation, the process of using fungi to degrade or remove environmental contaminants, has emerged as a promising, sustainable solution for large-scale pollution challenges. This abstract summarizes the current state and future potential of mycoremediation in large-scale applications, highlighting its efficacy, scalability, and environmental benefits. Mycoremediation utilizes the natural metabolic pathways of various fungal species to break down complex pollutants, including hydrocarbons, pesticides, heavy metals, and industrial wastes. Large-scale deployment strategies, such as mycofiltration, mycoaugmentation, and the integration of fungi into bioremediation frameworks, have demonstrated significant pollutant reductions in contaminated soils, water bodies, and industrial sites. This review also addresses the challenges faced in scaling up, such as environmental variables, fungal strain selection, and regulatory considerations. Advances in biotechnology, including genetic engineering and optimized cultivation methods, are expanding the capabilities and efficiency of fungal remediation. The abstract concludes by emphasizing the need for interdis-

D. R. Devi (✉)
Department of Pharmacognosy, Vignan Institute of Pharmaceutical Technology, Duvvada, Visakhapatnam, Andhra Pradesh, India

K. Sivaji
Department of Pharmacology, Dr.Yellapragada Subbarao Research Center, Dr. Samuel George Institute of Pharmaceutical Sciences, Markapuram, Prakasam district, Markapur, Andhra Pradesh, India

K. Basavaiah
Department of Chemistry, Andhra University, Visakhapatnam, Andhra Pradesh, India

A. Anjali · P. S. S. P. Kumar
Department of Pharmaceutics, A.K.R.G College of Pharmacy, Nallajerla, Andhra Pradesh, India

A. Roy et al. (eds.), *Mycoremediation of Xenobiotics*, Emerging Paradigms in Pharmaceutical Research, https://doi.org/10.1007/978-981-95-5743-1_5

ciplinary research, policy support, and public engagement to realize the full potential of mycoremediation as a key tool in large-scale environmental restoration and pollution management.

Keywords
Mycoremediation · Xenobiotics · Peroxidases · Large-scale application · Case studies · Myco-nanotechnology · Genetic engineering and integrated remediation systems

5.1 Introduction

Mycoremediation is the procedure that involves the application of fungi in order to remove toxins from different environmental elements, either dead or alive (Jeykumar et al. 2022). Mycoremediation is an inexpensive process that does not produce any toxic waste. Mycoremediation efficiency relies on choosing and applying the right fungus species to be used with the intended heavy metal or other contaminant. This might be an efficient and reasonable solution for the emerging issue of water and soil pollution (Akhtar and Mannan 2020). Mycoremediation involves a bioremediation procedure using fungi to destroy toxic substances; it could be conducted in the presence of filamentous fungi (molds) and macro fungi (mushrooms). Thus, mycoremediation can serve as a bio-tool for the transformation, degradation, or immobilization of pollutants (Bosco and Mollea 2019).

5.1.1 Importance of Addressing Xenobiotic Contamination

The contamination of the environment by xenobiotic compounds represents a global concern attributable to anthropogenic activities, such as urban expansion and population increase. The substantial volumes of hazardous substances released into the ecosystem result in significant environmental pollution. Prominent pollutants, including heavy metal ions, polycyclic aromatic hydrocarbons (PAHs), agricultural pesticides, fertilizers, and petroleum derivatives, are detected in deposits of soil and aqueous environments (Embrandiri et al. 2016).

Significant xenobiotic substances that are harmful to humans, animals, plants, and the environment were discussed in Table 5.1, Fig. 5.1.

5.1.1.1 Impact of Xenobiotics on Soil

Xenobiotics like dioxins, 1,1,1-trichloro-2,2-bis (4-chlorophenyl) ethane (DDT), polychlorinated biphenyls (PCBs), chlordane, polycyclic aromatic hydrocarbons (PAHs), and nitroaromatics are the major threat to the soil ecosystem of developed countries. Yet, some other pollutants like benzene, nitrobenzene, toluene, xylene, aniline, ethylbenzene, trinitrotoluene/dibenzofurans, and chlorinated solvents are reported to be xenobiotic, particularly within the soil ecosystem (Ramel et al. 2012). Personal care and cosmetic products are also xenobiotic contaminants, particularly

Table 5.1 Major xenobiotic compounds and their effects

Xenobiotic compounds	Possible effects	Consequences observed	References
Polychlorinated biphenyls (PCBs) PCB-156,180,194	Neurological diseases in pediatrics	Extensive neural abnormalities (abnormal reflexes and neural tissue damage)	Alkorta and Garbisu (2001), Samanta et al. (2002)
HalocarbonsCFCs, H(C) FCs, CH_3CCl_3, CCl_4, CFC-12 HFC-134a	Global warming and climate change	Loss of biodiversity and environmental species destruction	Borpatragohain et al. (2019)
Synthetic polymers	PVC and PP or related products accumulation	Food chain alteration and food webs, pollution of aquatic and soil habitats	Hayat et al. (2010), Strauch et al. (2008)
Pharmaceuticals Analgesics, antibiotics, antiepileptics, antiseptics, beta-blockers, estrogenic drugs	Cellular and tissue damage	Aquatic, terrestrial, and arboreal animals' reproductive potential may decrease. Increase the effect of mortality on scavengers	Mishra et al. (2019)
Polycyclic aromatic hydrocarbonsPAHs	Aquatic and avian ecosystem toxicity	Genotoxicity, stress due to oxidation, immunosuppression, and hormonal ailments	Essumang (2010)
Polybrominated biphenyls PBBs	Adverse effects on hormone T3 and T4 secretion	Thyroid gland and related hormone disorders	Ibor et al. (2019)
PesticidesHerbicides, fungicides, and insecticides	Biomagnification and bioaccumulation hazards	Endocrinal irregularities, embryonic cell toxicity in aquatic animals	Burgos-Aceves et al. (2018)
Heavy metals	Nephro- and hepatotoxicity, contamination of water tables, aquatic water	Metabolic disorder, cellular and organ damage, and a variety of carcinogenic effects	Verma and Singh (2006)

parabens in air and soil (Kumar et al. 2019) and azo dyes present in soil, because of one or more aromatic rings with azo bonds (Pande et al. 2022).

Anthropogenic processes that induce these soil chemical compounds include military transport, fuel burning, industrial processes, application of pesticides, fertilizers, and soil amendments in intensive agriculture that generate adverse impacts (Abdelkader and Nadjia 2012). Chemical properties of xenobiotics and conditions at a site affect the bioavailability and distribution of xenobiotics in soil, and SOM is a significant factor in this process (Zhang et al. 2011). Pesticides, such as herbicides, algaecides, insecticides, fungicides, bactericides, are agrotechnical and management chemicals that have been widespread toxins to the environment in the past century.

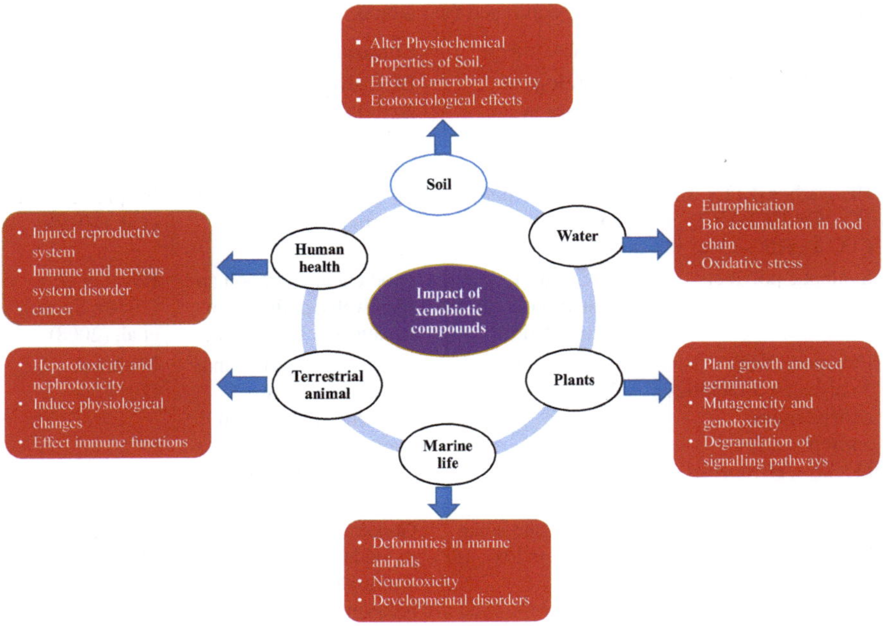

Fig. 5.1 Hazardous effects are caused by direct or indirect exposure to xenobiotics

5.1.1.2 Impact of Xenobiotics on Water

The human activities that pollute surface and ground water in the cities include urban industrial production, transportation, building, and home activities. The issue of the urban water systems' chemical pollution has been studied (Khan and Rampal 2014). The xenobiotic sensors will be subjected to the municipal wastewater in sewage treatment plants and then discharged into the wastewater. Water bodies also have traces of metals, xenobiotics, and synthetic organics, like PAHs, phthalates, and pesticides (Kobzev et al. 2020). The compounds of the xenobiotics end up in the water table due to leaching and affect aquatic ecosystems (Priyanka et al. 2022).

5.1.1.3 Impact of Xenobiotics on Plants

Xenobiotics have multiple effects on the physiological and morphological characteristics of the plant; for example, the photosynthetic pigments, protein, cysteine content, leaf area, and foliar surface of the plant are changed by the particulate matter produced by the transportation industry (Nag et al. 2021). The broad range of xenobiotics that have different structures and configurations from one another introduces some alterations in signal transduction, gene expression, and regulation in the higher plants. Phytohormones of xenobiotics, such as analogs, have essential relationships with plant hormone receptors and signal pathways (Tirkey et al. 2022). The metal ions needed to grow plants, such as Cu, Fe, Mo, and Zn, are toxic in large amounts, but the metal ions that do not affect plants, such as As, Cd, Hg, and Pb, are all toxic in low amounts in plant growth (Li et al. 2022).

5.1.1.4 Xenobiotic Effects on Marine Organisms

The marine life metabolic processes are having a high negative effect through xenobiotics, especially in developing fish embryos, causing structural deformities, growth retardation, and death. Changes in body shape, deformities, delayed hatching, and death occur in fishes (Srinivasan and Sadasivam 2021). Dyes and pigments are xenobiotics that inhibit sunlight penetration and gas exchange in trace amounts (Feller et al. 2021). Pesticides and herbicides cause xenobiotic contamination in marine life. Chemicals like organophosphorus, nitrophenols, morpholine, synthetic pyrethroids, and carbamates used in agriculture enter water bodies like seas and oceans. Insecticide β-cypermethrin severely threatens marine life and invertebrates (Benguenab and Chibani 2021).

5.1.1.5 Effects of Xenobiotics on Human Health

Ingestion, inhalation, percutaneous absorption, or any other form of intravenous contact would cause a health risk to humans, as they are exposing humans to an entire spectrum of xenobiotics, including drugs and nonessential foreign substances (Russell et al. 2021). The xenobiotics may disrupt the microbiota of the gut of human beings, leading to dysbiosis, indirectly linked to other undesirable health problems, including chronic illnesses, and damage to the human body DNA may occur. Other chemicals have a direct and indirect toxic and irritating effect on various organs and systems of the human body (Rhodes 2014).

5.2 Mechanisms of Fungal Detoxification

Fungi possess a high ability to break down a large variety of substances by releasing various extracellular enzymes, which assist in the breakdown of two important constituents: lignin and cellulose. Fungi are among the most reliable terrestrial organisms with the capacity to decay wood. The principal constituent of the fungi is the mycelium (vegetative portion of the fungus, which normally appears as thin white threads), in the majority of instances, responsible for breaking down lignin and Cellulose.

5.2.1 Mycoremediation by White-Rot Fungi

Fungi-based bioremediation is referred to as mycoremediation. These methods employ fungi that secrete enzymes that have the ability to degrade various pollutants and have shown promising results in the removal of contaminants from a location. Bioremediation processes are mostly reliant on bacteria and limited attempts at utilizing fungi. Fungi break down or transform organic and inorganic matter, playing a crucial role in the element cycle. They are valuable by decomposing organic material and eradicating the metal contamination risks. Fungi oxidize various important chemicals and environmental resilience. *P. chrysosporium* was the first fungus to be linked with the breakdown of organic pollution and has the ability to bioremediate pesticides, carbon tetrachloride, dioxins, and many other pollutants. *P. chrysosporia* fungus has become a model of bioremediation among fungi. Other species of white-rot fungi are *Trametes versicolor* and *Pleurotus ostreatus*.

5.2.1.1 Microbe Role in Xenobiotic Degradation

Biological organisms can be utilized to clean up chemical pollution through bioremediation. Microorganisms are greatly dependent upon the biotransformation of xenobiotics in soil, sediments, and aquatic environments. Bioremediation is based on the biological processes of living organisms (bacteria, fungi, and plants) and enzymes (Godoy et al. 2016; Ijoma and Tekere 2017). Microorganisms possess a remarkable capacity for catabolism with the assistance of numerous genes, enzymes, and pathways of biodegradation. Several microbes like *Alcaligenes* sp., *Cellulosimicrobium* sp., *Microbacterium* sp., *Micrococcus* sp., *Methanospirillum* sp., *Aeromonas* sp., *Sphingobium* sp., *Flavobacterium* sp., *Rhodococcus*, *Aspergillus* sp., *Penicillium*, *Trichoderma* sp., *Streptomyces*, *Rhodotorula*, *Candida*, and *Aureobasidium* have been isolated and identified and have shown a very good potential to biodegrade a wide range of xenobiotic pollutants in soil/water environments (Mishra et al. 2021). Yet, only a handful of representative microbial enzymes participate in xenobiotic detoxification, such as cytochrome P450s, cellulase, laccases, phytase, proteases, and lipases. Such enzymes can hydrolyze aromatic hydrocarbons, dyes, and halogenated compounds by different mechanisms (Fig. 5.2).

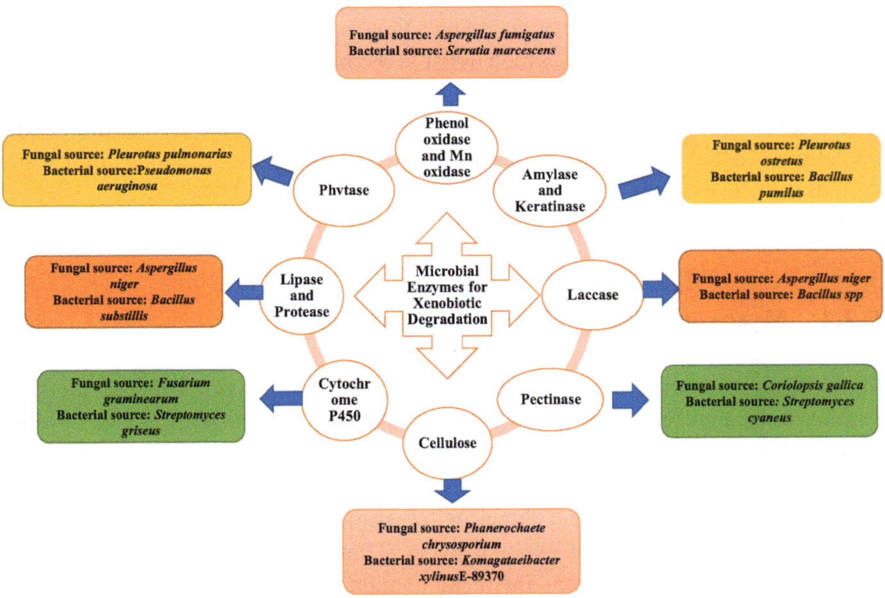

Fig. 5.2 Microbial enzymes for xenobiotic degradation

5.2.2 Mycorrhizal Associations

A fungus and plant root form a symbiotic relationship known as a mycorrhiza wherein both parties benefit: the fungus provides water and minerals, and the plant provides sugars (Smith and Read 2008). According to Gutierrez et al. the fungal hyphae can be either intracellular (found inside plant cortex cells or arbuscular mycorrhizas) or intercellular (found surrounding plant cortex cells or ectomycorrhizas). *Arbuscular mycorrhizae*-forming fungi are found in *Glomeromycota*, whereas ectomycorrhizae-forming fungi are found in *Ascomycota* or *Basidiomycota*. Ascomycota fungi that are mycorrhizal include truffle (Smith and Read 2008). Ectomycorrhizas with little intracellular structures and ectendomycorrhizas with intracellular and intercellular hyphae develop upon responses to mineral levels by ectomycorrhizal fungi (Fortas and Chevalier 1992) this is seen in the fungal mutants (Gea et al. 1994; Gay et al. 1994). In the semiarid and arid areas, desert truffle is a type of mycorrhizal fungi (Kagan-Zur et al. 1999), which belongs to six families of Pezizales. The *Terfezia* and *Tirmania* genera, which infect the Helianthemum species, are members of the *Pezizaceae* family. *Pezizaceae* is another family that has a representation of *Mattirolomyces terfezioides* and *Kalaharituber* and *Eremiomyces*, both are specific to acacia trees in the Kalahari Desert, and no *Cistaceae* is found there.

5.2.3 Physiological Adaptations in Response to Pollutants

Mycoremediation may be a cost-effective, sustainable, and efficient method to tackle the continuous growing issue of soil and water pollution. Fungi possessing vigorous growth, leading to the formation of extensive hyphal networks, diverse ligninolytic enzyme secretion, a high ratio of surface area to volume, tolerance to heavy metals, external environment stress adaptation in pH fluctuation and temperature, and pollutant remediation by metal protein binding, are ideal candidate for pollution neutralization (Kapahi and Sachdeva 2017; Bhattacharya et al. 2011). It can be utilized for the in situ cleanup of several pollutants like dyes, herbicides, and pharmaceutical drugs discharged by different industries, also used in bioreactors. Bioreactors refer to systems with controlled physicochemical conditions that are developed to favor microbial growth (Aragão et al. 2020). Controlled biomass of fungi and regulated utilization of metabolites in the bioreactor can be utilized to facilitate the decomposition of the pollutants (Tekere 2019). Up to now, bioreactors are constructed for the waste treatment eluted from the sugar industry and as pharmaceutical industry waste (Rodarte-Morales et al. 2012). They can also be employed ex situ in the remediation of soil contaminated with polycyclic aromatic hydrocarbons, herbicides, pesticides, tars, chlorinated solvents, and explosives (Tekere 2019). Various classes of mycoremediation of environmental pollutants:

- Poly-cyclic aromatic hydrocarbon (PAH)
- Heavy metal

- Agricultural waste (pesticides and herbicides)
- Phthalates, dyes, detergents
- Pharmaceutical wastes
- Mycoremediation of antibiotics

5.3 Fungal Species in Mycoremediation

Microorganisms are essential to the detoxification of soils and ecosystems. A promising technology called mycoremediation uses fungi to break down or change toxic compounds into less toxic or nontoxic forms. *Pleurotus ostreatus, Rhizopus arrhizus, Phanerochaete chrysosporium, Phanerochaete sordida, Trametes hirsuta, Trametes versicolor, Lentinus tigrinus,* and *Lentinus edodes* are among the fungal species used as mycoremediators. A variety of fungi, such as those illustrated in Fig. 5.3.

5.3.1 Focus on White-Rot Fungi and Their Enzymatic Capabilities

The term "white-rot" refers to the white surface of lumber that has been infiltrated by white-rot fungus, where the evacuation of lignin leaves a faded impression.

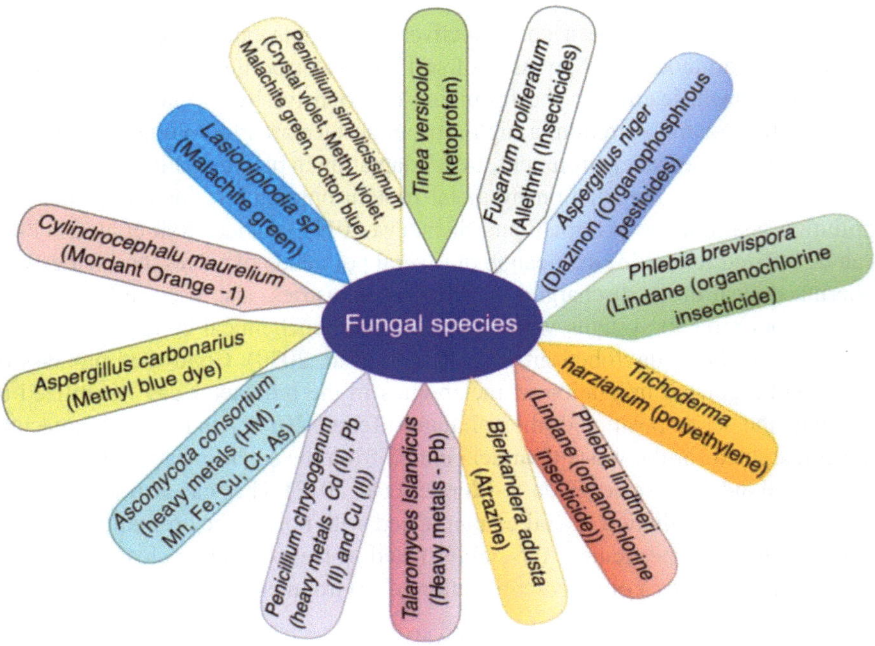

Fig. 5.3 Bioremediation of various pollutants and their specific fungal species

White-rot fungi have the amazing capacity to biodegrade lignin (Pointing 2001). White-rot fungi are categorized as distinct basidiomycetes, and only a small number of ascomycete families, such as Xylariaceae, are linked to white-rot decay (Eaton and Hale 1993). When the white-rot fungus *Phanerochaete chrysosporium* was discovered to decompose a wide variety of natural pollutants in the mid-1980s, the use of fungi for cleaning contaminated soil was first proven (Bumpus and Aust 1987). This ability was later exhibited for diverse species, such as *Trametes versicolor* and *Pleurotus ostreatus*, *Lentinus subnudus*, *Psathyrella candolleana* LCJ 178, and *Myrothecium gramineum* LCJ 177, *Porostereum spadiceum*, *Pleurotus floridanus* LCJ 155, Leucocoprinus cretaceous LCJ 164, and Agaricus sp. LCJ 169, *Ganoderma lucidum* and *Bjerkandera adusta* (Dhiman et al. 2020). All components of wood are deteriorated by white-rot fungi, including cellulose and lignin, and hemicellulose and lignin are also affected (Dashtban et al. 2010).

5.4 Large-Scale Mycoremediation Techniques

5.4.1 Myco-Nanotechnology

Nanomaterials can be produced using fungi, bacteria, and plant extracts. Compared to using bulk plants, applying mycology to the biosynthesis of nanoparticles has significant advantages, because it removes the requirement for full purification of the nanoparticles following synthesis (Fatima et al. 2015). The biogenesis of nanoparticles based on biosorption is suggested as a means of eliminating heavy metals from soils and wastewater. This approach may also facilitate the synthesis of heavy metal nanoparticles that may find use in technology (Gericke and Pinches 2006). The highly organized physical and biochemical activities of microbial cells (NPs) have been used to create metal nanoparticles in a novel way (Chan et al. 2023; Kumar et al. 2019). A reliable and affordable method of addressing the growing problem of marine and terrestrial pollution is mycoremediation. Strong growth, resistance to complex pollutants, increased surface area to volume ratio, multifunctional extracellular enzyme production, extensive hyphal network, and flexibility in responding to temperature, metal-binding protein, and pH variations are the main reasons for the advantages of fungi (Kumar et al. 2019; Deng et al. 2011).

5.4.2 Molecular Approach to Improve Mycoremediation

Fungi are extremely efficient in changing chemicals and degrading toxic substances in the environment. A number of fungi, which are filamentous, such as *Aspergillus*, *Mucor*, *Trichoderma*, and *Penicillium*, are resistant to organic and inorganic pollutants (Thenmozhi et al. 2013). Mushrooms may accumulate toxic metals in the soils in contaminated soils (Kacprzak et al. 2014). *Penicillium*, *Trichoderma*, *Aspergillus*, and *Mortierella* species have a high bioremediation ability (Garon et al. 2004). Wild *Pinus pinaster* genotypes containing ectomycorrhizal amassed 30 mg g^{-1} Cd more in shoots compared to other *Pinus pinaster* genotypes. Depending on Cd levels, *P. pinaster* genotypes exhibit divergent relationships with mycorrhizal relationships, as stated by Sousa et al. (Pundir et al. 2018). The accumulation of Cr (4.33 g^{-1} mg) and Cd (2.72 mg g^{-1}) in *Rhizopus* sp. and *Aspergillus* sp.1 was 1.20 mg g^{-1} and 2.27 mg g^{-1}, respectively. Studies indicate that *Trichoderma* enhances bioconcentration parameters of Zn, Cr, and Cd when it grows on Salix, Ni, and Cr in the presence of *Phalaris arundinacea* (Deng et al. 2011). Mycorrhiza can also be applied to mycorrhizal mediation, because revegetation can generate promising species in polluted areas through a change in fungus-host interactions. Penicillium italicum helped in the breakdown of fluorine with cyclodextrins. Mycoremediation potential can be enhanced by factors, such as soil characteristics, plant species, fungal species, concentration of pollutants, and soil properties. At pH 4, *Aspergillus* species exhibited improved copper and nickel extraction. According to Pundir et al. (2018), the removal of metal decreases with an increase in pH, because of the formation of metal hydroxide. The next-generation sequencing method is a solution to microarray problems, as it generates accurate sequence reads. It has an advantage in stem cell and cancer cell research in that it determines regulatory marks and protein-binding sites. Next-generation sequencing is cost-effective but consumes large quantities of computational resources, which requires the creation of bioinformatics techniques.

5.4.3 Mycofiltration

The filtration element of the mycelia, called mycofiltration, has been thought of as a way of stormwater disinfection. One review also demonstrated that *S. rugosoannulata* expelled up to 20% of *E. coli* particles in tempest water. Even though the majority of *E. coli* strains are harmless, certain strains are pathogenic. In addition to the fact that it leads to sickness, *E. coli* has been regarded as an indicator of fecal water pollution. Removal of these germs is beneficial to communities since all people require sterile water. To carry the contaminated tempest water, tests were installed to pass it through channels. The parasite was put together with two substrates. The organism was put into three birch wood chip holders after vivacious testing (when exposed to extreme temperatures) and one without growth and one with the parasite. One had an organism and the other one had an untested parasite. The compartments consisted of 75 birch chips and 25 rice straw. These samples were cloned thrice, making a total of 15 compartments. Experiments were conducted at intervals when tempest water went through the canals. The best was a 20% reduction in *E. coli* in specimens having non-vivaciously tried growth on wood chips. They were comparatively critical, as compared to the control. The above findings indicate that *S. rugosoannulata* mycelium could eliminate *E. coli* in adverse environments, owing to successful experimental growth. The cultivating technique and recovery power in extreme conditions are suggested. Also, the analysis of the

Pleurotus ostreatus and *Irpex lacteus* species' capacity to eliminate polycyclic aromatic hydrocarbons (PAHs) in soil showed their resilience. PAHs are industry spread carcinogens. They are harmful to the skin and immune system and enhance the development of tumors, and thus, are severe health issues.

5.4.4 Mycoaugmentation

Currently, bioaugmentation with fungi is being studied. The idea behind bioaugmentation is to improve the organisms' capacity and catabolic significance in order to improve the bioremediation process. Numerous studies have used bioaugmentation on a variety of matrices, including heavy metals (Xu et al. 2017; Garima and Singh 2014), diesel oil (Jayanthi 2016; Yang et al. 2017), activated sludge (Jayanthi et al. 2017), benzene, toluene, ethylbenzene, and xylene (BTEX) (Jayanthi et al. 2017). Naturally, the data gathered will help environmental engineers create and implement consortiums or consortia for the treatment and decontamination of soils contaminated by metals or metalloids. The effective removal of metals and metalloid pollutants from the soil is hoped to be facilitated by interactions that may improve metabolic competency within the fungal variety. Bioaugmentation in order to clean up landfill soil that was contaminated with metals and metalloids.

5.4.5 Fungal Bioreactors

The main concern of bioreactors, which are basically multiphase systems, is the transfer of mass (and occasionally heat). Stirred tank reactors make up the largest group of reactors used for cultivating fungus (the most common type). Although it is widely acknowledged that agitation causes a decrease in particle size, excessive agitation can have two main effects: it can cause hyphal fragments to chip off the pellet surface or even rupture the pellets directly, and it can negatively impact enzymatic secretion. Additionally, the size and size distribution of pellets have a significant impact on how easily particles maintain optimal performance in fluidized-bed bioreactors. Generally speaking, to preserve filtration, small particles with a narrow size distribution are preferred. However, inadequate filtration with bumping, gushing, and slugging of the bed is frequently the result of particle aggregation (Kunii and Levenspiel 1969). The ability to regulate pellet growth is one way to provide bioreactors using pellets with a longer lifespan. A technique that allows for the efficient "shaving" of fungal biomass beyond a certain thickness while maintaining an active film surface is crucial. Some examples of these bioreactor types are pulsating bioreactors and reciprocating jet bioreactors (Brauer 1991).

5.5 Case Studies of Successful Large-Scale Mycoremediation Projects

The contaminated soil with petroleum was detoxified by taking advantage of the natural capacity of fungi to degrade hydrocarbons by means of bioaccumulation and biosorption. Nevertheless, its application in the business sector remains immature because it is difficult to scale up (Dickson et al. 2019). Fungi, e.g., *Trametes versicolor* and *Pleurotus ostreatus*, have been effective in degrading pharmaceutical pollutants, with degradation rates reaching a maximum of 90%. It is based on the extracellular enzyme, which has been enhanced by the advancement of enzyme engineering and bioreactors, and scale-up has been problematic (Khan 2025). Strains of fungi, such as *Aspergillus niger*, can be utilized in order to eliminate cadmium (Cd) and chromium (Cr), which are heavy metal pollutants found in the soils. This method of bioremediation has a high degree of bioaccumulation, which makes it practical in in situ and ex situ (Khan et al. 2019). Mycoremediation has been explored to promote sustainable agricultural development, where different types of fungi are used to remove pollutants within the contaminated soils, thus enhancing the health and sustainability of the soils. In situ techniques treat the soil on the site, which causes minimal disturbance (Purohit et al. 2018). Mycoremediation strategies of PAH-contaminated environments rely on using fungal enzymes to break down the pollutants, which is an underresearched area that needs more studies, particularly in the fields of genomics and molecules (Park and Choi 2020). Fungal bioremediation can deal with a large variety of emerging pollutants. The study of fungal enzymes and fungal enzyme control is intended to close the gaps in knowledge and maximize the pollutant breakdown (Akpasi et al. 2023) presented in Table 5.2.

5.6 Challenges and Limitations in Scale-Up

The process of scaling up mycoremediation that employs fungi in the remediation of environmental pollutants has a number of challenges and limitations. The following are some of the main problems that are mentioned in the literature:

5.6.1 Environmental Considerations, Ecological Limitations, and Dietary Needs

Certain environmental, edaphic, and climatic factors are very crucial in mycoremediation work. When remediation is done, it has to be done in the best conditions as close to natural conditions by ensuring proper levels of moisture, pH, and temperature. The controlled conditions in the laboratory to the variable field conditions are difficult to transition to and slow down the scaling activity. To promote fungus growth and metabolism, the addition of organic substrates (e.g., straw, wood chips) is usually needed to tutor fungal remediation. The logistics and expenses that come

Table 5.2 Case studies of mycoremediation projects' scale and outcome

Project location	Pollutant(s)	Fungal species	Scale and outcome	Reference
Bhopal, India	Petroleum hydrocarbons	*Pleurotus ostreatus*	10,000 m^2 contaminated soil; 70–80% degradation	Singh et al. (2018)
United States (Industrial site)	Polycyclic aromatic hydrocarbons (PAHs)	*Phanerochaete chrysosporium*	5 acres; 60–85% decrease in PAHs	Lamar et al. (2015)
China (E-waste site)	Heavy metals (Pb, Cd)	*Aspergillus niger*	500 tons of soil treated; 50–70% metal removal	Wang et al. (2020)
Brazil (agricultural land)	Pesticide residues (atrazine)	*Trametes versicolor*	200 ha; 80% of atrazine degradation	Silva and Durrant (2019)
Italy (industrial contaminated site)	Chlorinated compounds	*Phanerochaete chrysosporium*	2000 m^3 soil; 70% elimination of contaminants	Federici et al. (2017)
Canada (oil spill site)	Petroleum hydrocarbons	*Pleurotus pulmonarius*	1000 m^2; 65–75% hydrocarbon degradation	Chiu et al. (2018)
Australia (diesel-contaminated soil)	Diesel	Native fungal species	70–85% decrease in diesel-range organics in 16 weeks	Harms et al. (2011)
Ecuador (oil-contaminated sites)	Petroleum hydrocarbons (TPH)	*Pleurotus ostreatus* (oyster mushroom)	1600 cubic meters of soil, and it was shown that 95% of TPH was degraded	Sánchez (2005)
Finland	Creosote, a source of persistent PAHs	(White-rot fungi) *Phanerochaete chrysosporium*	The project involved applying fungal inoculum to windrows of polluted soil and showed the scalability and efficiency of mycoremediation at industrial scale	Steffen et al. (2002)

with the sourcing and distribution of appropriate substrates within expansive territories may be heavy (Dickson et al. 2019).

5.6.2 Production of Mycotoxins

Although filamentous fungi can be useful in degrading pollutants, most of the fungi also produce mycotoxins. The capacity of bioremediation can be influenced by regulatory factors that control the production of mycotoxins, which makes the treatment of pollutants and mycotoxins a two-fold task (Chanda et al. 2015).

5.6.3 Enzyme Specificity and Pharmaceuticals

The enzyme-based mechanisms of degradation of pharmaceutical pollutants are complex. Although fungi can be used to degrade different pharmaceuticals with extracellular enzymes, scaling of these activities is still a challenging aspect given that enzyme specificity, stability, and optimal activities are considered (Khan 2025).

5.6.4 Surfactant Use and Soil Interactions

Surfactants have proven to be an effective way to improve mycoremediation, but they have shortcomings. The mobilization of hydrophobic pollutants, such as polycyclic PAHs, can be facilitated by surfactants, but it can also cause toxicity and influence the bioremediation. The mechanism of comprehending surfactant-soil contact is essential to scaffolding (Rathankumar et al. 2021).

5.6.5 Genetically Modified Organisms (GMOs)

The exploitation of GMOs in mycoremediation has the possible advantages of better degradation properties. Nonetheless, the ecological implications and regulatory advantages of releasing GMOs into the environment are severe, which makes it harder to apply them at large scale (Saxena et al. 2019).

5.6.6 Poor Fungal Resistance to Bearing Big Contaminant Burdens

High concentrations of poisonous contaminant are toxic to some fungal species. In highly contaminated areas, the growth of fungi might be suppressed, and gradual or watered-down methods of treatment need to be used (Pointing 2001). Introduction of foreign fungi can be resisted by native soil microbes competing with them for nutrients or through antagonistic activity. This competition tends to cause less or irregular remediation results on broader levels (Baldrian 2003). Most pollutants, especially hydrophobic ones, such as polycyclic aromatic hydrocarbons (PAHs), will be trapped in soil matrices, where their bioreactive properties are restricted to fungi. This lowered the degradation efficiency, and it may require pretreatment or addition of surfactants, which makes field operations complicated.

5.6.7 Issues of Regulatory and Public Acceptance

Commercial use of non-native or genetically modified fungi has regulatory issues and popular worries. It may take a long time or restrict the extent of project implementation by raising the required approvals and dealing with ecological risk assessment.

5.6.8 Difficulties in Monitoring and Standardization

It is not easy to gauge the success and effectiveness of mycoremediation in complicated settings. No standard protocols exist in terms of tracking the activity of fungus and contaminant degradation in the field, and comparing the outcomes of different projects is challenging (Zafra et al. 2015).

The above challenges indicate that a complex interplay between environmental, biological, and chemical factors can be scaled in mycoremediation efforts. The solution to such problems involves innovations in biotechnological technology, environmental knowledge, and control structures.

5.7 Developments in Mycoremediation Biotechnology

Increased efficiency and potential of mycoremediation have become significantly high due to recent developments in biotechnology. Such developments break the historical barriers and make a way to large-scale restoration of the environment.

5.7.1 Fungi Genetic Engineering

Genetic modification has enabled the development of better strains of fungi that have enhanced pollutant-degrading properties. Fungus can be genetically engineered to express crucial ligninolytic enzymes, such as laccases and peroxidases, to enhance their ability to break down hydrocarbons, dyes, and persistent organic pollutants. Genetically modified strains of *Trametes versicolor* and *Phanerochaete chrysosporium* have been shown to be more effective in the degradation of chlorinated chemicals and polycyclic aromatic hydrocarbons (PAHs).

5.7.2 Innovations in Systems Biology and Omics

High-throughput omics methods, such as genomics, transcriptomics, proteomics, and metabolomics, have illuminated the metabolic pathways and regulatory networks that contribute to the fungal degradation of pollutants (Janusz et al. 2017). Important genes, enzymes, and metabolites that facilitate the degradation of pollutants are identified using these technologies, and the methods of remediation can be optimized and rational strain improvements made (Zhang et al. 2016).

5.7.3 Biofilm Engineering and Immobilization

Immobilization of fungi on various carriers (e.g., charcoal, alginate beads or natural fibers) has provided improvements in stability and reusability of fungi in contaminated environments (Levin et al. 2004). Two advantages of immobilized fungal systems in reaction or soil beds are better access of fungus to contaminants and greater control of process parameters.

5.7.4 Synthetic Ecology and Fungal Consortia

Synthetic consortia (designed communities of bacteria and fungi) have become promising in order to enhance the number of toxins that can be degraded and enhance efficacy in cleaning up. To decompose pollutants more effectively, these consortia exploit synergistic interactions, where fungi and bacteria complement each other with respect to their metabolic activity.

5.7.5 Application of Nanobiotechnology

The combination of nano materials and fungi has led to the creation of nano biocatalysts with better catalytic performance and pollutant selectivity (Bilal et al. 2017). An example is the successful elimination of colors and phenolic chemicals in wastewater by the use of laccase-nanoparticle conjugates prepared with fungi.

5.8 Environmental and Economic Benefits

Mycoremediation would be an appealing alternative of environmental and monetary benefits as compared to the traditional remediation technologies in various environments.

5.8.1 Environmental Benefits

Mycoremediation degrades, transforms, or sequesters contaminants using naturally occurring fungi, eliminating harsh chemicals, incineration, or landfill disposal. Fungal extracellular enzymes are capable of breaking down a wide variety of chemical and inorganic pollutants, such as heavy metals, hydrocarbons, dyes, and pesticides, in a manner that is environmentally safe (Harms et al. 2011; Singh 2006). Mycoremediation improves soil fertility and soil structure by increasing organic matter and microbial diversity, unlike most conventional methods, which may eliminate or sterilize soil. When the contaminants are decomposed by fungi, humic compounds are also formed, thus facilitating plant growth and the long-term stability of the soil (Gadd 2008). The chances of dangerous byproducts or secondary contamination are minimized as fungal breakdown often results in complete mineralization of contaminants. Nevertheless, certain physicochemical methods might yield dangerous byproducts that might require further treatment.

5.8.2 Profitable Effects

Mycoremediation is relatively cheap as compared to other traditional cleanup procedures, like excavation, incineration, or carbonous chemical treatment. Fungi could be grown on untreated material, or low-carbon agricultural waste, and fungal energy demands are insignificant (Stamets 2005). Most mycoremediation remedies can be in situ, and therefore, less disposal and transport costs are incurred. It is possible to make fungal inoculation and substrate application adaptable to varying circumstances, such as remote or low-resource cases, because field-scale utilization has been proven (Sánchez 2005). Targeted mycoremediation techniques can be used to produce possible valuable by-products, i.e., mushroom edibles or compost, on treated lands. As a consequence, remediation practice is encouraged and provided with the greater economic reward (Eggen and Majcherczyk 1998).

5.9 Aspects of Regulation and Policy

Despite the possible economic and environmental benefits of mycoremediation, there are several legal and policy considerations that should be made before this approach can be applied on a large scale. The elements affect the viability, pace, and general acceptance of mycoremediation projects worldwide.

5.9.1 Evaluation of Environmental Risk and Biosafety

Regulatory authorities typically require extensive risk assessments before offering field deployment to investigate any environmental implications. This involves consideration of the possibility of unexpected ecological consequences, the risk of introducing alien or genetically modified strains of fungi, and the fate of by-products that are degradable (Harms et al. 2011; Singh 2006). Regulatory frameworks highlight the need for biosafety procedures so as to prevent the proliferation of potentially invasive species as well as to ensure that the transformation products or secondary metabolites pose no new risks.

5.9.2 Procedures of Approval and Uniformity

Mycoremediation projects on a large scale can involve complex, multistage review processes to obtain licenses. These may be required to be conducted through environmental impact studies, pilot-scale efficacy demonstration, and compliance with national and international laws that control the application of living things in environmental technology (3). Moreover, the efficacy and safety of mycoremediation have yet to be determined as the established procedures and standards, making it challenging to compare projects and delay the granting of regulatory authorization.

5.9.3 Stakeholder Attitude and Involvement

The focus of the regulatory environment of environmental biotechnologies is the acceptance of the people. There should be open dialogue and stakeholder consultation that can help in the issues related to the application of genetically modified fungi, potential impacts on biodiversity, and food safety (when edible mushrooms are involved), which can influence the policies (Gentry et al. 2004). Mycoremediation projects would need to be successful and create trust, which would depend on the community involvement and education.

5.9.4 International Policy and International Harmonization

The policies and regulations should be aligned since there are many contaminated locations on both national borders. International agencies, such as the Organization for Economic Co-operation and Development (OECD) and the United Nations Environment Programme (UNEP), have insisted on the best practices and principles to be used in the bioremediation methods, such as mycoremediation, to facilitate technology transfer and ensure environmental safety (UNEP 2009).

5.9.5 Rewards and Financial Resources

Some of the ways through which regulatory regimes can promote mycoremediation are incentives such as grants, tax credits, and research and development funding. To address the hindrances of finance and facilitate a wider implementation, the policy instruments that assist in sustainable cleanup technology would be beneficial (Sánchez 2005).

5.10 Prospects for the Future and Research Paths

Mycoremediation has proven to be very promising as a cost-effective and eco-friendly way of destroying various environmental pollutants. Nevertheless, to maximally make it work on a large scale, several research directions and future perspectives are necessary.

To create fungal strains that are better in degrading, more resilient, and selective to complex contaminants, future research will increasingly adopt state-of-the-art genomes, transcriptomics, and synthetic biology (Harms et al. 2011). By increasing the expression of enzymes, uptake of pollutants, and resistance to stressors using CRISPR/Cas9 and other gene editing technologies, the design of customized so-called super fungi to achieve certain remediation goals can be achieved. Recent studies have suggested that fungal-bacterial consortia have the potential to degrade long-term pollutants more efficiently than single species consortia. To develop multi organism repair communities that are resilient, future research will focus on understanding and optimizing these synergistic relationships that encompass quorum sensing, metabolite exchange, and niche specialization (Singh 2006). The incorporation of fungi with nanomaterials is one of the interesting directions. However, although new carrier materials could enhance the fungal immobilization, dispersal, and survival in harsh field conditions, nanoparticle-fungal hybrids would potentially enhance the pollutant adsorption and catalysis (Gentry et al. 2004). The use of these innovative materials in a sustainable and secure way will also be examined. Despite laboratory findings, field experiments are in a large scale. Future research should focus on long-term and multi-site experiments across a variety of soil and climate conditions in order to authenticate lab findings and develop standardized protocols to track, evaluate, and adhere to environmental regulations (UNEP 2009).

It is expected that more research on mycoremediation will be based on big data, machine learning, and predictive models. These technologies may be applied to direct site-specific remedial measures, optimize the treatment environment, and predict the effectiveness of fungi (Stamets 2005). To determine market feasibility, regulatory procedures, and social acceptability, interdisciplinary studies, which incorporate economics, policy, and social sciences, are necessary. The research on the source of financing and encouraging incentives and community participation and practices will facilitate its mainstream acceptance (8). Studies need to focus on the ability of mycoremediation methods to withstand extreme weather, temperature variations, and changing pollutants under different environmental conditions due to changes in climate (Zafra et al. 2015).

5.11 Conclusion

The mycoremediation approach is a promising, cost-effective, and sustainable approach to the treatment of large-scale environmental pollution. Natural metabolic process of fungi offers unique advantages to the decomposition and removal of many contaminants, including heavy metals and hydrocarbons as well as persistent chemical compounds. Comparing mycoremediation with traditional forms of remediation, when used on a large scale, mycoremediation has proven to be highly promising in cleaning contaminated soils, waterways, and industrial locations—with minimal or no environmental disturbance. Nevertheless, the choice of fungi species, climate, and the system of complementing biotechnologies are among the factors that should be considered thoroughly in order to make mycoremediation an effective intervention on a large scale. The main tool of the global effort to restore and protect our own ecosystems, mycoremediation, is distinctive as we seek effective and sustainable means to deal with pollution.

References

Abdelkader E, Nadjia L (2012) Degradation study of phenazin neutral red from aqueous suspension by paper sludge. J Chem Eng Process Technol 02(03). https://doi.org/10.4172/2157-7048.1000109

Akhtar N, Mannan MA-U (2020) Mycoremediation: expunging environmental pollutants. Biotechnol Rep (Amst) 26:e00452. https://doi.org/10.1016/j.btre.2020.e00452

Akpasi SO, Anekwe IMS, Tetteh EK, Amune UO, Shoyiga HO, Mahlangu TP, Kiambi SL (2023) Mycoremediation as a potentially promising technology: current status and prospects—a review. Appl Sci 13(8):4978. https://doi.org/10.3390/app13084978

Alkorta I, Garbisu C (2001) Phytoremediation of organic contaminants in soils. Bioresour Technol 79(3):273–276. https://doi.org/10.1016/s0960-8524(01)00016-5

Aragão MS, Menezes DB, Ramos LC, Oliveira HS, Bharagava RN, Romanholo Ferreira LF, Teixeira JA, Ruzene DS, Silva DP (2020) Mycoremediation of vinasse by surface response methodology and preliminary studies in air-lift bioreactors. Chemosphere 244:125432. https://doi.org/10.1016/j.chemosphere.2019.125432

Baldrian P (2003) Interactions of heavy metals with white-rot fungi. Enzym Microb Technol 32(1):78–91. https://doi.org/10.1016/s0141-0229(02)00245-4

Benguenab A, Chibani A (2021) Biodegradation of petroleum hydrocarbons by filamentous fungi (Aspergillus ustus and Purpureocillium lilacinum) isolated from used engine oil-contaminated soil. Acta Ecol Sin 41:416–423

Bhattacharya S, Das A, Mangai G, Vignesh K, Sangeetha J (2011) Mycoremediation of Congo red dye by filamentous fungi. Braz J Microbiol 42(4):1526–1536. https://doi.org/10.1590/S1517-838220110004000040

Bilal M, Asgher M, Parra-Saldivar R, Hu H, Wang W, Zhang X, Iqbal HMN (2017) Immobilized ligninolytic enzymes: an innovative and environmental responsive technology to tackle dye-based industrial pollutants—a review. Sci Total Environ 576:646–659. https://doi.org/10.1016/j.scitotenv.2016.10.13

Borpatragohain B, Sahoo S, Rai A (2019) An overview on the impact, interaction and fate of xenobiotic-soil organic matter complexes. Int J Chem Stud 7(3):4935–4941

Bosco F, Mollea C (2019) Mycoremediation in soil. In: Environmental chemistry and recent pollution control approaches. IntechOpen, pp 1–16

Brauer H (1991) Growth of fungi and bacteria in the reciprocating jet bioreactor. Bioprocess Eng 6:1–5

Bumpus JA, Aust SD (1987) Biodegradation of environmental pollutants by the white rot fungus Phanerochaete chrysosporium: involvement of the lignin degrading system. BioEssays 6:166–170

Burgos-Aceves MA, Cohen A, Smith Y, Faggio C (2018) MicroRNAs and their role on fish oxidative stress during xenobiotic environmental exposures. Ecotoxicol Environ Saf 148:995–1000. https://doi.org/10.1016/j.ecoenv.2017.12.001

Chan EA, Rendón-Barraza C, Wang B, Pu T, Ou J-Y, Wei H, Adamo G, An B, Zheludev NI (2023) Counting and mapping of subwavelength nanoparticles from a single shot scattering pattern. Nanophotonics 12(14):2807–2812. https://doi.org/10.1515/nanoph-2022-0612

Chanda A, Gummadidala PM, Gomaa OM (2015) Mycoremediation with mycotoxin producers: a critical perspective. Appl Microbiol Biotechnol 100(1):17–29. https://doi.org/10.1007/s00253-015-7032-0

Chiu SW, Gao T, Wong MH (2018) Mycoremediation of petroleum hydrocarbons by _Pleurotus pulmonarius_ in contaminated soil. Environ Pollut 233:1155–1163. https://doi.org/10.1016/j.envpol.2017.10.040

Dashtban M, Schraft H, Syed TA, Qin W (2010) Fungal biodegradation and enzymatic modification of lignin. Int J Biochem Mol Biol 1(1):36–50

Deng Z, Cao L, Huang H, Jiang X, Wang W, Shi Y, Zhang R (2011) Characterization of Cd-and Pb-resistant fungal endophyte Mucor sp. CBRF59 isolated from rapes (Brassica chinensis) in a metal-contaminated soil. J Hazard Mater 185:717–724

Dhiman N, Jasrotia T, Sharma P, Negi S, Chaudhary S, Kumar R, Mahnashi MH, Umar A, Kumar R (2020) Immobilization interaction between xenobiotic and Bjerkandera adusta for the biodegradation of atrazine. Chemosphere 257:127060. https://doi.org/10.1016/j.chemosphere.2020.127060

Dickson UJ, Coffey M, Di Bonito M, Mortimer RJG, Ray N (2019) Mycoremediation of petroleum contaminated soils: progress, prospects and perspectives. Environ Sci Process Impacts 21(9):1446–1458. https://doi.org/10.1039/c9em00101h

Eaton RA, Hale MD (1993) Wood: decay, pests and protection. Open J Forestry 3(2):1–520

Eggen T, Majcherczyk A (1998) Removal of polycyclic aromatic hydrocarbons (PAHs) in contaminated soil by white rot fungus Pleurotus ostreatus. Int Biodeterior Biodegrad 41(2):111–117

Embrandiri A, Kiyasudeen SK, Rupani PF, Ibrahim MH (2016) Environmental xenobiotics and its effects on natural ecosystem. In: Plant responses to xenobiotics. Springer, Singapore, pp 1–18

Essumang DK (2010) Distribution, levels, and risk assessment of polycyclic aromatic hydrocarbons (PAHs) in some water bodies along the coastal belt of Ghana. TheScientificWorldJ 10:972–985. https://doi.org/10.1100/tsw.2010.96

Fatima F, Bajpai P, Pathak N, Singh S, Priya S, Verma SR (2015) Antimicrobial and immunomodulatory efficacy of extracellularly synthesized silver and gold nanoparticles by a novel phosphate solubilizing fungus Bipolaris tetramera. BMC Microbiol 15(1):52. https://doi.org/10.1186/s12866-015-0391-y

Federici E, Giubilei M, D'Annibale A (2017) Mycoremediation of chlorinated compounds in soil: a case study in Italy. J Hazard Mater 332:267–276. https://doi.org/10.1016/j.jhazmat.2017.03.017

Feller FM, Eilebrecht S, Nedielkov R, Yücel O, Alvincz J, Salinas G, Ludwig KC, Möller H, Philipp B (2021) Investigations on the degradation of the bile salt cholate via the 9,10-Seco-pathway reveals the formation of a novel recalcitrant steroid compound by a side reaction in Sphingobium sp. Strain Chol11. Microorganisms 9(10):2146. https://doi.org/10.3390/microorganisms9102146

Fortas Z, Chevalier G (1992) Effet des conditions de culture sur la mycorhization del'Helianthemum guttatum par trois espe'ces de terfez des genres Terfezia et Tirmania d'Alge'rie. Can J Bot 70:2453–2460

Gadd GM (ed) (2008) British mycological society symposia: fungi in bioremediation series number 23. Cambridge University Press

Garima T, Singh SP (2014) Application of bioremediation on solid waste management: a review. J Bioremed Biodegr 05(06). https://doi.org/10.4172/2155-6199.1000248

Garon D, Sage L, Seigle-Murandi F (2004) Effects of fungal bioaugmentation and cyclodextrin amendment on fluorene degradation in soil slurry. Biodegradation 15(1):1–8. https://doi.org/10.1023/b:biod.0000009934.87627.91

Gay L, Normand L, Marmeisse B, Sotta B, Debaud JC (1994) Structural aspects of ectomycorrhiza of Pinus pinaster(Ait.) Sol. formed by an IAA-overproducer mutant of Hebeloma cylindrosporum Romagnesi. New Phytol 128:651–670

Gea I, Normand L, Vian B, Gay G (1994) Structural aspects of ectomycorrhiza of Pinus pinaster(Ait.) Sol. formed by an IAA-overproducer mutant of Hebeloma cylindrosporum Romagnesi. New Phytol 128:651–670

Gentry T, Rensing C, Pepper I (2004) New approaches for bioaugmentation as a remediation technology. Crit Rev Environ Sci Technol 34(5):447–494. https://doi.org/10.1080/10643380490452362

Gericke M, Pinches A (2006) Biological synthesis of metal nanoparticles. Hydrometallurgy 83(1–4):132–140. https://doi.org/10.1016/j.hydromet.2006.03.019

Godoy P, Reina R, Calderón A, Wittich R-M, García-Romera I, Aranda E (2016) Exploring the potential of fungi isolated from PAH-polluted soil as a source of xenobiotics-degrading fungi. Environ Sci Pollut Res Int 23(20):20985–20996. https://doi.org/10.1007/s11356-016-7257-1

Harms H, Schlosser D, Wick LY (2011) Untapped potential: exploiting fungi in bioremediation of hazardous chemicals. Nat Rev Microbiol 9(3):177–192. https://doi.org/10.1038/nrmicroEmuh
Hayat MT, Xu J, Ding N, Mahmood T (2010) Dynamic behavior of persistent organic pollutants in soil and their interaction with organic matter. In: Molecular environmental soil science at the interfaces in the earth's critical zone. Springer
Ibor OR, Adeogun AO, Regoli F, Arukwe A (2019) Xenobiotic biotransformation, oxidative stress and obesogenic molecular biomarker responses in Tilapia guineensis from Eleyele Lake, Nigeria. Ecotoxicol Environ Saf 169:255–265. https://doi.org/10.1016/j.ecoenv.2018.11.021
Ijoma GN, Tekere M (2017) Potential microbial applications of co-cultures involving ligninolytic fungi in the bioremediation of recalcitrant xenobiotic compounds. Int J Environ Sci Technol: IJEST 14(8):1787–1806. https://doi.org/10.1007/s13762-017-1269-3
Janusz G, Pawlik A, Sulej J, Swiderska-Burek U, Jarosz-Wilkolazka A, Paszczynski A (2017) Lignin degradation: microorganisms, enzymes involved, genomes analysis and evolution. FEMS Microbiol Rev 41(6):941–962. https://doi.org/10.1093/femsre/fux049
Jayanthi B (2016) Selected microbial diversity of contaminated landfill soil of Peninsular Malaysia and the behavior towards heavy metal exposure. Catena 147:25–31
Jayanthi B, Emenike CU, Auta SH, Agamuthu P, Fauziah SH (2017) Characterization of induced metal responses of bacteria isolates from active non-sanitary landfill in Malaysia. Int Biodeterior Biodegrad 119:467–475. https://doi.org/10.1016/j.ibiod.2016.10.053
Jeykumar P, Debnath C, Vijayaraghavan R, Muthuraj M (2022) Trends in bioremediation of heavy metal contaminations. Environ Eng Res 28(4)
Kacprzak MJ, Rosikon K, Fijalkowski K, Grobelak A (2014) The effect of Trichodermaon heavy metal mobility and uptake by Miscanthus giganteus, Salixsp., Phalaris arundinacea, and Panicum virgatum. Appl Environ Soil Sci 2014:1–10. https://doi.org/10.1155/2014/506142
Kagan-Zur V, Kuang J, Tabak S, Taylor FW, Roth-Bejerano N (1999) Potential verification of a host plant for the desert truffle Terfezia pfeilii by molecular methods. Mycol 103:1270–1274
Kapahi M, Sachdeva S (2017) Mycoremediation potential of Pleurotus species for heavy metals: a review. Bioresour Bioprocess 4(1). https://doi.org/10.1186/s40643-017-0162-8
Khan MF (2025) Fungi for sustainable pharmaceutical remediation: enzymatic innovations, challenges, and applications—a review. Processes 13(4):1034. https://doi.org/10.3390/pr13041034
Khan AM, Rampal S (2014) Effects of repeated oral administration of pazufloxacin mesylate and meloxicam on the antioxidant status in rabbits. J Am Assoc Lab Anim Sci: JAALAS 53(4):399–403
Khan I, Aftab M, Shakir S, Ali M, Qayyum S, Rehman MU, Haleem KS, Touseef I (2019) Mycoremediation of heavy metal (Cd and Cr)-polluted soil through indigenous metallotolerant fungal isolates. Environ Monit Assess 191(9):585. https://doi.org/10.1007/s10661-019-7769-5
Kobzev K, Kobzeva N, Chegge V, Balinskaya M, Bozhenko E, Saakian S, Sarkisian D (2020) Ecological problems in The Russian Federation. Impact on the health of people and the country's economy. E3S Web of Conferences 217:11001. https://doi.org/10.1051/e3sconf/202021711001
Kumar A, Chaturvedi A, Yadav K, Arunkumar K, Malyan S, Raja P, Kumar R, Khan S, Yadav K, Rana K (2019) Fungal phytoremediation of heavy metal-contaminated resources: current scenario and future prospects. In: Recent advancement in white biotechnology through fungi. Springer, Cham, pp 437–461
Kunii D, Levenspiel O (1969) Fluidization engineering. Wiley
Lamar RT, White RB, Dietrich DM (2015) Bioremediation of PAH-contaminated soils using white-rot fungi. Appl Microbiol Biotechnol 99(19):8111–8123. https://doi.org/10.1007/s00253-015-6829-1
Levin L, Papinutti L, Forchiassin F (2004) Evaluation of Argentinean white rot fungi for their ability to produce lignin-modifying enzymes and decolorize industrial dyes. Bioresour Technol 94(2):169–176. https://doi.org/10.1016/j.biortech.2003.12.002
Li H, Ma Y, Yao T, Ma L, Zhang J, Li C (2022) Biodegradation pathway and detoxification of β-cyfluthrin by the bacterial consortium and its bacterial community structure. J Agric Food Chem 70(25):7626–7635. https://doi.org/10.1021/acs.jafc.2c00574

Mishra VK, Singh G, Shukla R (2019) Impact of xenobiotics under a changing climate scenario. In: Climate change and agricultural ecosystems. Woodhead Publishing

Mishra S, Lin Z, Pang S, Zhang W, Bhatt P, Chen S (2021) Recent advanced technologies for the characterization of xenobiotic degrading microorganisms and microbial communities. Front Bioeng Biotechnol 9:632059

Nag M, Lahiri D, Dutta B, Jadav G, Ray RR (2021) Biodegradation of used polyethene bags by a new marine strain of Alcaligenes faecalis LNDR-1. Environ Sci Pollut Res 28:41365–41379

Pande V, Pandey SC, Sati D, Bhatt P, Samant M (2022) Microbial interventions in bioremediation of heavy metal contaminants in agroecosystem. Front Microbiol 13:824084. https://doi.org/10.3389/fmicb.2022.824084

Park H, Choi I-G (2020) Genomic and transcriptomic perspectives on mycoremediation of polycyclic aromatic hydrocarbons. Appl Microbiol Biotechnol 104(16):6919–6928. https://doi.org/10.1007/s00253-020-10746-1

Pointing SB (2001) Feasibility of bioremediation by white-rot fungi. Appl Microbiol Biotechnol 57(1–2):20–33. https://doi.org/10.1007/s002530100745

Priyanka JV, Rajalakshmi S, Senthil Kumar P, Krishnaswamy VG, Al Farraj DA, Elshikh MS, Abdel Gawwad MR (2022) Bioremediation of soil contaminated with toxic mixed reactive azo dyes by co-cultured cells of Enterobacter cloacae and Bacillus subtilis. Environ Res 204(Pt B):112136. https://doi.org/10.1016/j.envres.2021.112136

Pundir R, Chary GHVC, Dastidar MG (2018) Application of Taguchi method for optimizing the process parameters for the removal of copper and nickel by growing Aspergillus sp. Water Resour Ind 20:83–92. https://doi.org/10.1016/j.wri.2016.05.001

Purohit J, Chattopadhyay A, Biswas MK, Singh NK (2018) Mycoremediation of agricultural soil: bioprospection for sustainable development. In: Mycoremediation and environmental sustainability. Fungal biology. Springer, pp 91–120. https://doi.org/10.1007/978-3-319-77386-5_4

Ramel F, Sulmon C, Serra A-A, Gouesbet G, Couée I (2012) Xenobiotic sensing and signalling in higher plants. J Exp Bot 63(11):3999–4014. https://doi.org/10.1093/jxb/ers102

Rathankumar AK, Kalita S, Varjani S, Saikia K, Kumar PS, George J, Bharadwaj N, Vaidyanathan VK (2021) Surfactant-aided mycoremediation of soil contaminated with polycyclic aromatic hydrocarbon (PAHs): progress, limitation, and countermeasures. J Chem Technol Biotechnol 97(2):391–408. https://doi.org/10.1002/jctb.6721

Rhodes CJ (2014) Mycoremediation (bioremediation with fungi)-growing mushrooms to clean the earth. Chem Speciat Bioavailab 26:196–198

Rodarte-Morales AI, Feijoo G, Moreira MT, Lema JM (2012) Biotransformation of three pharmaceutical active compounds by the fungus Phanerochaete chrysosporium in a fed batch stirred reactor under air and oxygen supply. Biodegradation 23(1):145–156. https://doi.org/10.1007/s10532-011-9494-9

Russell JN, Perry BJ, Bergsveinson J, Freeman CN, Sheedy C, Nilsson D, Braul L, Yost CK (2021) Metagenomic and metatranscriptomic analysis reveals enrichment for xenobiotic-degrading bacterial specialists and xenobiotic-degrading genes in a Canadian Prairie two-cell biobed system. Environ Microbiol Rep 13(5):720–727. https://doi.org/10.1111/1758-2229.12990

Samanta SK, Singh OV, Jain RK (2002) Polycyclic aromatic hydrocarbons: environmental pollution and bioremediation. Trends Biotechnol 20(6):243–248. https://doi.org/10.1016/s0167-7799(02)01943-1

Sánchez C (2005) Bioremediation of oil-contaminated soil by the white-rot fungus Pleurotus ostreatus in the Ecuadorian Amazon: a field study. J Environ Sci Health A Tox Hazard Subst Environ Eng 40(4):885–897

Saxena G, Kishor R, Saratale GD, Bharagava RN (2019) Genetically modified organisms (GMOs) and their potential in environmental management: constraints, prospects, and challenges. Springer Singapore, pp 1–19. https://doi.org/10.1007/978-981-13-3426-9_1

Silva CM, Durrant LR (2019) Degradation of atrazine by _Trametes versicolor_ in agricultural soils of Brazil. J Environ Sci Health B 54(2):93–102. https://doi.org/10.1080/03601234.2018.1546111

Singh H (2006) Mycoremediation. Wiley

Singh S, Singh P, Singh SK (2018) Mycoremediation of petroleum hydrocarbon contaminated soil: a case study from Bhopal, India. J Environ Manag 217:415–424. https://doi.org/10.1016/j.jenvman.2018.04.017

Smith SE, Read JD (2008) Mycorrhizal symbiosis, 3rd edn. Academic Press

Srinivasan S, Sadasivam SK (2021) Biodegradation of textile azo dyes by textile effluent non-adapted and adapted Aeromonas hydrophila. Environ Res 194:110643. https://doi.org/10.1016/j.envres.2020.110643

Stamets P (2005) Mycelium running: how mushrooms can help save the world. Ten Speed Press

Steffen KT, Hatakka A, Hofrichter M (2002) Degradation of polycyclic aromatic hydrocarbons (PAHs) by the litter-decomposing basidiomycete Stropharia coronilla and other white rot fungi. FEMS Microbiol Ecol 41(1):1–7

Strauch G, Möder M, Wennrich R, Osenbrück K, Gläser H-R, Schladitz T, Müller C, Schirmer K, Reinstorf F, Schirmer M (2008) Indicators for assessing anthropogenic impact on urban surface and groundwater. J Soils Sediments 8(1):23–33. https://doi.org/10.1065/jss2007.06.234

Tekere M (2019) Microbial bioremediation and different bioreactors designs applied. In: Biotechnology and bioengineering. IntechOpen

Thenmozhi R, Arumugam K, Nagasathya A, Thajuddin N, Paneerselvam A (2013) Studies on Mycoremediation of used engine oil contaminated soil samples. Adv Appl Sci Res 4(2):110–118

Tirkey SR, Ram S, Mitra M, Mishra S (2022) Performance analysis of Pseudomonas sp. strain SA3 in naphthalene degradation using phytotoxicity and microcosm studies. Biodegradation 33(2):169–180. https://doi.org/10.1007/s10532-022-09972-3

United Nations Environment Programme (2009) Guidelines for the application of bioremediation in environmental management

Verma A, Singh SN (2006) Biochemical and ultrastructural changes in plant foliage exposed to auto-pollution. Environ Monit Assess 120(1–3):585–602. https://doi.org/10.1007/s10661-005-9105-5

Wang X, Chen M, Wang J (2020) Mycoremediation of heavy metals in e-waste contaminated soil using _Aspergillus niger. Environ Sci Pollut Res 27(14):16513–16524. https://doi.org/10.1007/s11356-020-08245-6

Xu Q, Li X, Ding R, Wang D, Liu Y, Wang Q, Zhao J, Chen F, Zeng G, Yang Q, Li H (2017) Understanding and mitigating the toxicity of cadmium to the anaerobic fermentation of waste activated sludge. Water Res 124:269–279. https://doi.org/10.1016/j.watres.2017.07.067

Yang Q, Zhao H, Du B (2017) Bacteria and bacteriophage communities in bulking and non-bulking activated sludge in full-scale municipal wastewater treatment systems. Biochem Eng J 119:101–111. https://doi.org/10.1016/j.bej.2016.12.017

Zafra G, Absalón ÁE, Cortés-Espinosa DV (2015) Morphological changes and growth of filamentous fungi in the presence of pollutants. Curr Opin Biotechnol 33:152–160

Zhang C, Wang S, Yan Y (2011) Isomerization and biodegradation of beta-cypermethrin by Pseudomonas aeruginosa CH7 with biosurfactant production. Bioresour Technol 102(14):7139–7146. https://doi.org/10.1016/j.biortech.2011.03.086

Zhang J, Presley GN, Hammel KE, Ryu J-S, Menke JR, Figueroa M, Hu D, Orr G, Schilling JS (2016) Localizing gene regulation reveals a staggered wood decay mechanism for the brown rot fungus Postia placenta. Proc Natl Acad Sci U S A 113(39):10968–10973. https://doi.org/10.1073/pnas.1608454113

Bioreactors Involved in Mycoremediation

6

Praveen Singh, Geetha S. J., Amit Saxena, Gaurav Pandey, Saurabh Shukla, and Sanket J. Joshi

Abstract

The escalating release of synthetic xenobiotics from industrial, pharmaceutical, and agricultural activities has created unprecedented environmental contamination challenges. Mycoremediation, utilizing fungi for environmental detoxification, offers a sustainable biotechnological solution to address persistent organic pollutants, pharmaceutical residues, and agricultural chemicals. White-rot fungi, particularly *Phanerochaete chrysosporium* and *Trametes versicolor*, possess unique extracellular enzymatic systems, including lignin peroxidases, manganese peroxidases, and laccases, that enable broad-spectrum xenobiotic degradation. These nonspecific oxidative enzymes, combined with intracellular

P. Singh · Geetha S. J.
Amity Institute of Microbial Technology, Amity University Rajasthan, Jaipur, Rajasthan, India

A. Saxena
Department of Petroleum Engineering and Geo-Engineering, Rajiv Gandhi Institute of Petroleum Technology, Jais, India

G. Pandey
Department of Petroleum Engineering and Earth Science, Energy Cluster, School of Advanced Engineering, UPES, Dehradun, Uttarakhand, India

Earth Cryosphere Institute, Tyumen Scientific Center SB, RAS, Tyumen, Russia

S. Shukla
Department of Civil Engineering and Architecture, University of Catania, Catania, Italy

S. J. Joshi (✉)
Amity Institute of Microbial Technology, Amity University Rajasthan, Jaipur, Rajasthan, India

Amity Microbial Culture Centre, Amity University Rajasthan, Jaipur, Rajasthan, India
e-mail: sjjoshi@jpr.amity.edu

A. Roy et al. (eds.), *Mycoremediation of Xenobiotics*, Emerging Paradigms in Pharmaceutical Research, https://doi.org/10.1007/978-981-95-5743-1_6

cytochrome P450 monooxygenases, facilitate the transformation of structurally diverse contaminants that resist conventional biological treatments. Field applications demonstrate mycoremediation's effectiveness across diverse contamination contexts, from European industrial sites to North American agricultural soils and Asian pharmaceutical effluents. While challenges remain in optimizing environmental conditions and scaling laboratory results to field applications, mycoremediation represents a cost-effective, ecologically sustainable approach that integrates pollution remediation with circular economy principles, positioning fungi as powerful allies in global environmental restoration efforts.

Keywords

Mycoremediation · White-rot fungi · Bioreactor · Field Applications

6.1 Introduction

The rapid expansion of industrialization, urbanization, and agricultural intensification over the past century has accelerated the release of synthetic chemicals into the environment, leading to an unprecedented accumulation of toxic and recalcitrant pollutants. These compounds, collectively referred to as xenobiotics, include pesticides, pharmaceuticals, petroleum hydrocarbons, dyes, industrial solvents, and plastic-derived additives. Unlike naturally occurring organic substances, xenobiotics are often anthropogenic in origin and characterized by complex molecular structures, halogen substitutions, or high aromaticity that render them resistant to conventional biological degradation (Amobonye et al. 2023). Their persistence compromises soil health, contaminates groundwater systems, and bioaccumulates in food webs, ultimately threatening ecosystem functions and human well-being. Traditional physicochemical remediation technologies like adsorption, thermal treatments, advanced oxidation, or chemical precipitation, while effective under specific conditions, frequently suffer from high operational costs, limited scalability, secondary waste generation, and unsuitability for mixed-contaminant environments (Ba et al. 2014). This has driven the search for alternative, eco-friendly, and economically feasible remediation approaches capable of addressing the limitations of existing methods.

Among emerging strategies, mycoremediation, the application of fungi for detoxifying polluted environments, has garnered significant attention as a sustainable biotechnological solution. Fungi, particularly wood-decaying species such as white-rot fungi, exhibit extraordinary catabolic versatility due to their unique extracellular enzymatic machinery. Unlike bacteria, which depend on substrate-specific metabolic pathways, fungi deploy nonspecific oxidative and hydrolytic enzymes, including lignin peroxidases, manganese peroxidases, and laccases, that evolved to degrade the complex polymeric structure of lignin in plant biomass (Bilal et al. 2018). Coincidentally, this enzymatic repertoire enables fungi to transform structurally diverse xenobiotics, many of which share chemical features with lignin, such as high aromatic content and complex bond arrangements. These oxidative systems can cleave stable carbon-chlorine bonds, disrupt polycyclic aromatic hydrocarbons (PAHs), and oxidize phenolic or heterocyclic compounds, thereby breaking down contaminants that often remain recalcitrant in conventional biological treatments (Birolli et al. 2020).

The promise of mycoremediation extends beyond laboratory demonstrations. Field studies have documented the successful deployment of fungi for degrading pesticides in agricultural soils, detoxifying industrial effluents, mineralizing hydrocarbons in oil-contaminated sites, and reducing pharmaceutical residues in wastewater treatment facilities (Bokade and Bajaj 2023). White-rot fungi such as *Phanerochaete chrysosporium* and *Trametes versicolor* are particularly noteworthy: their ligninolytic systems have been shown to degrade dyes, polychlorinated biphenyls (PCBs), pentachlorophenol, and even microplastic additives. The ability of fungal enzymes to function extracellularly provides flexibility in environmental applications, since pollutants need not be transported into microbial cells for transformation (Bonatti et al. 2023). Furthermore, fungi possess robust growth forms, extensive mycelial networks that enable them to penetrate contaminated matrices, bridging air, soil, water, and sediment interfaces that often pose barriers for microbial remediation.

Despite these advantages, the transition from laboratory efficacy to large-scale environmental deployment remains complex. Mycoremediation performance is strongly influenced by environmental conditions, including pH, temperature, oxygen availability, nutrient supply, and the presence of co-contaminants (Cabana et al. 2007). Fungal degradation often requires co-metabolism, in which enzymatic activity is enhanced or sustained by the simultaneous presence of natural substrates, such as lignocellulose. In contaminated sites with heavy metal co-pollution, fungal growth and enzymatic activity may be inhibited, complicating remediation outcomes. Additionally, achieving consistent results under variable field conditions requires careful optimization of inoculation strategies, substrate amendments, and reactor design for wastewater or soil treatment (Camerano Spelta Rapini et al. 2024).

Beyond the biological and environmental dimensions, the viability of mycoremediation must also be considered within economic and regulatory frameworks. Relative to conventional physicochemical techniques, fungal remediation is generally cost-effective, relying on inexpensive substrates (e.g., agricultural residues or sawdust) as carriers that stimulate fungal growth and enzyme production (Cao et al. 2021). Moreover, mycoremediation aligns well with circular economy principles, as degradation byproducts can potentially be valorized, for instance, ligninolytic fungi generate biosorbents, biopolymers, or secondary metabolites with industrial relevance (Champramary et al. 2023). Nevertheless, standardization and regulatory acceptance remain challenges. Wide heterogeneity in test conditions, strain selection, and pollutant types makes it difficult to establish universal protocols for fungal remediation. Regulators require consistent evidence regarding not only pollutant removal but also ecotoxicological safety of degradation intermediates before large-scale adoption can be realized (Chane et al. 2024).

The future of mycoremediation research is increasingly intertwined with advances in biotechnology and systems-level ecology. Genetic engineering tools, such as CRISPR-Cas and synthetic biology approaches, offer opportunities to enhance the secretion of ligninolytic enzymes, develop fungal strains with expanded pollutant tolerance, and engineer consortia of fungi and bacteria with complementary biodegradative capacities (Chen et al. 2023). Omics-based technologies, including genomics, transcriptomics, and proteomics, are illuminating the regulatory networks governing xenobiotic degradation, enabling the rational design of optimized fungal phenotypes (Chmelová et al. 2024). Concurrently, ecological research is exploring the potential of mixed fungal consortia, as natural microbial communities often outperform monocultures by fostering synergistic interactions and resource sharing (Christian et al. 2005). Developing standardized testing frameworks for distinct classes of xenobiotics will further accelerate the translation of fungal remediation into scalable, real-world applications.

In sum, mycoremediation offers a paradigm shift in environmental detoxification, combining ecological sustainability with remarkable degradative potential. By harnessing the versatile enzymatic systems and adaptive growth capacities of fungi, this approach addresses limitations inherent to conventional remediation methods (Cruz-Morató et al. 2014). While technical, ecological, and regulatory challenges remain, ongoing research continues to strengthen the foundations for large-scale deployment (Curiel-Alegre et al. 2024). Mycoremediation not only mitigates the risks posed by persistent xenobiotics but also exemplifies how nature-inspired biotechnologies can simultaneously remediate pollution and contribute to sustainable resource cycles. As the global demand for environmentally sound remediation strategies intensifies, harnessing fungal systems for xenobiotic detoxification represents both an urgent necessity and a promising frontier in environmental biotechnology (D'Annibale et al. 2000).

6.2 Fungal Mechanisms for Xenobiotic Degradation

Fungi possess remarkable biochemical versatility that allows them to colonize and thrive in diverse ecological niches, including highly polluted environments. Their ability to degrade xenobiotic complex synthetic compounds resistant to natural biodegradation stems from an array of enzymatic systems that evolved primarily for the decomposition of lignocellulosic biomass, particularly lignin (Dey et al. 2024). While bacteria often rely on specific catabolic pathways tailored to narrow classes of compounds, fungi employ broad-spectrum oxidative and hydrolytic mechanisms (Fig. 6.1). This metabolic flexibility enables them to attack xenobiotics with diverse chemical structures, such as polycyclic aromatic hydrocarbons (PAHs), chlorinated pesticides, pharmaceuticals, phenolic dyes, and industrial solvents (Dey et al. 2020).

The degradative machinery of fungi can be categorized into two major systems:

(i) Extracellular enzyme systems, dominated by peroxidases and laccases that initiate oxidative cleavage outside the fungal cell (Fonseka et al. 2025).

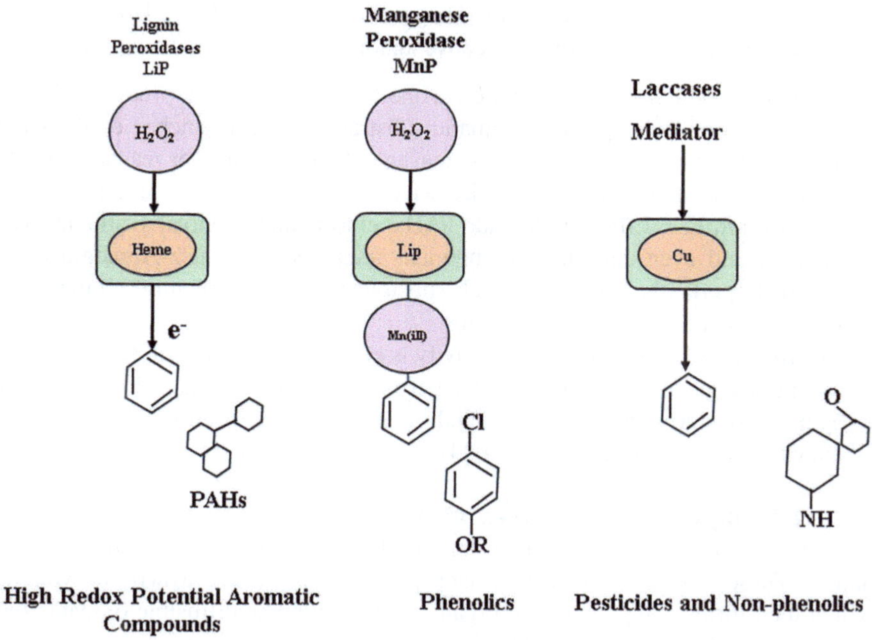

Fig. 6.1 Schematic representation of the white-rot-fungal enzyme system for xenobiotic degradation

(ii) Intracellular xenobiotic-transforming systems, most notably cytochrome P450 monooxygenases that catalyze phase-I biotransformations. Together, these enzymatic networks underpin the efficiency and ecological relevance of fungal mycoremediation (Flórez Córdoba et al. 2024).

6.2.1 Extracellular Enzyme Systems

Extracellular oxidative enzymes are central to the fungal strategy for xenobiotic degradation. These enzymes, secreted into the extracellular environment, enable fungi to circumvent the requirement for cellular uptake of hydrophobic or complex pollutants, a major advantage when tackling poorly soluble or high-molecular-weight compounds (Fragoeiro and Magan 2008). The most studied and environmentally significant extracellular enzymes include lignin peroxidases (LiPs), manganese peroxidases (MnPs), and laccases.

6.2.1.1 Lignin Peroxidases (LiPs)

Lignin peroxidases, first characterized in *Phanerochaete chrysosporium*, are heme-containing enzymes with exceptionally high redox potential. Their primary ecological role is to depolymerize lignin, one of the most chemically complex and recalcitrant natural polymers (García-Delgado et al. 2015, 2021). However, their oxidative power also makes them effective against a wide range of xenobiotics.

- *Mechanism of Action*: LiPs operate via one-electron oxidation, producing highly reactive cation radicals from aromatic substrates (García-Sánchez et al. 2018). These radicals undergo subsequent cleavage, demethylation, or rearrangements, leading to destabilization and breakdown of otherwise persistent compounds.
- *Substrate Range*: LiPs can degrade PAHs, chlorinated phenols, synthetic dyes, dioxins, and even certain petrochemical fractions (González-Abradelo et al. 2019). Importantly, they can attack substrates with no structural similarity to lignin, reflecting their non-specificity.
- *Environmental Relevance*: Their activity is effective under natural soil and water conditions, given sufficient hydrogen peroxide as a co-substrate. However, management of peroxide levels is crucial, as excessive concentrations can lead to enzyme inactivation (Guo et al. 2024).

6.2.1.2 Manganese Peroxidases (MnPs)

Manganese peroxidases are another defining feature of fungal extracellular degradation. These enzymes utilize hydrogen peroxide to oxidize Mn(II) to Mn(III), which then forms complexes with organic acids, acting as diffusible oxidants that extend the oxidative reach beyond the enzyme's immediate site of secretion.

- *Mechanism of Action*: The Mn(III)-organic acid chelates function as redox mediators, diffusing into substrates otherwise inaccessible to the enzyme (Harms et al. 2011). This indirect oxidation expands fungal activity into soil matrices and aqueous environments where xenobiotic pollutants are dispersed.
- *Target Compounds*: MnPs are particularly effective against phenolic substrates, chlorophenols, and certain classes of pesticides. They also contribute significantly to the oxidation of PAHs and textile dyes (Hu et al. 2021).
- *Ecological Significance*: Because manganese is abundant in soils worldwide, MnPs represent a naturally advantageous system for in situ mycoremediation (Imam et al. 2022). Their reliance on diffusible redox agents allows oxidative activity to occur at considerable distances from fungal hyphae, enhancing environmental penetration.

6.2.1.3 Laccases

Laccases are copper-dependent oxidases with a broad ecological distribution across wood-decaying fungi, soil fungi, and even some marine fungal species (Jaiswal and Shukla 2020). Their ability to catalyze the oxidation of both phenolic and non-phenolic compounds by reducing molecular oxygen to water makes them valuable for xenobiotic degradation.

- *Mechanism of Action*: Laccases catalyze electron transfer from a donor substrate (e.g., phenol, amine, polyaromatic compound) to molecular oxygen. While their redox potential is generally lower than that of peroxidases, their activity can be dramatically expanded when used in conjunction with "mediators" (Kathhivaran and Gnanadoss 2021). These small, diffusible molecules, such as ABTS (2,2′-azino-bis) or natural phenolics, shuttle electrons between the enzyme and otherwise recalcitrant substrates, effectively broadening the substrate range.

- *Applications in Xenobiotic Degradation*: Laccases are highly useful for degrading industrial effluents containing phenolic compounds, textile dyes, and pharmaceutical residues (Košnář et al. 2018). Their involvement in decolorization processes makes them particularly attractive for treating dye wastewater.
- *Advantages for Mycoremediation*: Laccases function efficiently under mild environmental conditions and often require only molecular oxygen, eliminating the need for costly co-substrates such as hydrogen peroxide. Their extracellular nature also allows broad environmental adaptability.

6.2.2 Cytochrome P450 Systems

While extracellular enzymes dominate initial xenobiotic depolymerization, intracellular systems, particularly cytochrome P450 monooxygenases (CYPs), are instrumental in metabolizing xenobiotics into more soluble and less toxic intermediates (Li et al. 2024). These enzymes represent a cornerstone of fungal phase I metabolism, comparable to liver P450s in mammals.

6.2.2.1 General Characteristics

Cytochrome P450s are heme-thiolate proteins that catalyze the insertion of an oxygen atom into hydrophobic or recalcitrant substrates. They are membrane-associated enzymes located in the endoplasmic reticulum or mitochondrial membranes. Their catalytic cycle involves electron transfer from NADPH through cytochrome P450 reductase, enabling oxidative modifications such as hydroxylation, epoxidation, N-dealkylation, or O-dealkylation (Li et al. 2022).

6.2.2.2 Role in Xenobiotic Detoxification

- *Hydroxylation*: Hydroxylation of aromatic hydrocarbons, alkanes, and chlorinated compounds improves solubility and facilitates further enzymatic attack (Lladó et al. 2013).
- *Epoxidation*: Conversion of alkenes to epoxides often serves as an intermediate step for subsequent ring cleavage (Mattila et al. 2022).
- *Dealkylation and Demethylation*: Many persistent pollutants, including pesticides and pharmaceuticals, undergo P450-mediated modifications that render them more susceptible to further degradation or conjugation (Migliore et al. 2022).

6.2.2.3 Examples of Xenobiotic Transformations

- *Agricultural Pesticides*: CYP enzymes hydroxylate organophosphates and carbamates, facilitating their breakdown (Mori et al. 2021).
- *Pharmaceuticals*: Antidepressants, antibiotics, and endocrine-disrupting compounds undergo P450-mediated metabolism in fungal cells, paralleling mammalian detoxification pathways (Muter 2023).

- *Aromatic Hydrocarbons*: PAHs such as benzopyrene are attacked by fungal P450s, producing hydroxylated derivatives that can subsequently be mineralized through extracellular peroxidases (Ortúzar et al. 2022).

6.2.2.4 Integration with Phase II Metabolism

Following P450-mediated reactions, fungi often employ phase-II conjugation mechanisms (such as glycosylation, sulfation, and glutathione conjugation) to detoxify xenobiotic intermediates. These modifications increase polarity and promote excretion or sequestration within fungal cells, preventing toxic accumulation (Park and Choi 2020). This two-step detoxification cycle mirrors higher organisms' defense systems, highlighting the evolutionary significance and ecological robustness of fungal metabolism.

6.2.3 Synergistic Interplay Between Extracellular and Intracellular Systems

The efficiency of fungal xenobiotic degradation arises not from isolated enzyme activities, but from the synergistic interplay between extracellular and intracellular enzyme systems. Extracellular enzymes initiate depolymerization and oxidative transformations of bulky or insoluble contaminants, generating smaller, partially oxidized intermediates (Park et al. 2019). These intermediates can then diffuse into fungal cells, where cytochrome P450s and other intracellular enzymes complete detoxification. Conversely, intracellular metabolism can produce metabolites that act as natural mediators, enhancing extracellular enzyme activity (Passarini et al. 2021). This two-tiered strategy reflects the ecological adaptability of fungi to persist in chemically hostile environments and underscores their potential for effective bioremediation strategies.

6.2.4 Environmental and Biotechnological Significance

The enzymatic mechanisms described above have profound implications for bioremediation:

- *Versatility*: Fungal enzymes can degrade structurally diverse compounds, reducing dependence on pollutant-specific pathways (Pozdnyakova et al. 2016).
- *Scalability*: Extracellular enzyme systems function without substrate uptake, enabling field-level applications such as mycoreactors, soil inoculation, and biofilters.
- *Sustainability*: The reliance on naturally occurring fungal systems and low-cost inducers (e.g., lignocellulosic wastes) offers an economically viable alternative to chemical treatments (Rathankumar et al. 2022).
- *Engineering Opportunities*: Advances in fungal genomics and metabolic engineering provide avenues to overexpress key enzymes, optimize P450 systems, and design fungal consortia with enhanced remediation efficiency (Romero-Reyes et al. 2023).

Fungi represent a uniquely powerful tool for xenobiotic detoxification due to their dual arsenal of extracellular oxidative enzymes and intracellular cytochrome P450 monooxygenases. Lignin peroxidases and manganese peroxidases generate radical-mediated oxidative pathways capable of dismantling recalcitrant pollutants, while laccases provide complementary oxygen-dependent transformations broadened by mediators (Rughöft et al. 2020). Intracellularly, cytochrome P450s perform hydroxylation, epoxidation, and dealkylation that metabolize xenobiotics into less toxic, more soluble derivatives. The integration of these enzymatic systems yields a highly versatile remediation platform, offering both practical efficiency and ecological sustainability (Satpati et al. 2023). A summary of some of the reported fungal species used in xenobiotic mycoremediation applications is highlighted in Table 6.1. As increasing volumes of xenobiotic pollutants challenge global environmental health, harnessing fungal mechanisms provides a forward-looking approach. Continued exploration of fungal enzyme dynamics, coupled with genetic engineering and process optimization, will enhance the scalability and predictability of mycoremediation strategies, placing fungi at the forefront of sustainable environmental biotechnology (Singh and Tripathi 2022).

6.3 Target Xenobiotic Compounds

The environmental burden of xenobiotic pollutants arises not only from their widespread use but also from their persistence, toxicity, and resistance to natural degradation. Unlike naturally occurring organic matter, xenobiotics are often anthropogenically synthesized molecules designed for chemical stability and biological potency. This resistance to degradation poses profound risks: persistence in soils and waters, bioaccumulation in living organisms, biomagnification through food webs, and long-term ecological toxicity (Singh et al. 2023). Mycoremediation has emerged as a promising solution to address these contaminants due to the broad substrate specificity of fungal enzyme systems, particularly those of white-rot fungi. The following sections highlight three major classes of xenobiotics that represent critical targets for fungal detoxification: persistent organic pollutants (POPs), pharmaceutical compounds, and agricultural chemicals (Soares et al. 2021).

6.3.1 Persistent Organic Pollutants (POPs)

6.3.1.1 Characteristics and Challenges

Persistent organic pollutants are among the most difficult contaminants to remediate because of their hydrophobicity, halogen substitutions, and resistance to biological attack. POPs, such as polychlorinated biphenyls (PCBs), dioxins, and organochlorine pesticides (e.g., DDT, aldrin, and chlordane), persist in the environment for decades (Sun et al. 2024). Their lipophilicity promotes accumulation in soil organic matter and aquatic sediments, from where they enter food chains and bioaccumulate in animal and human tissues. Chronic exposure is linked to carcinogenicity,

Table 6.1 Some of the reported fungal species used in xenobiotic mycoremediation applications

Fungal species	Primary enzyme systems	Optimal environmental conditions	Target contaminant classes	Typical degradation rates (lab/field)	Latest evidence
Phanerochaete chrysosporium	LiP, MnP, laccases	Temp: 25–35 °C; pH 5.5–7.5; moisture 50–70%	PAHs, chlorinated hydrocarbons, pesticides	60–90% (PAHs); 50–80% (PCBs) (Tien and Kirk 1983; Pointing 2001)	New enzymatic pathways for phenanthrene degradation identified (flavoprotein monooxygenase and intradiol dioxygenases), expanding known ligninolytic and non-ligninolytic routes (Zhang et al. 2023; Fonseka et al. 2025). Recent reviews/analyses reaffirm *P. chrysosporium* as a model white-rot degrader and discuss process variables that improve PAH/PCB removal (Torres-Farradá et al. 2024).
Trametes versicolor	Laccases, MnP	Temp: 20–30 °C; pH 4.5–6.5	Dyes, pharmaceuticals (e.g., BPA, estradiols), pesticides	70–85% (dyes); 60–75% (estradiols) (Bumpus and Brock 1988; Rodríguez-Couto 2009)	Recent work shows highly effective immobilized *T. versicolor* laccase formulations with improved thermal/pH stability and higher BPA removal (Li et al. 2024; Guo et al. 2024). Genomic profiling studies explain strong benzopyrene degradation potential (Chmelová et al. 2024). Reviews in 2024–2025 update laccase applications to pharmaceutical removal (Ortúzar et al. 2022; Sun et al. 2024).

Pleurotus ostreatus	LiP, laccases	Temp: 20–28 °C; pH 5–6.5	Organophosphates, carbamates, petroleum hydrocarbons	50–80% (pesticides) (Pointing 2001; Kulshreshtha et al. 2014)	Field/soil trials and pilot biopile studies (2024–2025) show spent mushroom substrate (SMS) and *P. ostreatus* amendments significantly boost petroleum hydrocarbon (TPH/PAH) removal in contaminated soils (Dey et al. 2024; Curiel-Alegre et al. 2024). Recent syntheses report high lab reductions up to 80–87% TPH in optimized setups (Zhao et al. 2022; Satpati et al. 2023)
Indigenous consortia	Mixed enzymatic profiles (fungal + bacterial)	Site-specific—adapt amendments to local soil/water	Mixed organics, pharmaceuticals, pesticides, petroleum hydrocarbons	Variable; site- and amendment-dependent (Harms et al. 2011; Singh 2006)	2024 reviews and experimental studies highlight engineered or prospectively selected fungal-bacterial consortia as especially effective for petroleum/PAH removal (Xiang et al. 2022; Champramary et al. 2023). Strategies include native-isolate consortia, designed synthetic consortia, and combinations with soil amendments to improve stability and degradation rates (Zou et al. 2024; Wang et al. 2023).

endocrine disruption, neurotoxicity, and reproductive disorders, making their removal an urgent environmental health concern (Torres-Farradá et al. 2024).

6.3.1.2 Fungal Degradation of POPs

White-rot fungi, especially *Phanerochaete chrysosporium*, have been extensively studied for their capacity to break down chlorinated aromatic hydrocarbons.

- *PCBs*: Experimental studies have demonstrated extensive PCB dechlorination and ring cleavage using lignin peroxidases and manganese peroxidases. The efficiency varies depending on the degree and position of chlorination, with lower-chlorinated congeners being more susceptible (Tortella et al. 2005). Unlike bacterial systems that frequently halt at partially dechlorinated intermediates, fungi can completely mineralize PCBs under optimized conditions.
- *Dioxins*: Fungal enzymes can initiate hydroxylation and ring cleavage reactions that destabilize the highly aromatic structures of dioxins (Wahla et al. 2023). *Trametes versicolor* has been reported to significantly reduce the toxicity of dioxin-contaminated soils.
- *Organochlorine pesticides*: Several fungi, including *Pleurotus ostreatus* and *Ganoderma lucidum*, have demonstrated the ability to degrade DDT and lindane through oxidative dechlorination mechanisms (Wahla et al. 2023).

6.3.1.3 Advantages Over Conventional Methods

Conventional methods for POP remediation, such as incineration and chemical oxidation, are energy-intensive, costly, and generate harmful byproducts. By contrast, fungal mycoremediation offers a biological system that not only reduces pollutant load but also detoxifies intermediates. Moreover, fungal mycelia can extend into sediment matrices, enabling biodegradation in otherwise inaccessible microenvironments (Wang et al. 2021).

6.3.2 Pharmaceutical Compounds

6.3.2.1 Emerging Contaminants of Concern

In recent decades, pharmaceutical residues have emerged as a new class of "contaminants of emerging concern" (CECs). Drugs, including antibiotics, hormones, analgesics, antidepressants, and anti-inflammatory agents, are increasingly detected in hospital effluents, municipal wastewater, and surface waters. Conventional wastewater treatment plants are often ill-equipped to remove these micropollutants because pharmaceuticals are designed to resist metabolic breakdown and maintain biological activity at low concentrations. Even low-dose environmental exposure can cause antimicrobial resistance (AMR), endocrine disruption, and ecotoxicological stress on aquatic organisms (Wang et al. 2023).

6.3.2.2 Mycoremediation Potentials

Fungi are particularly effective against pharmaceuticals due to their nonspecific enzyme systems, which can oxidize diverse xenobiotics across different therapeutic categories.

- *Hormones and Endocrine-Disrupting Compounds (EDCs): Trametes versicolor* has demonstrated complete mineralization of hormones, such as 17β-estradiol and ethinylestradiol, both potent endocrine disruptors. This ability to cleave steroid ring structures represents a significant advantage over bacterial pathways that typically yield only partially transformed, biologically active metabolites.
- *Antibiotics*: Research has shown that fungi, including *Aspergillus spp.* and *Pleurotus ostreatus*, can degrade tetracyclines, sulfonamides, and fluoroquinolones. This capacity helps mitigate the development of antimicrobial resistance in aquatic environments (Xiang et al. 2022).
- *Anti-Inflammatory Drugs*: Nonsteroidal anti-inflammatory drugs (NSAIDs) such as diclofenac, ibuprofen, and naproxen persist in municipal wastewater effluents. *Phanerochaete chrysosporium* and *Coriolus versicolor* have been reported to degrade these compounds, often via hydroxylation and subsequent ring cleavage.

6.3.2.3 Mechanistic Insights

Laccase and peroxidase enzyme systems oxidize aromatic rings and phenolic structures found in many drugs, while cytochrome P450 monooxygenases hydroxylate or epoxidize xenobiotics within fungal cells (Xu et al. 2023). This dual-action extracellular breakdown and intracellular transformation enables complete detoxification rather than partial modification, a significant improvement compared to bacterial treatments.

6.3.2.4 Environmental Implications

The fungal degradation of pharmaceuticals is not merely about pollutant removal it directly contributes to public health protection by reducing the risk of hormone disruption in aquatic organisms, mitigating antibacterial resistance, and lowering toxicological risks in drinking water sources. Importantly, fungal remediation technologies can be integrated into biofilters for hospital effluent treatment and tertiary wastewater treatment units, making them viable at scale (Xu et al. 2023).

6.3.3 Agricultural Chemicals

6.3.3.1 Pesticide Residues as Global Pollutants

Agricultural intensification has led to heavy reliance on pesticides, herbicides, and fungicides, many of which are xenobiotic in nature and persist in soils and groundwater. Runoff from agricultural fields contaminates rivers and lakes, while residues often persist long-term in soils, posing risks to pollinators, soil biodiversity, and human health (Ye et al. 2024).

Key xenobiotic pesticide categories include:

- Organophosphates (e.g., chlorpyrifos, malathion, parathion)
- Carbamates (e.g., carbaryl, aldicarb)
- Triazine herbicides (e.g., atrazine, simazine)
- Chlorinated insecticides (e.g., aldrin, heptachlor, endosulfan)

6.3.3.2 Fungal Biodegradation of Pesticides

Mycoremediation offers an effective and economical approach for the simultaneous treatment of mixed pesticide contamination.

- *Organophosphates*: *Pleurotus ostreatus* degrades chlorpyrifos and malathion by hydrolytic cleavage of ester bonds, reducing toxicity substantially.
- *Triazine Herbicides*: *Fusarium oxysporum* and *Trametes versicolor* have been reported to cometabolize and degrade atrazine, one of the most recalcitrant herbicides, via hydroxylation and oxidative ring cleavage.
- *Carbamates*: Enzymes from *Pleurotus* spp. catalyze hydrolysis and demethylation, reducing potency and toxicity (Zhang et al. 2023).

6.3.3.3 Mixed Pesticide Remediation

A major advantage of fungal systems is their ability to attack multiple chemical classes simultaneously. While bacterial consortia often require specific strains for each pesticide, a single fungal inoculation, whether using white-rot fungi or edible mushrooms, can significantly reduce multiple contaminants (Zhao et al. 2022). This versatility enhances the economic feasibility of large-scale field remediation projects.

6.3.4 Comparative Advantages and Challenges

The degradation of these three xenobiotic categories—POPs, pharmaceuticals, and pesticides—demonstrates the breadth of fungal catabolism. However, challenges remain:

- Environmental Variability: Field conditions (pH, temperature, oxygen, co-contaminants) significantly influence degradation efficiency.
- Partial Mineralization: While fungi excel at initiating degradation, ensuring complete detoxification requires careful monitoring of transformation products.
- Scaling Issues: Laboratory studies provide promising results, but translation to real-world contaminated soils and wastewater systems often requires bioreactor optimization, inoculation methods, and substrate amendment strategies.

Despite these limitations, the potential is enormous (Zhuo and Fan 2021). Mycoremediation not only directly reduces pollutant levels but also integrates well with circular economic principles. Agricultural residues can serve as fungal substrates, and fungal biomass produced during remediation can be valorized into animal feed, biofertilizers, or biopolymers, coupling waste treatment with resource recovery (Zou et al. 2024).

Xenobiotics, whether arising from industrial production, medical systems, or agricultural intensification, represent one of the greatest environmental challenges of the twenty-first century. Among these, persistent organic pollutants, pharmaceutical residues, and agricultural pesticides stand out as particularly harmful due to their toxicity, persistence, and ubiquity (Jaiswal and Shukla 2020). Fungal mycoremediation offers unique advantages: non-specific enzymatic systems that degrade structurally diverse compounds, extracellular oxidative machinery that circumvents constraints of cellular uptake, and versatile metabolic pathways that enable mineralization of toxic intermediates (Kathhivaran and Gnanadoss 2021).

By degrading PCBs, dioxins, and organochlorine pesticides, fungi address some of the most intractable legacy pollutants. Through the removal of hormones, antibiotics, and anti-inflammatory drugs, fungi help mitigate the emerging crisis of pharmaceutical pollution and antimicrobial resistance (Mattila et al. 2022). By managing pesticide residues in soils and water, fungi contribute to safer, more sustainable agricultural practices.

In all cases, fungal systems demonstrate both environmental relevance and technological promise. When properly harnessed and integrated with existing waste management and remediation infrastructures, mycoremediation has potential not only to detoxify heavily contaminated environments but also to enable sustainable cycles of waste valorization. Table 6.2 highlights prevailing regulatory frameworks and approval requirements for mycoremediation implementation across different parts of the world. Future research should continue to refine fungal strain selection, engineer more pollutant-tolerant consortia, and develop scalable treatment frameworks (Migliore et al. 2022). Ultimately, fungi stand as powerful allies in global efforts to tackle xenobiotic pollution while advancing sustainable and circular strategies for environmental protection.

6.4 Application Strategies and Technologies

The field of mycoremediation has evolved significantly over the past three decades, moving from laboratory demonstrations of fungal xenobiotic degradation to applied strategies designed for field-scale implementation. Central to this translation is the choice of application strategy, whether fungal inoculation and biodegradation processes are carried out directly at contaminated sites (in situ) or within engineered systems under controlled conditions (ex situ) (Passarini et al. 2021). Each approach involves distinct benefits, limitations, and technological considerations, and the choice often depends on site characteristics, contaminant type, cost constraints, and regulatory frameworks.

6.4.1 In Situ Approaches

In situ *mycoremediation* refers to the application of fungi directly to contaminated sites such as soils, sediments, or groundwater systems without the need for

Table 6.2 Regulatory frameworks and approval requirements for mycoremediation implementation across major jurisdictions

Jurisdiction	Regulatory body	Required approvals	Safety assessments	Monitoring protocols	Approval timelines
United States	EPA (Environmental Protection Agency)	Bioremediation permit, ecological risk assessment (EPA 2023)	Introduced organism risk evaluation, NPDES permits (EPA 2023)	Quarterly sampling over 3–5 years	6–12 months
European Union	ECHA (European Chemicals Agency)	REACH compliance, GMO considerations (ECHA 2022)	Ecotoxicology, soil/water quality monitoring (ECHA 2022)	Long-term post-remediation monitoring	12–18 months
Canada	Canada - Environment and Climate Change Canada (CECCC 2017)	Site-specific remediation plan approval (ECCC 2023)	Indigenous species preference, ecological impact (ECCC 2023)	Annual reporting and groundwater monitoring	9–15 months
Japan	Ministry of Environment, Japan (2015)	Environmental impact assessment, GMO usage permits (MOE Japan 2022)	Biosafety evaluation, native microbial community impact (MOE Japan 2022)	Real-time monitoring with adaptive management	12 months
Australia	Australia - Department of Agriculture, Water and Environment (2019)	Remediation permit, risk-based site management (DAWE 2021)	Risk assessments for introduced species (DAWE 2021)	Scheduled monitoring and reporting	6–12 months
India	MoEFCC, CPCB, SPCBs	CTE/CTO permits, contaminated site remediation per draft hazardous and other wastes (Management and Transboundary Movement) Rules 2024 (India MoEFCC 2024)	Environmental impact assessment, biosafety evaluation for introduced organisms (India MoEFCC 2024)	Quarterly water/ soil monitoring, annual reporting	6–18 months

excavation, transportation, or extensive preprocessing. This strategy has gained increasing interest due to its potential to treat pollutants in place with minimal ecological disturbance (Rathankumar et al. 2022). Unlike conventional remediation methods such as soil washing or incineration, in situ fungal treatment emphasizes long-term, naturally mediated processes that combine biodegradation with ecosystem restoration.

6.4.1.1 Bioaugmentation

One of the most common in situ techniques is bioaugmentation, which involves inoculating contaminated environments with specific fungal strains known for strong xenobiotic-degrading capabilities. For example, wood-decaying white-rot fungi such as *Phanerochaete chrysosporium*, *Trametes versicolor*, and *Pleurotus ostreatus* are widely applied due to their extracellular enzymatic arsenal (Satpati et al. 2023).

- *Implementation*: Bioaugmentation requires the preparation of fungal biomass (spores or mycelium), often immobilized on carrier materials such as sawdust, straw, corn cobs, or alginate beads, to enhance survival and dispersion. Carriers also provide sustained nutrient sources, enabling fungal colonization.
- *Nutrient Amendments*: To improve fungal establishment, organic amendments (e.g., lignocellulosic residues or molasses) are introduced to stimulate enzyme production. These aid the co-metabolic degradation of xenobiotics by inducing ligninolytic pathways.
- *Advantages*: Bioaugmentation introduces robust degrader strains that can accelerate pollutant breakdown even in harsh or contaminated conditions where indigenous fungi are insufficient (Singh and Tripathi 2022).
- *Challenges*: Success depends heavily on matching strains with environmental conditions. Soil chemistry, pH, oxygen levels, and temperature fluctuations can inhibit introduced fungi. Additionally, competition with indigenous microbial communities may reduce colonization efficiency (Singh et al. 2023).

6.4.1.2 Biostimulation

An alternative strategy is biostimulation, which enhances the activity of indigenous fungal populations rather than introducing non-native organisms.

- *Mode of Action*: This involves optimizing environmental parameters or adding low-cost stimulants (organic matter, trace nutrients, or electron donors) that encourage local fungi to express degradative enzymes. For example, sawdust amendments in pesticide-contaminated agricultural fields can stimulate ligninolytic fungi already present within soil ecosystems (Soares et al. 2021).
- *Sustainability Advantage*: Biostimulation is often more sustainable because it avoids risks associated with ecological imbalance or genetic drift that may accompany the introduction of external strains. It also reduces regulatory hurdles, as no alien species are introduced into the environment.

- *Limitations*: Indigenous fungi may lack the required catabolic pathways for certain pollutants, and stimulants may only partly enhance xenobiotic degradative potential. Moreover, results can be inconsistent across different soil or climatic conditions (Sun et al. 2024).

6.4.1.3 Hybrid and Supportive Approaches

Practical field applications often employ hybrid strategies combining bioaugmentation with biostimulation, introducing fungal inoculum alongside supplementation with organic residues or natural mediators. For instance, in oil spill remediation, fungi inoculated into contaminated soils may be co-cultivated with lignocellulose to stimulate enzyme induction (Wahla et al. 2023). Additionally, mycorrhizal fungi may be integrated into phytoremediation efforts, where plants and fungi establish synergistic associations that enhance pollutant uptake and biodegradation.

6.4.1.4 Real-World Applications

Several case studies illustrate the promise of in situ fungal remediation:

- PCB-contaminated soils treated with *P. chrysosporium* showed notable dechlorination and ring cleavage within months when supplemented with straw.
- Petroleum hydrocarbons in oil-polluted soils have been degraded by *P. ostreatus*, with efficiency enhanced by cocultivation with crop residues.
- Atrazine-contaminated agricultural soils showed significant reductions when fungal inoculum was applied under controlled irrigation and nutrient input.

These successes highlight the adaptability and ecological compatibility of in situ fungal treatment, though consistent performance across sites remains a continuing research challenge (Ye et al. 2024).

6.4.2 Ex Situ Systems

Ex situ *mycoremediation* refers to the treatment of contaminated soils, sediments, or effluents outside their original location, within engineered facilities where conditions can be carefully controlled. This approach is particularly useful for heavily polluted sites with complex contaminant mixtures, high toxicity, or acute risk to health, where immediate containment and effective biodegradation are necessary.

6.4.2.1 Solid-Phase Bioreactor Systems

Solid-phase bioreactors are one of the most advanced ex situ fungal treatment technologies. Excavated contaminated soils or solid waste materials are introduced into reactors inoculated with fungi, along with bulking agents such as straw or woodchips.

- Process Control: Bioreactors allow fine-tuned regulation of key parameters moisture, aeration, temperature, and pH, enabling optimal fungal growth and enzyme secretion.

- Advantages: Enhances efficiency of degradation, reduces variability, and minimizes risks associated with environmental fluctuations (Guo et al. 2024).
- Applications: Used for soils contaminated with polycyclic aromatic hydrocarbons (PAHs), chlorophenols, pharmaceutical residues, and dye-contaminated effluents. *Trametes versicolor* and *Coriolus hirsutus* have demonstrated success in semi-pilot-scale bioreactors handling mixed contaminants.

6.4.2.2 Composting and Landfarming Systems

Composting systems represent a cost-effective *ex situ* application in which contaminated soils are mixed with organic amendments (compost, manure, straw) to stimulate fungal activity. The heat generated during composting also enhances pollutant volatilization and microbial community synergy.

- *Advantages*: Simple, low-cost design that can be implemented on a medium scale.
- *Limitations*: Less controlled than reactor systems; efficiency varies with pollutant type (better for pesticides and PAHs than halogenated compounds).

Landfarming variants involve spreading contaminated soils over large, prepared surfaces inoculated with fungi, periodically aerated, irrigated, and amended with nutrients. This semi-controlled strategy offers cost efficiency for large volumes of moderately contaminated soils (Dey et al. 2024).

6.4.2.3 Slurry and Suspension Bioreactors

For aqueous or semisolid wastes (e.g., sludge, wastewater containing pharmaceuticals or dyes), slurry-phase systems provide high fungal contact with pollutants. These systems allow direct mixing, agitation, and aeration for uniform treatment (Dey et al. 2020). *Phanerochaete chrysosporium* slurry reactors have successfully degraded textile dyes and antibiotics under optimized redox conditions.

Advantages of Ex Situ Systems

- High control over conditions → reproducibility and efficiency
- Ability to monitor pollutant degradation, enzyme activity, and toxic intermediates
- Applicability to highly contaminated or mixed waste streams

Limitations

- Costs associated with excavation, transport, and engineering infrastructure can be prohibitive.
- Treatment capacity may be limited depending on reactor size and throughput.
- Potential risk of recontamination during soil removal and transport.

6.4.3 Comparative Perspectives

Both in situ and ex situ methods highlight the versatility of fungal remediation:

- *In situ approaches* are most suited for diffuse, moderately contaminated sites where low-cost, sustainable long-term treatment is desired with minimal distur-

bance. They rely heavily on ecological adaptation and synergy with indigenous organisms (Bilal et al. 2018).
- *Ex situ systems* excel with hotspots of severe pollution, industrial waste, or complex mixtures where precision, speed, and containment are priorities. These systems are more expensive but provide reliable control, monitoring, and compliance with strict remediation targets (Birolli et al. 2020).

Hybrid strategies are evolving toward integrated frameworks, where initial containment and treatment using ex situ reactors are followed by in situ bioaugmentation to restore ecological functionality.

6.4.4 Future Directions in Application Technologies

Looking ahead, technological innovations are likely to enhance the feasibility and efficiency of fungal remediation strategies:

1. *Immobilization Technologies*: Use of biochar, alginate beads, nanomaterials, or synthetic carriers to immobilize fungi for enhanced resilience in soil and water.
2. *Process Integration*: Coupling mycoremediation with phytoremediation, bacterial consortia, and physicochemical pretreatments (e.g., advanced oxidation) to enhance pollutant degradation.
3. *Automation and Monitoring*: Biosensors and molecular tools for real-time monitoring of fungal activity and contaminant breakdown.
4. *Circular Economy Applications*: Valorizing fungal biomass after remediation into biofertilizers, animal feed additives, or biopolymers.
5. *Standardized Protocols*: Developing guidelines that specify fungal strain selection, inoculum preparation, and treatment conditions tailored to different xenobiotic classes.

The practical application of mycoremediation depends on carefully selecting between in situ and *ex situ* strategies based on contamination characteristics, cost constraints, and site management priorities. In situ bioaugmentation and biostimulation provide sustainable, low-cost options for long-term ecological restoration, while ex situ bioreactors and composting systems offer superior control and faster pollutant removal for highly contaminated environments (Champramary et al. 2023). Integration of these approaches, supported by technological innovations and ecological insights, will be critical for scaling mycoremediation from promising research to widespread environmental practice. By leveraging fungi's inherent catabolic versatility through these application strategies, mycoremediation can transition from isolated case studies to robust, field-ready solutions for xenobiotic detoxification across diverse environmental settings (Chane et al. 2024).

6.5 Case Studies and Field Applications

Successful real-world applications of mycoremediation provide valuable insights into the opportunities and challenges of scaling laboratory findings to the field. While mechanistic studies demonstrate the potential of individual fungi to degrade xenobiotics under controlled conditions, case studies illustrate the importance of site-specific adaptation, project design, and environmental management (Akhtar and Mannan 2020).

6.5.1 European Industrial Site Remediation

One of the most well-documented large-scale projects occurred at a former chemical manufacturing facility in Germany, where soils and sediments were contaminated with polycyclic aromatic hydrocarbons (PAHs) and chlorinated solvents following decades of industrial operations (Akerman-Sanchez and Rojas-Jimenez 2021). Recognizing that excavation and high-temperature incineration would be prohibitively expensive and disruptive, regulators approved a pilot *ex situ* biotreatment system employing fungi.

- *Implementation*: Treatment beds were constructed with excavated soils, incorporating bulking agents such as straw and woodchips to improve aeration. *Phanerochaete chrysosporium* and *Trametes versicolor* were co-inoculated to exploit their complementary enzymatic profiles.
- *Management*: Moisture levels were maintained between 50% and 60% of field capacity using irrigation systems, while nutrient supplementation (molasses and nitrogen-rich amendments) was applied every 8 weeks to sustain fungal activity. Weather shelters were installed to maintain consistent conditions and prevent runoff.
- *Outcomes*: Over 18 months, contaminant concentrations were reduced by 85%, with notable decreases in both high-molecular-weight PAHs (benzopyrene, chrysene) and chlorinated solvent residues. Microtox assays confirmed reduced toxicity.

This case demonstrates the feasibility of mycoremediation for industrial legacy pollutants and highlights how careful environmental management (pH, aeration, moisture) plays as significant a role as fungal inoculation itself (Amobonye et al. 2023).

6.5.2 North American Agricultural Site Treatment

In California's Central Valley, an agricultural site heavily contaminated with pesticide residues posed significant risks to groundwater and soil fertility. The pesticides included organophosphate insecticides (chlorpyrifos, malathion), triazine herbicides (atrazine, simazine), and carbamates.

- *Strategy*: The remediation team opted for an in situ *bioaugmentation* strategy using locally isolated fungal strains rather than laboratory reference cultures. The decision was based on the recognition that indigenous fungi are often better adapted to local soil and climate conditions, providing greater resilience to environmental fluctuations (Bokade and Bajaj 2023).
- *Implementation*: Straw mulching and irrigation were applied to maintain soil conditions. Periodic additions of compost supported a favorable carbon-to-nitrogen ratio, encouraging ligninolytic activity.
- *Observations*: Effectiveness varied with season; degradation was most efficient in spring and autumn, when moderate temperatures and moisture supported optimal fungal growth.
- *Results*: Within 12 months, organophosphate residues were reduced by 70%, triazines by 55%, and carbamates by 60%.

The case underscores the importance of strain selection and ecological adaptation. Locally adapted fungi often outperform “model strains” in the field, suggesting the value of bioprospecting indigenous fungal isolates for regional contamination challenges (Hu et al. 2021).

6.5.3 Asian Pharmaceutical Manufacturing Waste

In India, a pharmaceutical manufacturing site discharged residues comprising antibiotics, solvents, and intermediate products. Standard biological treatment in wastewater plants proved inadequate for these xenobiotics. Researchers implemented an ex situ consortium-based mycoreactor system.

- *Approach*: A fungal consortium including *Aspergillus niger*, *Trametes versicolor*, and *Pleurotus ostreatus* was designed to exploit complementary degradative pathways. *A. niger* tolerated high solvent levels, *T. versicolor* produced high levels of laccases targeting aromatic residues, while *P. ostreatus* metabolized antibiotics.
- *Bioreactor Design*: A slurry-phase bioreactor with controlled aeration and nutrient dosing was constructed. Wastewater was continuously circulated, ensuring consistent fungal-pollutant contact.
- *Outcome*: Antibiotic residues (fluoroquinolones, macrolides) were reduced by 80–90% within 30 days, and overall chemical oxygen demand (COD) was reduced by 70%. Toxicity to aquatic microalgae was markedly diminished after treatment (Satpati et al. 2023).

This case illustrates the power of fungal consortia and reactor-based controlled systems for high-strength industrial effluents with diverse chemical burdens.

6.6 Factors Affecting Remediation Efficiency

The success of fungal remediation depends on a range of abiotic and biotic factors that interact with fungal physiology, enzyme production, and contaminant bioavailability.

6.6.1 Environmental Parameters

- *Temperature*: Most white-rot fungi operate optimally between 25 and 35 °C. Seasonal variations strongly influence treatment efficiency; protective measures such as shading, insulation, or thermoregulated composting may be necessary during extremes.
- *Moisture*: Fungal mycelial growth requires adequate soil moisture. Studies show that the most effective degradation occurs at 40–70% of field capacity. Values above 80% risk creating anaerobic conditions that suppress ligninolytic enzyme activity (Soares et al. 2021).
- *pH*: Optimal fungal growth and enzyme activity generally occur in slightly acidic to neutral soils (pH 5.5–7.5). Highly acidic or alkaline conditions may require lime or sulfur amendments.
- *Oxygen Availability*: Aerobic conditions are critical for oxidative enzymes such as laccases and peroxidases; poorly aerated soils or oversaturated substrates hinder enzymatic expression.

6.6.2 Contamination Characteristics

- *Concentration*: While fungi tolerate moderate pollutant loads, extremely high contaminant levels can be toxic to fungal cells, mandating staged treatment or dilution. For example, PAH concentrations above 5000 mg/kg often require pretreatment.
- *Bioavailability*: Hydrophobic compounds bound strongly to soil organic matter may resist enzymatic attack. Surfactants, amendments (biochar, humic acids), or fungal surfactants (biosurfactants) can enhance desorption (Wahla et al. 2023).
- *Co-contamination*: Pollutant mixtures can interact antagonistically or synergistically. For instance, heavy metals often inhibit fungal enzyme expression, while certain aromatic hydrocarbons induce ligninolytic enzymes, stimulating degradation. Understanding these interactions is crucial for predicting outcomes.

6.7 Economic and Regulatory Considerations

6.7.1 Cost–Benefit Analysis

Economic feasibility remains a driver for technological adoption.

- *Cost Advantage*: Mycoremediation is typically 50–75% cheaper than thermal or chemical remediation. Costs are primarily associated with inoculum production, substrate carriers, nutrient supplementation, and monitoring.
- *Lifecycle Perspective*: Although treatment timeframes extend to 6–24 months, the low energy inputs and minimal secondary waste generation result in favorable lifecycle analyses compared to energy-intensive technologies.
- *Constraints*: Delays may affect sites requiring urgent remediation. Decision-makers must weigh slower kinetics against sustainability and cost savings (Wang et al. 2021).

6.7.2 Regulatory Framework

Adoption of mycoremediation is influenced by regulatory and safety considerations.

- *Jurisdictional Variation*: Europe and North America have emerging frameworks for approving bioremediation projects, while other regions lack specific guidelines.
- *Risk Assessments*: Authorities require evidence that introduced fungi will not adversely alter local ecosystems. Use of indigenous strains often reduces regulatory hurdles.
- *Monitoring*: Regulators often mandate long-term monitoring beyond active remediation to confirm pollutant stability and ecological safety (Satpati et al. 2023).

6.8 Integration with Other Technologies

Fungal remediation rarely exists in isolation; instead, hybrid and enhanced systems often maximize efficiencies.

6.8.1 Hybrid Remediation Systems

- *Fungi + Bacteria*: Sequential fungal pretreatment to oxidize persistent compounds, followed by bacterial mineralization of intermediates, has proven effective for chlorinated organics.
- *Fungi + Plants (Phytoremediation):* Plant roots support fungal colonization, while fungi improve rhizosphere pollutant turnover. For instance, willows and poplars associated with white-rot fungi show enhanced removal of PAHs (Mori et al. 2021).

6.8.2 Enhanced Remediation Approaches

- *Bioelectrochemical Systems*: Low-voltage electrical stimulation can enhance ligninolytic enzyme activity, accelerating degradation rates.
- *Nanotechnology in Mycoremediation*: Fungal enzymes immobilized on nanocarriers exhibit greater stability; nano-structured supports can also deliver nutrients or mediators directly to polluted micro-sites. Risks associated with nanomaterial toxicity must, however, be managed (Mori et al. 2021).

6.9 Future Research Directions

As with other biotechnologies, research and innovation play key roles in expanding the scope and scalability of mycoremediation.

6.9.1 Genetic Engineering Approaches

- *CRISPR-based Editing*: Targeted modifications of ligninolytic enzymes (LiP, MnP, laccases) could improve redox potentials and substrate ranges (Mori et al. 2021).
- *Synthetic Biology*: Introduction of non-native degradation pathways may allow fungi to metabolize emerging contaminants such as PFAS or microplastics additives.
- *Challenges*: Public perception, GMO regulations, and ecological safety testing will be critical before field deployment (Mori et al. 2021).

6.9.2 Fungal Consortia Development

- *Multi-Species Consortia*: By assembling species with complementary enzymatic profiles, researchers can target complex pollutant mixtures, as shown in the Indian pharmaceutical case study.
- *Fungi-Bacteria Synergy*: Coculture systems leverage bacterial pathways for mineralization while fungi initiate oxidative transformation (Mori et al. 2021).

6.9.3 Advanced Monitoring and Control

- Biosensors and Omics Tools: Molecular methods (qPCR, metatranscriptomics) and biosensors can provide real-time monitoring of enzyme expression and pollutant breakdown.

- Predictive Modelling: Machine learning-based algorithms are emerging tools to predict success rates and optimize remediation parameters, reducing trial-and-error designs (Fonseka et al. 2025).

6.10 Conclusions and Future Prospects

Mycoremediation has advanced from conceptual promise to practical reality through field demonstrations across diverse contamination contexts, from industrial soils in Europe to agricultural pesticide residues in North America and pharmaceutical effluents in Asia. These successes confirm fungi's broad substrate specificity, ecological resilience, and cost-effectiveness.

- Ecological Relevance: Fungi not only degrade pollutants but also contribute to soil restoration and ecological balance, unlike many physicochemical methods.
- Economic Advantage: Cost reductions of up to 75% compared with conventional methods make mycoremediation attractive, particularly in developing regions facing widespread contamination.
- Integration Potential: Synergy with phytoremediation, bacterial systems, and advanced technologies enhances scope and efficiency.
- Future Vision: Genetic engineering, synthetic biology, advanced monitoring, and fungal consortia design will likely extend fungal treatment to pollutants currently recalcitrant to biodegradation.

As regulatory frameworks mature and environmental pressures intensify, mycoremediation is expected to emerge as a mainstream remediation technology. Furthermore, integration within circular economy principles, such as valorizing remediated fungal biomass into biofertilizers, biopolymers, or feedstock, aligns environmental cleanup with resource recovery. This dual benefit strengthens the case for fungi not simply as "nature's recyclers" but as cornerstones of sustainable environmental biotechnology.

Acknowledgment PS, GSJ, and SJJ kindly acknowledge DST-FIST grant to AIMT, AUR (Project No.: SR/FST/LS-I/2019/502).

References

Akerman-Sanchez G, Rojas-Jimenez K (2021) Fungi for the bioremediation of pharmaceutical-derived pollutants: a bioengineering approach to water treatment. Environ Adv 4:100071. https://doi.org/10.1016/j.envadv.2021.100071

Akhtar N, Mannan MA (2020) Mycoremediation: expunging environmental pollutants. Biotechnol Rep 26:e00452. https://doi.org/10.1016/j.btre.2020.e00452

Amobonye A, Aruwa CE, Aransiola S, Omame J, Alabi TD, Lalung J (2023) The potential of fungi in the bioremediation of pharmaceutical active compounds: a comprehensive review. Front Microbiol 14:1207792. https://doi.org/10.3389/fmicb.2023.1207792

Australia- Department of Agriculture, Water and the Environment (2019) National environment protection (assessment of site contamination) measure. Australian Government. https://www.dcceew.gov.au/environment/protection/nepc

Ba S, Haroune L, Cruz-Morató C, Jacquet C, Touahar IE, Bellenger JP et al (2014) Synthesis and characterization of combined cross-linked laccase and tyrosinase aggregates transforming acetaminophen as a model phenolic compound in wastewaters. Sci Total Environ 487:748–755. https://doi.org/10.1016/j.scitotenv.2014.04.056

Bilal M, Asgher M, Parra-Saldivar R, Hu H, Wang W, Zhang X, Iqbal HMN (2018) Immobilized ligninolytic enzymes: an innovative and environmental responsive technology to tackle dye-based industrial pollutants—a review. Sci Total Environ 640–641:836–851. https://doi.org/10.1016/j.scitotenv.2018.05.396

Birolli WG, Santos DA, Alvarenga N, Garcia ACFS, Romão LPC, Porto ALM (2020) A review on the use of fungi for the bioremediation of pharmaceutical compounds. J Hazard Mater 399:122906. https://doi.org/10.1016/j.jhazmat.2020.122906

Bokade P, Bajaj A (2023) Molecular advances in mycoremediation of polycyclic aromatic hydrocarbons: exploring fungal bacterial interactions. J Basic Microbiol 63(3–4):239–256. https://doi.org/10.1002/jobm.202200499

Bonatti E, dos Santos A, Birolli WG, Rodrigues-Filho E (2023) Biotransformation of steroids by fungi: a review of steroid metabolism and enzyme systems. World J Microbiol Biotechnol 39(6):152. https://doi.org/10.1007/s11274-023-03590-8

Bumpus JA, Brock BJ (1988) Biodegradation of crystal violet by the white rot fungus *Phanerochaete chrysosporium*. Appl Environ Microbiol 54(5):1143–1150

Cabana H, Jones JP, Agathos SN (2007) Elimination of endocrine disrupting chemicals using white rot fungi and their lignin modifying enzymes: a review. Eng Life Sci 7(5):429–456. https://doi.org/10.1002/elsc.200620206

Camerano Spelta Rapini C, Di Berardino C, Peserico A, Capacchietti G, Barboni B (2024) Can mammalian reproductive health withstand massive exposure to polystyrene micro- and nanoplastic derivatives? A systematic review. Int J Mol Sci 25(22):12166

Canada- Environment and Climate Change Canada (2017) Federal contaminated sites action plan (FCSAP): guidance manual. Government of Canada. https://www.canada.ca/en/environment-climate-change/services/federal-contaminated-sites.html

Cao H, Zhang X, Wang S, Liu J, Han D, Zhao B, Wang H (2021) Time-course transcriptome analysis reveals the mechanisms of Burkholderia sp. responding to Benzopyrene stress. Front Microbiol 12:761216. https://doi.org/10.3389/fmicb.2021.761216

Champramary S, Indic B, Szűcs A, Tyagi C, Languar O, Hasan KMF et al (2023) The mycoremediation potential of the armillarioids: a comparative genomics analysis. Front Bioeng Biotechnol 11:1189640. https://doi.org/10.3389/fbioe.2023.1189640

Chane AD, Košnář Z, Hřebečková T, Jozífek M, Doležal P, Tlustoš P (2024) Persistent polycyclic aromatic hydrocarbons removal from sewage sludge-amended soil through phytoremediation combined with solid-state ligninolytic fungal cultures. Fungal Biol 128(2):1675–1683. https://doi.org/10.1016/j.funbio.2024.01.007

Chen S, Zhu M, Guo X, Yang B, Zhuo R (2023) Coupling of Fenton reaction and white rot fungi for the degradation of organic pollutants. Ecotoxicol Environ Saf 254:114697. https://doi.org/10.1016/j.ecoenv.2023.114697

Chmelová D, Ondrejovič M, Miertuš S (2024) Laccases as effective tools in the removal of pharmaceutical products from aquatic systems. Life 14(2):230. https://doi.org/10.3390/life14020230

Christian V, Shrivastava R, Shukla D, Modi HA, Vyas BR (2005) Degradation of xenobiotic compounds by lignin-degrading white-rot fungi: enzymology and mechanisms involved. Indian J Exp Biol 43(4):301–312

Cruz-Morató C, Lucas D, Llorca M, Rodriguez-Mozaz S, Gorga M, Petrovic M et al (2014) Hospital wastewater treatment by fungal bioreactor: removal efficiency for pharmaceuticals and endocrine disruptor compounds. Sci Total Environ 493:365–376. https://doi.org/10.1016/j.scitotenv.2014.05.117

Curiel-Alegre S, de la Fuente-Vivas D, Khan AHA, García-Tojal J, Velasco-Arroyo B, Rumbo C et al (2024) Advances in biochar and PGPR engineering system for hydrocarbon degradation: a promising strategy for environmental remediation. Environ Res 252(2):118880. https://doi.org/10.1016/j.envres.2024.118880

D'Annibale A, Stazi SR, Vinciguerra V, Di Mattia E, Sermanni GG (2000) Characterization of immobilized laccase from Lentinula edodes and its use in olive mill wastewater treatment. Process Biochem 36(1–2):51–58. https://doi.org/10.1016/S0032-9592(00)00180-5

Department of Agriculture, Water and the Environment (2021) National environment protection (assessment of site contamination) measure. Australian Government. https://www.dcceew.gov.au/environment/protection/nepc

Dey P, Malik A, Mishra A, Singh DK, Patel N, Khan E (2020) Mechanistic insight to mycoremediation potential of an indigenous white-rot fungus Ganoderma lucidum for degradation of persistent organic pollutants. Ecotoxicol Environ Saf 196:110557. https://doi.org/10.1016/j.ecoenv.2020.110557

Dey P, Gola D, Mishra A, Malik A, Singh DK, Patel N (2024) Unveiling fungal strategies: mycoremediation in multi-metal pesticide environment using proteomics. Sci Rep 14:23120. https://doi.org/10.1038/s41598-024-74517-y

Environment and Climate Change Canada (2023) Federal contaminated sites action plan (FCSAP): guidance manual. Government of Canada. https://www.canada.ca/en/environment-climate-change/services/federal-contaminated-sites.html

European Chemicals Agency (2022) Registration, evaluation, authorisation and restriction of chemicals (REACH). ECHA. https://echa.europa.eu/regulations/reach

Flórez Córdoba M, Correa Álvarez M, Hidalgo López S, Granda Ramírez F, Osorno Bedoya L (2024) Biodegradación de polietileno de baja densidad en suelo con hongos del género Aspergillus sp. Ciencia e Ingeniería Neogranadina 34(1):57–66. https://doi.org/10.18359/rcin.6841

Fonseka HWL, Senarath MDAH, Daranagama DADA, Ratnayake RMCS (2025) Mechanisms and potential for fungal bioremediation of polycyclic aromatic hydrocarbons and other environmental contaminants: a comprehensive review. Asian J Mycol 8(1):86–128. https://doi.org/10.5943/ajm/8/1/6

Fragoeiro S, Magan N (2008) Environmental factors and bioremediation of xenobiotics using white rot fungi. Mycobiology 36(4):238–246. https://doi.org/10.4489/MYCO.2008.36.4.238

García-Delgado C, Alfaro-Barta I, Eymar E (2015) Combination of biochar amendment and mycoremediation for polycyclic aromatic hydrocarbons immobilization and biodegradation in creosote-contaminated soil. J Hazard Mater 285:259–266. https://doi.org/10.1016/j.jhazmat.2014.12.002

García-Delgado C, Alfaro-Barta I, Eymar E (2021) Combination of biochar amendment and mycoremediation for polycyclic aromatic hydrocarbons immobilization and biodegradation in creosote-contaminated soil. J Hazard Mater 390:122157. https://doi.org/10.1016/j.jhazmat.2020.122157

García-Sánchez M, Košnář Z, Mercl F, Aranda E, Tlustoš P (2018) A comparative study to evaluate natural attenuation, mycoaugmentation, phytoremediation, and microbial-assisted phytoremediation strategies for the bioremediation of an aged PAH-polluted soil. Ecotoxicol Environ Saf 147:165–174. https://doi.org/10.1016/j.ecoenv.2017.08.012

González-Abradelo D, Pérez-Llano Y, Peidro-Guzmán H, Sánchez-Carbente MR, Folch-Mallol JL, Aranda E, Gunde-Cimerman N (2019) First demonstration that ascomycetous halophilic fungi (Aspergillus sydowii and Aspergillus destruens) are useful in xenobiotic mycoremediation under high salinity conditions. Bioresour Technol 279:287–296. https://doi.org/10.1016/j.biortech.2019.02.002

Guo E, Zhao L, Li Z, Chen L, Li J, Lu F et al (2024) Biodegradation of bisphenol A by a Pichia pastoris whole-cell biocatalyst with overexpression of laccase from Bacillus pumilus and investigation of its potential degradation pathways. J Hazard Mater 474:134779. https://doi.org/10.1016/j.jhazmat.2024.134779

Harms H, Schlosser D, Wick LY (2011) Untapped potential: exploiting fungi in bioremediation of hazardous chemicals. Nat Rev Microbiol 9(3):177–192. https://doi.org/10.1038/nrmicro2519

Hu X, Zhou Q, Luo Y (2021) Occurrence and source analysis of typical veterinary antibiotics in manure, soil, vegetables and groundwater from organic vegetable bases, northern China. Environ Pollut 258:113829. https://doi.org/10.1016/j.envpol.2019.113829

Imam A, Suman SK, Vempatapu BP (2022) Pyrene remediation by Trametes maxima: an insight into secretome response and degradation pathway. Environ Sci Pollut Res 29:44135–44147. https://doi.org/10.1007/s11356-022-18888-7

India- Ministry of Environment, Forest and Climate Change (2024) Draft Remediation of Contaminated Sites Rules, 2024. Government of India. https://moef.gov.in

Jaiswal S, Shukla P (2020) Alternative strategies for microbial remediation of pollutants via synthetic biology. Front Microbiol 11:808. https://doi.org/10.3389/fmicb.2020.00808

Ministry of the Environment, Japan (2015) Basic environment plan. Ministry of the Environment. https://www.env.go.jp/en/

Kathhivaran A, Gnanadoss JJ (2021) White rot fungi-mediated bioremediation as a sustainable method for xenobiotic degradation. Environ Exp Biol 19:103–119. https://doi.org/10.22364/eeb.19.10

Košnář Z, Mercl F, Tlustoš P (2018) Ability of natural attenuation and phytoremediation using maize (Zea mays L.) to decrease soil contents of polycyclic aromatic hydrocarbons (PAHs) derived from biomass fly ash in comparison with PAHs-spiked soil. Ecotoxicol Environ Saf 153:16–22. https://doi.org/10.1016/j.ecoenv.2018.01.049

Kulshreshtha S, Mathur N, Bhatnagar P (2014) Bioremediation of pesticides using white rot fungi. Pestic Biochem Physiol 113:79–84. https://doi.org/10.1016/j.pestbp.2014.05.003

Li S, Sun K, Latif A, Si Y, Gao Y, Huang Q (2022) Insights into the applications of extracellular laccase-aided humification in livestock manure composting. Environ Sci Technol 56(12):7412–7425. https://doi.org/10.1021/acs.est.1c08042

Li J, Liu J, Tang S, Yan K, Zhang J (2024) Biophotoelectrocatalytic detoxification of bisphenol A using bismuth-doped TiO2 nanotubes photoanode coupled with laccase-immobilized cathode. J Environ Chem Eng 12(6):114910. https://doi.org/10.1016/j.jece.2024.114910

Lladó S, Covino S, Solanas AM, Viñas M, Petruccioli M, D'annibale A (2013) Comparative assessment of bioremediation approaches to highly recalcitrant PAH degradation in a real industrial polluted soil. J Hazard Mater 248–249:407–414. https://doi.org/10.1016/j.jhazmat.2013.01.020

Mattila H, Österman-Udd J, Mali T, Lundell T (2022) Label-free differentially proteomic analysis of interspecific interaction between white-rot fungi highlights oxidative stress response and high metabolic activity. Front Fungal Biol 3:837605. https://doi.org/10.3389/ffunb.2022.837605

Migliore L, Fiori M, Spadoni A, Galli E (2022) Biodegradation of oxytetracycline by Pleurotus ostreatus mycelium: a mycoremediation technique. J Hazard Mater 215–216:227–232. https://doi.org/10.1016/j.jhazmat.2012.02.056

Ministry of Environment, Forest and Climate Change (2024) Draft hazardous and other wastes (management and transboundary movement) rules 2024. Government of India. https://moef.gov.in

Ministry of Environment, Japan (2022) Basic environment plan. Ministry of the Environment. https://www.env.go.jp/en/

Mori T, Kitamura S, Araki Y, Kamakura M, Miyakozawa I, Tateishi N (2021) Cytochrome P450-mediated neonicotinoid metabolism in the white-rot fungus *Phanerochaete chrysosporium*. Appl Environ Microbiol 87(8):e02714–e02720. https://doi.org/10.1128/AEM.02714-20

Muter O (2023) Biodegradation of persistent organic pollutants: current status and future prospects. Microorganisms 11(3):710. https://doi.org/10.3390/microorganisms11030710

Ortúzar M, Esterhuizen M, Olicón-Hernández DR, González-López J, Aranda E (2022) Pharmaceutical pollution in aquatic environments: a concise review of environmental impacts and bioremediation systems. Front Microbiol 13:869332. https://doi.org/10.3389/fmicb.2022.869332

Park H, Choi IG (2020) Genomic and transcriptomic perspectives on mycoremediation of polycyclic aromatic hydrocarbons. Appl Microbiol Biotechnol 104(16):6919–6928. https://doi.org/10.1007/s00253-020-10746-1

Park H, Min B, Jang Y, Kim J, Lipzen A, Sharma A et al (2019) Comprehensive genomic and transcriptomic analysis of polycyclic aromatic hydrocarbon degradation by a mycoremediation fungus, Dentipellis sp. KUC8613. Appl Microbiol Biotechnol 103(19):8145–8155. https://doi.org/10.1007/s00253-019-10089-6

Passarini MRZ, Rodrigues MVN, da Silva M, Sette LD (2021) Marine-derived filamentous fungi and their potential application for polycyclic aromatic hydrocarbon bioremediation. Mar Pollut Bull 169:112558. https://doi.org/10.1016/j.marpolbul.2021.112558

Pointing SB (2001) Feasibility of bioremediation by white-rot fungi. Appl Microbiol Biotechnol 57(1–2):20–33. https://doi.org/10.1007/s002530100745

Pozdnyakova NN, Chernyshovaa MP, Grineva VS, Landesmanb EO, Turkovskayaa V (2016) Degradation of fluorene and fluoranthene by the basidiomycete Pleurotus ostreatus. Appl Biochem Microbiol 52:621–628. https://doi.org/10.1134/S0003683816060132

Rathankumar AK, Saikia K, Ramachandran K, Batista RA, Cabana H, Vaidyanathan VK (2022) Surfactant-aided mycoremediation of soil contaminated with polycyclic aromatic hydrocarbons. Environ Res 209:112926. https://doi.org/10.1016/j.envres.2022.112926

Rodríguez-Couto S (2009) Dye removal by immobilised fungi. Biotechnol Adv 27(3):227–235. https://doi.org/10.1016/j.biotechadv.2008.12.001

Romero-Reyes MA, Patterson KN, Heemstra JM (2023) Co-immobilization of enzymes and aptamers to create self-regenerating ultrafiltration membranes for toxin removal. ACS Mater Lett 5(6):1565–1569. https://doi.org/10.1021/acsmaterialslett.3c00055

Rughöft S, Jehmlich N, Gutierrez T, Kleindienst S (2020) Comparative proteomics of Marinobacter sp. TT1 reveals corexit impacts on hydrocarbon metabolism, chemotactic motility, and biofilm formation. Microorganisms 9(1):3. https://doi.org/10.3390/microorganisms9010003

Satpati GG, Gupta S, Biswas RK, Choudhury AK, Kim JW, Davoodbasha M (2023) Microalgae mediated bioremediation of polycyclic aromatic hydrocarbons: strategies, advancement and regulations. Chemosphere 344:140337. https://doi.org/10.1016/j.chemosphere.2023.140337

Singh H (2006) Mycoremediation: fungal bioremediation. Wiley-Interscience, Hoboken

Singh R, Tripathi A (2022) Mycoremediation of pesticide residues in agricultural soil: current status, challenges, and future prospects. J Environ Manag 312:114901. https://doi.org/10.1016/j.jenvman.2022.114901

Singh R, Glick BR, Rathore D (2023) Mycoremediation as a potentially promising technology: current status and prospects—a review. Appl Sci 13(8):4978. https://doi.org/10.3390/app13084978

Soares SRP, Silva BF, Barbosa DP, Sette LD (2021) Mycoremediation of sewage sludge and manure with marine fungi for the removal of organic pollutants. Front Mar Sci 9:946220. https://doi.org/10.3389/fmars.2022.946220

Sun K, Wu ZH, Liu J, Cheng ZH, Liu DF, Lin H, Yu HQ (2024) Unappreciated role of secondary metabolism-derived small mediators in degrading bisphenol A and antibiotics by a laccase-expressing fungus. Environ Pollut 363:125219. https://doi.org/10.1016/j.envpol.2024.125219

Tien M, Kirk TK (1983) Lignin-degrading enzyme from Phanerochaete chrysosporium: purification, characterization, and catalytic properties of a unique H_2O_2-requiring oxygenase. Proc Natl Acad Sci 80(24):7495–7499

Torres-Farradá G, Thijs S, Rineau F, Guerra G, Vangronsveld J (2024) White rot fungi as tools for the bioremediation of xenobiotics: a review. J Fungi 10(3):167. https://doi.org/10.3390/jof10030167

Tortella GR, Diez MC, Duran N (2005) Fungal diversity and use in decomposition of environmental pollutants. Crit Rev Microbiol 31(4):197–212. https://doi.org/10.1080/10408410500304066

U.S. Environmental Protection Agency (2023) Ecological risk assessment. U.S. EPA. https://www.epa.gov/risk/ecological-risk-assessment

Wahla AQ, Anwar S, Fareed MI, Ikram W, Ali L, Alharby HF et al (2023) Immobilization of metribuzin-degrading bacteria on biochar: enhanced soil remediation and bacterial community restoration. Front Microbiol 13:1027284. https://doi.org/10.3389/fmicb.2022.1027284

Wang L, Luo X, Pan Y, Zheng Z, Yin R, Tian X, Zhang L (2021) Mechanism of laccase induction via emodin in Trametes versicolor. Front Bioeng Biotechnol 9:653800. https://doi.org/10.3389/fbioe.2021.653800

Wang M, Zhang W, He T, Rong L, Yang Q (2023) Degradation of polycyclic aromatic hydrocarbons in aquatic environments by a symbiotic system consisting of algae and bacteria: green and sustainable technology. Arch Microbiol 206(1):10. https://doi.org/10.1007/s00203-023-03734-2

Xiang L, Harindintwali JD, Wang F, Redmile-Gordon M, Chang SX, Fu Y et al (2022) Integrating biochar, bacteria, and plants for sustainable remediation of soils contaminated with organic pollutants. Environ Sci Technol 56(23):16546–16566. https://doi.org/10.1021/acs.est.2c02976

Xu S, Li M, Hu Z, Shao Y, Ying J, Zhang H (2023) The potential use of fungal co-culture strategy for discovery of new secondary metabolites. Microorganisms 11(2):464. https://doi.org/10.3390/microorganisms11020464

Ye Y, Liu H, Wang Z, Qi Q, Du J, Tian S (2024) A cellulosomal yeast reaction system of lignin-degrading enzymes for cellulosic ethanol fermentation. Biotechnol Lett 46(4):531–543. https://doi.org/10.1007/s10529-024-03485-0

Zhang W, Li Q, Wang J, Wang Z, Zhan H, Yu X et al (2023) Biodegradation of Benzo[a]pyrene by a white-rot fungus Phlebia acerina: surfactant-enhanced degradation and possible genes involved. J Fungi 9(10):978. https://doi.org/10.3390/jof9100978

Zhao X, Miao R, Guo M, Shang X, Zhou Y, Zhu J (2022) Biochar enhanced polycyclic aromatic hydrocarbons degradation in soil planted with ryegrass: bacterial community and degradation gene expression mechanisms. Sci Total Environ 838(2):156076. https://doi.org/10.1016/j.scitotenv.2022.156076

Zhuo R, Fan F (2021) A comprehensive insight into the application of white rot fungi and their lignocellulolytic enzymes in the removal of organic pollutants. Sci Total Environ 778:146132. https://doi.org/10.1016/j.scitotenv.2021.146132

Zou Q, Zhao L, Guan L, Chen P, Zhao J, Zhao Y et al (2024) The synergistic interaction effect between biochar and plant growth-promoting rhizobacteria on beneficial microbial communities in soil. Front Plant Sci 15:1501400. https://doi.org/10.3389/fpls.2024.1501400

Mycoremediation and Soil Microbiome Synergies

7

Lakshay Virmani, Simran Kumari, and Kriti Bhandari

Abstract

This chapter explores how the integration of mycoremediation with the soil microbiome can enhance the breakdown of xenobiotic pollutants. It accentuates the synergistic roles of fungi and bacteria in cooperative degradation, nutrient cycling, and community stability. Special attention is given to mycorrhizal associations and rhizosphere interactions, which reinforce pollutant tolerance and uptake in plants. The chapter outlines all the practical strategies for designing and deploying microbial consortia, optimizing soil conditions, and using molecular tools to monitor ecological dynamics. This integrative approach not only increases the effectiveness of remediation but also promotes long-term soil health and ecological durability. The future insights talk about an understanding of how fungal-microbial networks can be harnessed in real-world settings to deliver more sustainable, efficient, and ecologically compatible solutions to environmental contamination. This integrative approach not only increases the effectiveness of remediation but also promotes long-term soil health and ecological durability by restoring biodiversity and improving soil structure. Furthermore, the chapter emphasizes scalability considerations, ensuring that laboratory findings can be translated to field-level interventions with minimal ecological disruption. The future insights focus on deepening our understanding of fungal–microbial networks in real-world contaminated sites, aiming to deliver more sustainable, efficient, and ecologically compatible solutions to environmental contamination. By bridging the gap between ecological theory and applied biotechnology, this synthesis offers a pathway toward next-generation remediation strategies that are resilient enough under changing environmental conditions.

L. Virmani · S. Kumari · K. Bhandari (✉)
Department of Biotechnology, Delhi Technological University, Shahbad Daulatpur, Delhi, India
e-mail: kritibhandari@dtu.ac.in

A. Roy et al. (eds.), *Mycoremediation of Xenobiotics*, Emerging Paradigms in Pharmaceutical Research, https://doi.org/10.1007/978-981-95-5743-1_7

Keywords

Soil microbiome · Fungal-bacterial synergy · Environmental remediation · Microbial · Consortia · Sustainable degradation

7.1 Introduction

Mycoremediation is the process by which fungi degrade, transform, or immobilize environmental contaminants, restoring balance to polluted ecosystems. Fungi accomplish this through secreted enzymes, organic acids, and biosorption mechanisms, making them particularly effective against a wide array of pollutants, such as heavy metals to complex organic compounds. The ability of fungi to survive in diverse and extreme environments, coupled with their metabolic versatility, makes mycoremediation a cost-effective and eco-friendly bioremediation.

The soil microbiome has a diverse community of microorganisms, like bacteria, fungi, archaea, and more, that inhabit the soil ecosystem. These microbial assemblies regulate key processes such as nutrient cycling, organic matter decomposition, and biological interactions that sustain soil health. Recent high-throughput sequencing studies have emphasized the inside networks between bacteria and fungi in soil ecosystems, which reveal how community complexity enhances functional potential (Gao et al. 2024a; Zhang et al. 2024).

7.1.1 Associative Remediation

Integrating fungal and broader microbial functions amplifies remediation outcomes by combining fungal enzymatic degradation with bacterial metabolic flexibility and ecological toughness. Modulating bacterial fungal networks, for example, through organic alterations like biochar, which enhances cross-domain interactions, increases contaminant breakdown, and promotes soil structural repair (Gao et al. 2024a; Zhang et al. 2024). Editorial perspectives also define how mycorrhizal fungi work in sync with bacterial communities to improve nutrient uptake under stress and support microbial-driven pollutant transformation (Mitra et al. 2024).

While single-fungus treatments can show pollutant degradation capabilities, they suffer from limited ecological adaptability and reduced functional area. Isolating fungi in monocultures may overlook the complementary metabolic pathways and stabilization effects offered by bacterial partners. Field-scale studies emphasize the importance of biodiversity: more complex microbial consortia will sustain long-term resilience, whereas monocultures often fail under environmental fluctuations (Fig. 7.1) (Mitra et al. 2024).

This chapter conceptualizes:

- Defines and contextualizes mycoremediation within soil restoration.
- Details about the ecological roles of soil microbiomes and fungal–bacterial networks.

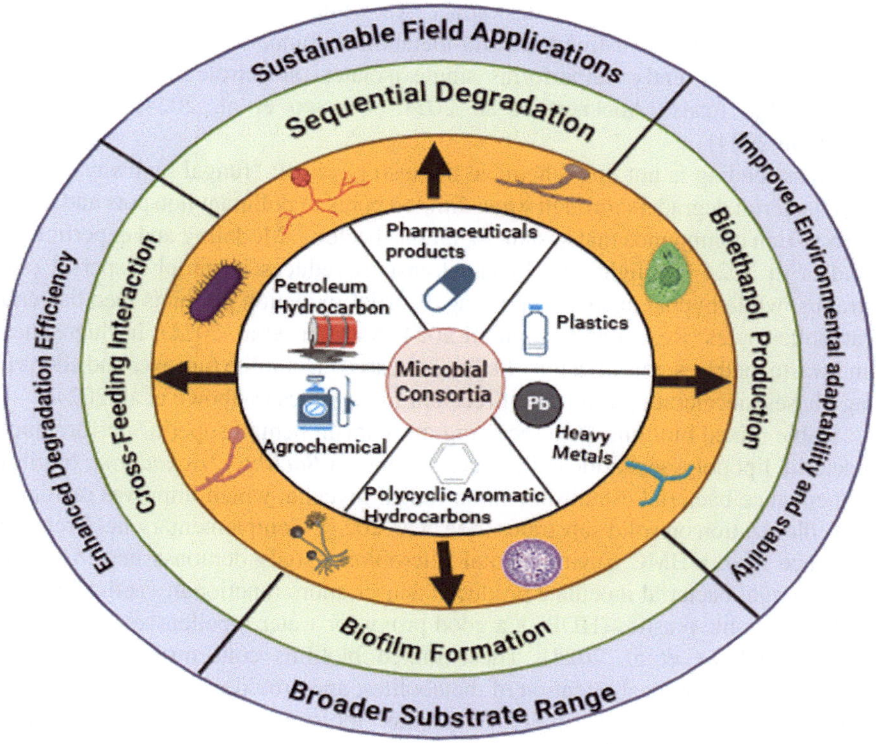

Fig. 7.1 Pictorial representation of broad field analysis of microbial consortia

- Explores the scientific rationale and emerging evidence for combined fungal–microbial bioremediation strategies.
- Critique the shortcomings of monoculture-based approaches.
- Proposes integrated remediation models that leverage the synergies of fungal and microbial interactions. Through synthesizing recent research and highlighting application opportunities, we seek to provide both theoretical underpinning and practical guidance for advancing next-generation bioremediation.

7.2 Fungal-Bacterial Synergies

7.2.1 Cooperative Degradation Pathways

Fungal–bacterial consortia often split complex xenobiotic catabolism into complementary steps, like fungal hyphae, which oxidize or solubilize hydrophobic substrates, like PAHs, generating smaller or more bioavailable intermediates that bacterial partners mineralize. Recent microcosm and review work shows that filamentous fungi, like white-rot, saprotrophs, and AMF, create "first-pass" oxidative products via extracellular peroxidases/laccases, while bacteria supply ring-cleavage dioxygenases and downstream central-metabolism funnels, which is an arrangement that consistently outperforms single isolates on petroleum pollutants and mixed PAH loads (Anckaert et al. 2024; Wibowo et al. 2024; Rezaei and Moghimi 2024).

Cross-feeding is not only chemical but also physical: "fungal highways" transport bacterial degraders through water-limited pores to pollutant hotspots and exude carbon-rich compounds that sustain fastidious bacteria. Modeling and experimental studies in 2024 highlight that hyphal transport reduces bacterial dispersal constraints, while hyphal exudates and fungal partial-oxidation products feed bacterial catabolic routes (Rezaei and Moghimi 2024; Anckaert et al. 2024). In rhizosphere and waste matrices, this division of labor increases metabolite turnover and shortens lag phases, accelerating removal of recalcitrant organics (Wibowo et al. 2024).

Stable mixed biofilms are a recurring mechanism behind superior performance. Bacterial lipopolysaccharide (LPS) cues strengthen *Bacillus–Trichoderma* biofilms and enhance bacterial establishment on fungal mycelia, which improves resilience and colonization on solid substrates (Xie et al. 2024). Complementary field-relevant evidence from a BMC Environmental Microbiome study demonstrated that combined fungal bacterial inoculate produced denser, more functionally active biofilms on hydrophobic plastics (HDPE), a good proxy for water-repellent xenobiotic surfaces (MacLean et al. 2024). These mixed biofilms concentrate extracellular enzymes, increase local retention of metabolites, and provide diffusion bridges that maintain catabolic fluxes under fluctuating moisture and nutrient conditions (Wibowo et al. 2024). Multiple 2024 reports connect the synergy to practical pollutant removal. A review on petroleum hydrocarbons documents higher degradation rates when fungi and bacteria are co-inoculated, attributing gains to sequential oxidation–cleavage pathways and enhanced mass transfer (Wibowo et al. 2024). Soil- and plant-assisted systems also benefit: a 2024 agronomic study combining degrading bacteria with maize achieved roughly a 59–79% removal of phenanthrene and naphthalene in 30 days, outperforming either component alone, which is consistent with rhizosphere-anchored, fungus-assisted community effects (Gao et al. 2024a, b). Mechanistically, 2024–2025 work further shows that mixed communities degrade hydrophobic substrates more efficiently when stabilized as biofilms on solid matrices and when hyphae shuttle bacteria to contaminants that would otherwise remain inaccessible (MacLean et al. 2024; Rezaei and Moghimi 2024).

7.2.2 Implications for Enhancing Degradation Rates In Situ

For in situ deployments, three design levers are present in the latest literature:

- Structure that co-formulates fungi that produce extensive hyphal networks with bacterial specialists to exploit the dispersal along fungal highways and mixed biofilms
- Function that select partners with complementary oxidative and ring-cleavage capacities to enable true sequential metabolism

- Habitat engineering that provides mild amendments like lignocellulosic carriers that encourage hyphal spread and biofilm stability while avoiding catabolic repression.

Systematic reviews in 2024–2025 recommend consortia over single strains for variable field conditions, citing greater functional redundancy and robustness to stress (Wibowo et al. 2024; Qattan 2025a, b). Where plants are part of the system, pairing fungal bacterial inoculate with appropriate hosts can amplify rhizodeposition and stabilize pH/moisture, further increasing removal efficiency (Gao et al. 2024a; Zhang et al. 2024). Overall, the 2024–2025 evidence base supports co-cultured, structure-aware designs that leverage hyphal transport, cross-feeding, and mixed biofilms to shorten start-up times and sustain high degradation rates in real soils and sediments (Table 7.1) (Anckaert et al. 2024; Xie et al. 2024; MacLean et al. 2024).

7.3 Plant–Fungal Interactions

Arbuscular mycorrhizal fungi expand the functional root system via extra-radical hyphae, shifting contaminant fate at the root–soil interface. By altering sorption equilibrium, precipitating metals on hyphal walls, and promoting sequestration through glomalin and other extracellular polymers, AMF can reduce bioavailable fractions and limit translocation to shoots with an outcome central to food-chain safety (Zhang et al. 2024; Shukla et al. 2025). These effects co-occur with AMF-driven remodeling of the rhizosphere redox and pH microgradients that favor immobilization or safe compartmentation of cadmium and other metals (Zhang et al. 2024; Zhao et al. 2025). Collectively, AMF improves plant performance on contaminated substrates while lowering contaminant mobility and exposure risk (Fig. 7.2).

7.3.1 Enhanced Nutrient Acquisition and Stress Tolerance

Beyond pollutant handling, AMF consistently enhances phosphorus and nitrogen capture and water use, buffering hosts against drought, salinity, and oxidative stress. Mechanistically, AMF modulates ion transporters, antioxidant systems, and osmolyte pathways while remodifying root architecture to optimize acquisition under stress (Shukla et al. 2025; Lethielleux-Juge 2025). These nutrient stress benefits are not required, like improved nutritional status, elevating detoxification capacity, like thiol metabolism and lignification, thereby indirectly accelerating xenobiotic tolerance and recovery of biomass on impacted soils (Lethielleux-Juge 2025).

Root exudates act as the chemical currency that coordinates plant fungus microbiome dialogues. Low-molecular-weight acids, phenolics, sugars, and lipids modulate AMF chemotropism and colonization while also recruiting degrader and metal-transforming bacteria that complement fungal functions (Ma et al. 2025;

Table 7.1 Synergistic fungal–bacterial consortia for pollutant degradation

Pollutant/matrix	Fungal partner(s)	Bacterial partner(s)	Mechanism/synergy	Evidence type	Performance highlights	References
LDPE microplastics (lab films)	*Sarocladium kiliense*	*Bacillus cereus*	Biofilm-mediated oxidation; sequential degradation	Lab co-culture	26.3% mass loss in 60 d; half-life reduced to 134 d	Alidoosti et al. (2025)
LDPE (landfill isolates)	*Alternaria* sp. + *Trametes* sp.	–	Mixed-fungal enzymatic action on PE films	Lab study	Reported PE degradation (2024 study)	Yang et al. (2024)
Multi-antibiotics in food waste medium	*Pichia pastoris* (laccase-+)	*Bacillus amyloliquefaciens*, *C. glutamate*, *Y. lipolytica*	Co-degradation of antibiotics; resource utilization to lipopeptides	Lab artificial consortium	~100% oxytetracycline removal in 48–72 h	Shang et al. (2024)
LDPE (landfill-sourced bacteria)	–	*Bacillus* sp. + *Priestia* sp.	Biofilm formation, esterase activity, weight loss	Lab mixed culture	Degradation parameters consolidated in 2025 study	Ting and Sallahudin (2025)
LLDPE (microcosm enrichment)	*Debaryomyces hansenii* + *Castellaniella denitrificans*	+ accessory *Rhodococcus*	Additive degradation + direct polymer attack	Microcosm enrichment	Effective LLDPE film degradation consortium	Salinas et al. (2023)

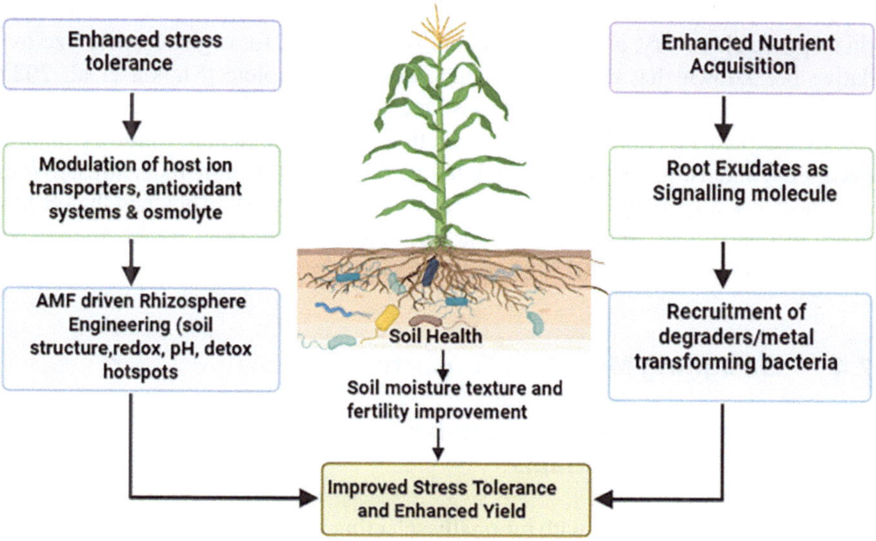

Fig. 7.2 Associative correlation with respect to plant fungal interactions

Wang et al. 2024a, b). Recent works showing specific exudate components can both suppress pathogens and enrich beneficial taxa, for example, methyl ferulate secreted by tobacco inhibited *Phytophthora nicotiana* and simultaneously recruited *Bacillus* spp., demonstrating dual defensive and community-engineering roles (Ma et al. 2025). Reviews in 2025 further highlight how exudation fluxes are plastic with environment and genotype, enabling a targeted watch of rhizosphere interactions relevant to remediation (Wang et al. 2024a, b). Emerging rhizosphere engineering integrates plant selection, AMF inoculation, and amendments like biochar, organic ligands to shape micro-biogeochemical hotspots that favor detoxification. Strategically chosen plants and AMF consortia can shift metal speciation toward less bioavailable pools, while amendments tune porosity, electron acceptors, and sorption sites to enhance immobilization or safe phytoextraction (Li et al. 2025a, b; Zhang et al. 2024). Field-relevant studies now resolve how root-induced fractionation of cadmium differs by soil type, knowledge that guides whether to favor stabilization in comparison with extraction in situ (Zheng et al. 2025). Such interventions work best when they deliberately couple exudate-mediated microbial recruitment with AMF-driven nutrient uplift, yielding stable communities capable of sustained pollutant transformation.

7.3.2 Integration of Phytoremediation and Mycoremediation Strategies

An integrated phyto-myco framework aligns plant traits like deep rooting, exudate chemistry with fungal capacities to deliver complementary removal pathways.

Plants provide carbon flow and hydraulic control, while fungi extend reach, adjust rhizosphere chemistry, and either immobilize metals or, for organics, catalyze oxidative breakdown that downstream microbes can complete (Shukla et al. 2025; Lethielleux-Juge 2025). Practically, this means pairing AMF compatible hosts with locally adapted inoculate, timing inoculation to developmental windows of high exudation, and using low-dose amendments to stabilize function under stress. When designed this way, integrated systems report higher removal efficiency, better biomass recovery, and improved soil health indices than plant-based approaches (Table 7.2) (Shukla et al. 2025; Li et al. 2025a, b).

7.4 Designing Microbial Consortia

7.4.1 Criteria for Selecting Compatible Fungal and Bacterial Strains

Successful consortia begin with rationally selecting the strain: choosing the partners whose metabolic capacities are complementary, like oxidative and ring-cleavage steps, biosorption, and mineralization, that share tolerance ranges for pH, temperature, and pollutant concentration and that do not carry antagonistic traits that would drive exclusion (Ruan et al. 2024; Efremenko et al. 2024).

Additional practical filters include:

- Evidence of pollutant-relevant enzyme systems like fungal ligninolytic enzymes, bacterial dioxygenases.
- Growth kinetics compatible with intended deployment, so one partner does not overrun the other.
- Amenability to formulation, like spore production and desiccation tolerance for field use (Efremenko et al. 2024; Ningombam et al. 2025). Computational tools and metabolic models (e.g., SuperCC) can help predict complementary pathways and identify keystone strains from complex source microbiomes prior to assembly (Ruan et al. 2024).

Laboratory screening proceeds in tiers: isolate characterization, pairwise co-culture assays, and small-scale microcosm trials. Initial screens should measure pollutant transformation, growth in co-culture, and production of key enzymes or metabolites; pairwise assays reveal antagonism or facilitation (Efremenko et al. 2024). Microcosm experiments that mimic soil or wastewater physicochemistry identify context-dependent failures (nutrient limitation, metal stress) early. Omics readouts like metatranscriptomics, proteomics, and simple activity assays such as enzyme assays and substrate disappearance inform whether communities operate via sequential metabolism or cross-feeding (Ruan et al. 2024). Immobilization on carriers like biochar and alginate beads during screening can also show whether physical co-location stabilizes interactions (Efremenko et al. 2024).

Co-cultivation methods shape community assembly and function. Bottom-up approaches mix pre-characterized strains at controlled ratios and monitor dynamics;

Table 7.2 Field- and near-field applications of fungal–bacterial consortia

Site/scale	Contaminant type	Inoculum/ strategy	Deployment and monitoring notes	Outcome highlights	References
EU Horizon "MySOIL" pilot (11 countries)	Mixed petroleum contaminants	Microbial consortia including fungal components	Standardized monitoring, decision-support protocols	Real-world applicability; protocol and KPI development	MySOIL Project (2025)
Cd-contaminated vegetable fields (China)	Cadmium (heavy metal)	AMF-focused amendments (concept includes PGPR)	Field plots; AMF community tracking via amplicon	AMF community shift supports phytoremediation framework	Li et al. (2024)
Chlorantraniliprole-spiked soil (lab mimic)	Insecticide (CAP)	Hexa-bacterial consortium (*Bacillus*, *Alcaligenes*, *Pseudomonas*)	Microcosm/soil incubation (20 d at 30 °C)	Up to 99.3% CAP removal; complete degradation in 20 d	Singh et al. (2024)
Sewage water + agricultural reuse (pots)	OM and metals (Cu, Ni, Cd, Pb)	Fungal consortium (*A. niger*, *Sordariomycetes*, *Coniochaetaceae*)	Pot trials measuring maize physiological responses	88% organic load removal; >80% heavy metals cleanup	Daâssi et al. (2024)
Integrated aquaculture wastewater	Aquaculture effluent pollutants	Photosynthetic bacteria + fungal community consortia	Aquaculture waste facility testing	Demonstrates synergy of bacterial and fungal pollutant removal	Yang et al. (2025)

top-down approaches enrich functional communities from contaminated matrices and subsequently isolate keystones for reconstruction (Ruan et al. 2024). Physical strategies like immobilization, carrier materials, and layered agar/slurry systems that help maintain spatial structure, enhance cross-feeding, and reduce competitive washout (Efremenko et al. 2024). Process controls such as staged inoculation help in introducing the fungus first to build hyphal networks, then bacterial degraders, cosubstrate dosing to prevent intermediate accumulation, and micro-aeration adjustments frequently improve stability and overall degradation in laboratory trials (Ruan et al. 2024; Qattan 2025a, b).

7.4.2 Inoculation Strategies for Field-Scale Deployment

Field deployment requires formulation and delivery considerations distinct from lab work. Practical strategies reported recently include carrier-based formulations like biochar, sawdust, and alginate beads that protect cells and facilitate slow release, seed or root coatings for rhizosphere targeting, and repeated high-dose inoculations to overcome indigenous community inertia (Efremenko et al. 2024; Ruan et al. 2024). Recent studies show that dose and repetition matter: higher and repeated inoculations produced stronger and more persistent reassembly of functional microbiomes in soil trials. Where possible, combine inoculation with mild biostimulation, like nutrient balancing and organic amendments, to prevent substrate limitation while avoiding catabolic repression (Qattan 2025a, b). Pilot testing under site-specific physicochemical regimes is essential before scaling.

Competition and competitive exclusion are common failure modes. Contemporary strategies for managing dominance include the following: selecting metabolic complementarity to reduce niche overlap, spatial structuring (immobilization or carrier partitioning) to create microhabitats, staged inoculation to allow slower growers to establish, and engineered cross-feeding circuits or quorum-sensing control for dynamic population regulation (Efremenko et al. 2024; Jaiswal et al. 2025). Modeling approaches, like genome-scale metabolic models, can forecast competitive outcomes and suggest strain ratios that minimize takeover while maximizing function (Ruan et al. 2024). Finally, monitoring, like qPCR, marker genes, and activity assays, is required post-deployment to detect unintended drift and guide corrective reinoculation or amendment (Fig. 7.3).

7.5 Environmental Optimization

7.5.1 Importance of Soil pH, Texture, and Organic Matter

Soil pH, texture, and organic matter content jointly set the chemical and physical stage on which fungi, bacteria, and plants interact to transform the xenobiotics. pH controls metal speciation, enzyme kinetics, and microbial community composition; neutral to slightly acidic microzones often have the capabilities to maximize

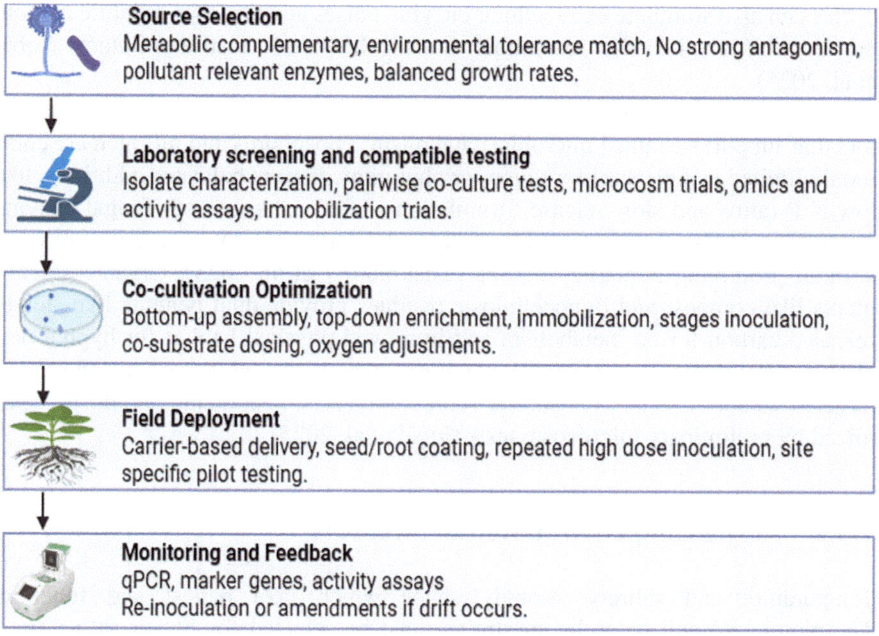

Fig. 7.3 Systematic workflow for designing microbial consortia

enzymatic degradation of many organics, while strongly alkaline or very acidic conditions can immobilize enzymes or shift speciation toward more toxic ionic forms (Iqbal et al. 2025; Chakraborty et al. 2025). Texture determines water retention and pore connectivity: clay-rich soils retain more organic matter and moisture and therefore support sustained microbial activity, whereas sandy soils promote faster drainage and sporadic substrate availability (Bogati et al. 2025). Soil organic matter provides the energy and sorption matrix for microbes and pollutants: higher SOM raises microbial biomass, supplies labile carbon for co-metabolic pathways, and forms organo-mineral complexes that can sequester metals or modulate bioavailability (Iqbal et al. 2025). Designing remediation systems, therefore, begins with baseline characterization and, where needed, targeted amendments to bring pH, texture-mediated water retention, and SOM into ranges that favor the intended microbial functions.

Water availability is often the proximate limiter of biodegradation in field settings. Fungi show greater drought resilience than many bacteria, maintaining hyphal growth and storage compounds under low-water regimes (Canarini et al. 2024), but overall biodegradation rates and bacterial activities decline sharply with soil drying (Bogati et al. 2025). Practical moisture management, therefore, balances persistence and aeration: controlled irrigation or moisture-retention amendments, like biochar and compost, sustain enzyme production and hyphal extension without creating long-term anoxia that inhibits aerobic degraders. Periodic wetting to drying cycles can also stimulate extracellular enzyme pulses and turnover of labile carbon, but must be tuned to site constraints to avoid losses of volatile pollutants (Iqbal et al. 2025).

Biostimulation has the potential for adding limiting nutrients or electron acceptors that support sustained microbial catabolism. Phosphorus and nitrogen are commonly limited in soils used for myco- or phytoremediation; balanced additions, like low N:P ratios and slow-release formulations, often enhance both fungal enzyme production of laccases, peroxidases, etc., and bacterial ring-cleavage pathways without promoting nontarget blooms (Chakraborty et al. 2025). Organic amendments like compost and lignocellulosic residues provide dual benefits like slowly released carbon for co-metabolism and improved physical habitat for hyphal networks and biofilms. Recent studies caution that excessive labile carbon can repress targeted catabolic genes (catabolite repression), so amendments should be optimized by preliminary microcosm tests (Iqbal et al. 2025).

7.5.2 Addressing Environmental Stressors

Temperature and salinity strongly shape which taxa persist and function. Thermotolerant and halotolerant strains must be selected for hot or saline sites; planting salt-tolerant species with adapted microbiomes has been shown to rapidly reduce and reshape microbial networks toward remediation-friendly assemblages (Cui et al. 2025). Likewise, pollutant co-stress of heavy metals with organics can inhibit degraders: integration of immobilizing agents like clays and metal-sequestering amendments helps relieve toxicity and allow enzymatic degradation to proceed (Chakraborty et al. 2025). Temperature modulation is largely operational seasonal timing and shallow mulches and should align inoculation with windows of optimal microbial activity identified in site trials.

Invasive plants, opportunistic pathogens, or strongly antagonistic microbes can derail remediation consortia. Early screening using small-scale mesocosms reveals opportunists that outcompete functional strains; physical containment, like carrier immobilization, microsite placement, and ecological tactics, which promotes native keystone taxa via tailored exudate-stimulating amendments, reduces takeover risk (Iqbal et al. 2025). Biocontrol strategies such as introducing benign antagonists to pathogens or using plant hosts that favor mutualists can stabilize desired communities. Continuous monitoring after deployment allows rapid corrective measures, like re-inoculation and amendment adjustments, to maintain functional balance.

7.6 Monitoring Community Dynamics

Monitoring microbial community dynamics is fundamental to understanding how ecosystems respond to environmental changes, pollutant exposure, and management interventions. A suite of complementary tools, including molecular methods, traditional techniques, functional assays, degradation tracking, and adaptive management frameworks, that somehow allows comprehensive and responsive ecosystem assessment.

7.6.1 Use of Molecular Tools

Modern molecular techniques revolutionize our ability to monitor community composition and function. Quantitative PCR (qPCR) facilitates precise, sensitive quantification of specific target genes or taxa (Smith and Jones 2020), enabling temporal tracking of pathogens, degraders, or keystone populations. Metabarcoding, which employs high-throughput sequencing of marker genes, e.g., 16S rRNA or ITS, delivers taxonomic profiles across communities, revealing shifts in diversity and relative abundance (Thompson et al. 2019). Metagenomics goes further by sequencing total community DNA, uncovering not just who is present but also their functional potential, such as pollutant-degrading pathways or antibiotic resistance genes (Brown et al. 2021). Collectively, these molecular tools allow rapid, culture-independent monitoring of community structure and predicted function.

Despite the dominance of molecular methods, culturing and microscopy remain indispensable. Culturing allows isolation of live organisms, which facilitates phenotypic characterization, physiological assays, and calibration of molecular data (Lee and Patel 2018). Meanwhile, microscopy, including epifluorescence or electron microscopy, permits direct visualization of microbes, informing on morphology, spatial relationships (e.g., biofilm architecture), and biomass estimation (Garcia-Ruiz et al. 2017). These traditional approaches validate molecular findings and can reveal disturbances such as cell size variation or morphological differentiation not captured by sequencing alone.

Assessing ecosystem function requires more than taxonomic snapshots; functional activity assays are crucial. Enzyme assays, such as dehydrogenase for general microbial activity or specific hydrolases for nutrient cycling, provide real-time indicators of metabolic potential (Martinez et al. 2015). Respiration rate measurements, often via CO_2 evolution or oxygen consumption, track whole-community metabolic activity (Harrison and Smith 2016). These proxies reflect vitality and ecological function, signaling how communities respond to stressors or resource inputs even when community composition remains stable.

For ecosystems confronting contamination, tracking pollutant degradation kinetics offers direct evidence of remediation. This involves measuring the disappearance of target pollutants, e.g., hydrocarbons, pesticides, and the appearance of metabolites over time. Approaches include chemical analytics of μg/L concentration over time, isotope-labeling, e.g., ^13C-labeled substrates, and modeling degradation rates, half-life constants, and first-order kinetics (Nguyen et al. 2022). Coupling chemical data with microbial community analyses, e.g., abundance of degradation pathway genes via qPCR or metagenomics, links microbial dynamics to functional outcomes (Nguyen et al. 2022). Such combined monitoring provides mechanistic insight into which community members contribute to pollutant removal and how rapid contaminant reduction occurs.

7.6.2 Adaptive Management Based on Monitoring Feedback

Effective ecosystem management must be adaptive: monitoring results should feed back into decision-making. Through adaptive management, observed changes in community composition, functional activity, or degradation rates guide adjustments in treatments such as nutrient amendments, bioaugmentation, aeration, or pollutant removal strategies (Walker et al. 2018). For instance, if metagenomic analysis indicates insufficient presence of degraders, managers may introduce specific strains or promote conditions for their growth; if enzyme assays or respiration data signal declining activity, moisture or oxygen levels might be adjusted (Walker et al. 2018). The iterative loop of monitoring leads to interpretation, which further leads to management, then monitoring, which enables responsive and optimized approaches to sustaining or restoring ecosystem health.

7.7 Case Studies

7.7.1 Field Studies Demonstrating Consortia-Based Remediation

Recent field and microcosm-scale studies provide strong, real-world evidence that tailored microbial consortia can outperform single strains for pollutant removal. In paddy-soil microcosms and soil bioaugmentation trials, indigenous consortia degraded the pesticides fipronil and thiobencarb rapidly (>80% removal in many treatments) and accelerated in-soil dissipation by 1.3–13× relative to uninoculated controls (Faridy et al. 2024). Similarly, an indigenous, biosurfactant-producing bacterial consortium applied in soil microcosms removed large fractions of crude-oil hydrocarbons (aqueous removal ≈96% in 9 days; soil removal ≈65% over 120 days), demonstrating the value of in situ biosurfactant producers for addressing hydrophobic contaminants (Hosseini et al. 2025). Field collections from polluted rivers have also yielded consortia that tolerate and remove high proportions of Cu, Zn and Ni in laboratory and near-field tests, illustrating potential for heavy-metal reduction using locally adapted mixes (Qurbani et al. 2025).

The literature from 2024 to 2025 documents consortia acting on a broad pollutant spectrum. Representative, recent examples include:

- *Pesticides (fipronil, thiobencarb)*: Enrichment-derived bacterial consortia achieved rapid degradation in paddy soils and yielded detailed metabolite pathways, supporting safe mineralization rather than mere transformation to persistent intermediates (Faridy et al. 2024).
- *Hydrocarbons (crude oil)*: Engineered and native biosurfactant-producing consortia achieved high removal of saturated hydrocarbons in aqueous tests and substantial removal in soil microcosms, underscoring synergy between degraders and surfactant producers to overcome bioavailability limits (Hosseini et al. 2025).

- *Heavy metals (Cu, Zn, Ni)*: Indigenous consortia displayed enhanced metal tolerance and sequestration/reduction efficiencies, reporting reductions of >80% for some metals in lab/near-field assays, often via combined biosorption, bioaccumulation, and biomineralization mechanisms (Qurbani et al. 2025; Qattan 2025a, b).
  Collectively, these studies show consortia can be tuned to organics via enzymatic degradation/co-metabolism and to metals via binding, redox transformation, or immobilization.

7.7.2 Implementation and Management

Field and microcosm experiences since 2024 show reoccurring practical lessons:

- *Use of indigenous or preadapted strains when possible.* Local isolates establish and persist better than exogenous strains, improving long-term efficacy (Hosseini et al. 2025; Faridy et al. 2024).
- *Address bioavailability up front.* For hydrophobic pollutants, pairing degraders with biosurfactant-producers or employing amendment strategies increases mass transfer and speeds degradation (Hosseini et al. 2025).
- *Monitor metabolites and ecotoxicity.* Rapid loss of parent compound is insufficient; pathway analysis is needed to ensure nontoxic end products (Faridy et al. 2024).
- *Environmental constraints matter.* High contaminant concentrations, extreme pH, salinity, or drought can suppress consortia activity; optimization of inoculum size, nutrient balance, and moisture is often required (Faridy et al. 2024; Qurbani et al. 2025).
- *Consortia design must balance complexity and manageability.* Very large, undefined mixes can be unpredictable; rationally assembled or well-characterized consortia like bottom-up design give better control and easier regulatory assessment (Qattan 2025a, b).

Recent reviews and case studies emphasize that consortia-based bioremediation can be cost-competitive versus excavation or chemical treatments when long-term environmental, social, and disposal costs are included. Key practical points:

- *Up-front costs* include isolation/characterization, scale-up (fermentation), and site conditioning (nutrient amendments, aeration).
- *Operational savings* accrue from avoiding heavy machinery, landfill disposal, and secondary chemical use; passive remediation also reduces community disruption (Qattan 2025a, b).
- *Scalability trade-offs*: lab and microcosm successes do not always translate directly to full site scale because of inoculum establishment and heterogeneity, with pilot plots and staged rollouts reducing risk (Hosseini et al. 2025).
- *Regulatory and monitoring costs* (ecotoxicity testing, long-term monitoring) must be budgeted; however, when coupled with adaptive monitoring frame-

works, iterative adjustments can optimize cost-effectiveness over the project lifetime (Faridy et al. 2024; Qattan 2025a, b).

In short, consortia strategies frequently offer favorable life-cycle economics for soils and waters with moderate to severe contamination, but success hinges on local adaptation, thoughtful pilot trials, and robust monitoring that ties microbial dynamics to chemical endpoints.

7.8 Challenges and Future Prospects

7.8.1 Technical Challenges in Establishing and Maintaining Consortia

Engineering and deploying microbial consortia in real-world environments face a host of technical hurdles. First, achieving stable assembly of multi-strain systems can be difficult due to interspecies competition, variable growth rates, and sensitivity to environmental conditions (Qattan 2025a, b). Even well-designed consortia may destabilize over time, losing key functional members or succumbing to invasive environmental microbes (Qurbani et al. 2025). Scale-up from laboratory microcosms to field applications often entails inoculum volume increases of several orders of magnitude; maintaining viability and activity during production, storage, and field delivery, e.g., with carriers like biochars or granules, is technically demanding and cost-intensive (Hosseini et al. 2025). Nutrient amendments designed to stimulate consortia can also unintentionally enrich unwanted species, undermining degrader performance (Faridy et al. 2024). Finally, environmental heterogeneity variability in pH, moisture, pollutant concentration, and oxygenation can create micro-niches, where consortia underperform, necessitating refined delivery methods, physical mixing, or adaptive dosing strategies (Hosseini et al. 2025).

Deploying microbial consortia, even indigenous ones, carries ecological risks. Horizontal gene transfer between introduced and native microbes can unintentionally spread antibiotic resistance or catabolic capacities to unintended recipients (Qattan 2025a, b). Metabolite byproducts from degradation pathways may be toxic or persistent, especially if intermediate products accumulate or microbial metabolism stalls (Faridy et al. 2024). Introduction of large microbial loads may alter community balance, leading to nutrient drawdown, oxygen depletion, or creation of anaerobic zones, potentially harming nontarget organisms (Hosseini et al. 2025). These risks demand rigorous monitoring not only of target pollutant concentrations but also of ecotoxicology, resistance gene abundance, and native community shifts over time before and after deployment (Qurbani et al. 2025).

Regulation of bioremediation using microbial consortia remains nascent and fragmented across jurisdictions. Even when using local isolates, regulatory authorities often demand detailed dossiers: strain identification, genomic profiles, containment strategies, ecotoxicology assays, and environmental fate modeling (Qattan 2025a, b). Guidelines for evaluating genetically modified or synthetic strains are even stricter, often requiring multiyear monitoring and contingency plans if risks emerge. Policymakers must set a balance between environmental benefit, public safety, and innovation while ensuring transparency and community engagement. Harmonizing regulations, clarifying standards for genomic and field monitoring, and creating pathways for pilot-level testing could accelerate adoption (Qattan 2025a, b).

Precision agriculture offers tailored, data-driven opportunities to enhance consortia deployment. By using spatial soil sensors (for moisture, nutrient gradients, pollutant hotspots), decision support systems can guide targeted applications of microbial consortia via variable-rate sprayers or drip irrigation (Qurbani et al. 2025). Combined mapping of microbial hotspots and real-time data augmented by remote sensing or soil health indicators, which can optimize inoculation timing, dose, and placement, reducing costs and improving uptake. Further, integrating consortia that degrade agrochemicals with biofertilizer capabilities, e.g., nitrogen-fixing or phosphate-solubilizing partners, aligns remediation with crop productivity, promoting sustainable agricultural loops (Hosseini et al. 2025).

7.8.2 Research Needs and Emerging Tools

Innovative tools are emerging to address these challenges. Synthetic biology enables the rational design of minimal consortia with defined roles, engineered stability, or fail-safe kill-switches to limit spread (Qattan 2025a, b). Machine learning approaches can predict compatibility among strains, optimize consortia composition, and forecast performance across environmental gradients so far more rapidly than iterative bench experiments (Qurbani et al. 2025). Recent studies leverage environmental 'omics data alongside ML to infer successful co-occurrence networks and functional guilds, streamlining candidate selection (Faridy et al. 2024). Advanced biofilm- or encapsulation-based delivery systems, informed by modeling of release kinetics, could protect consortia in the field and extend functional lifespan (Hosseini et al. 2025). Increasingly, multi-omics monitoring frameworks allow real-time, adaptive feedback control loops that are used in sequencing-based monitoring, metabolite sensors, and remote chemical analysis feed into adjustment of inoculum, nutrient, aeration, or irrigation strategies for iterative optimization (Qattan 2025a, b).

7.9 Conclusion

Over the past decade, and particularly in recent years, microbial consortia-based bioremediation has evolved from a largely experimental concept to a promising applied technology. The studies reviewed in this chapter illustrate that diverse microbial partnerships were often being drawn from indigenous environments, which can outperform single-strain inoculants in degrading a wide range of pollutants, including pesticides, hydrocarbons, and heavy metals (Faridy et al. 2024; Hosseini et al. 2025; Qurbani et al. 2025). Field and microcosm trials demonstrate that consortia succeed by combining complementary metabolic pathways, enhancing pollutant bioavailability, and providing functional redundancy that buffers against environmental fluctuations (Qattan 2025a, b). The technical toolkit for monitoring community dynamics, ranging from qPCR and metagenomics to enzyme assays and degradation kinetics, enables practitioners to track not only pollutant disappearance but also microbial community shifts and functional stability. Despite these advances, substantial technical and ecological challenges remain. Achieving stable consortia establishment, preventing loss of functional members, avoiding ecological risks such as unwanted gene transfer, and ensuring that intermediate metabolites are nontoxic all require careful design, rigorous testing, and long-term monitoring (Faridy et al. 2024; Qattan 2025a, b). On the policy front, regulatory frameworks are still catching up, especially for engineered or synthetic consortia. Recent developments in precision agriculture, synthetic biology, and machine learning open opportunities to tailor consortia for site-specific needs, improve deployment efficiency, and predict field performance before large-scale trials. These technologies, coupled with adaptive management, can transform consortia-based remediation from an opportunistic intervention into an integrated part of environmental and agricultural management strategies (Hosseini et al. 2025; Qurbani et al. 2025).

7.9.1 Emphasis on the Need for Holistic, Systems-Based Approaches

The collective evidence underscores the importance of treating bioremediation not as a single-step intervention but as a complex system requiring integrated consideration of microbial ecology, environmental chemistry, site management, and socio-economic context. In practice, this means moving away from "deploy-and-forget" inoculation strategies toward adaptive, feedback-driven frameworks, where microbial performance is continuously monitored, interpreted, and adjusted (Qattan 2025a, b). A systems-based perspective recognizes that pollutants rarely exist in isolation; they often co-occur with other stressors, such as nutrient imbalances, salinity, or metal contamination. Similarly, microbial communities operate within broader ecological networks that include plants, soil fauna, and abiotic factors. Designing effective consortia thus requires multilevel thinking, identifying synergies between microbial degraders and plant hosts, coordinating remediation with soil fertility improvement, and aligning interventions with crop cycles in agricultural landscapes (Qurbani et al. 2025). Novel approaches also demand cross-disciplinary collaboration. Microbiologists, soil scientists, engineers, chemists, environmental economists, and policymakers each bring essential perspectives. For example, engineers can design delivery systems that protect microbes from environmental stress, while economists can quantify life cycle benefits and social return on investment. Without such integration, even technically promising consortia may fail due to logistical, regulatory, or social barriers.

7.9.2 Call for Decision-Making

The promise of microbial consortia in remediation is clear, but realizing its full potential requires decisive action from all stakeholders:

- *Researchers* must prioritize the development of predictive models linking consortia composition, environmental variables, and pollutant degradation outcomes. Investment in high-throughput screening, omics-based functional profiling, and machine learning-driven design platforms will accelerate the identification of robust, adaptable microbial partnerships. More longitudinal field studies, especially in diverse climatic and soil conditions, are needed to understand long-term persistence, ecological integration, and resilience to perturbations (Faridy et al. 2024).
- *Practitioners* should adopt adaptive management as standard practice. This involves implementing cost-effective monitoring tools, piloting at small scales before full deployment, and combining bioremediation with other land management strategies, such as phytoremediation or biochar amendment. Documentation of both successes and failures in field applications will be invaluable for refining best practices.
- *Policymakers* must develop clear, evidence-based regulatory frameworks that safeguard environmental and public health without hindering innovation. Provisions should account for the differences between native, non-modified consortia and engineered strains, with rough monitoring requirements. Policy incentives, such as subsidies, credits, or public-private partnerships, can lower the economic barriers to adoption, especially for agricultural or industrial sectors with large-scale contamination challenges (Qattan 2025a, b).

Ultimately, consortia-based remediation should not be seen solely as an emergency response to pollution events. Instead, it can become a proactive component of sustainable land and water management, embedded within long-term strategies to maintain ecological integrity, safeguard food security, and meet environmental quality standards. Achieving this vision will require sustained investment in research, demonstration of projects, regulatory innovation, and public engagement.

References

Anckaert CE, Moore JC, Johnson D (2024) Mutualism re-sequenced. Curr Biol. https://doi.org/10.1016/j.cub.2023.12.026

Bogati KA, Sewerniak P, Walczak M (2025) Unraveling the effect of soil moisture on microbial diversity and enzymatic activity in agricultural soils. Microorganisms 13(6):1245. https://doi.org/10.3390/microorganisms13061245

Brown AL, Nguyen PT, Ramirez M (2021) Metagenomic insights into microbial functional potential in contaminated environments. J Environ Microbiol 13(2):145–158

Canarini A, Fuchslueger L, Schnecker J, Metze D, Nelson DB, Kahmen A et al (2024) Soil fungi remain active and invest in storage compounds during drought independent of future climate conditions. Nat Commun 15:10410. https://doi.org/10.1038/s41467-024-54537-y

Chakraborty S, Talukdar A, Dey S, Bhattacharya S (2025) Role of fungi, bacteria and microalgae in bioremediation of emerging pollutants with special reference to pesticides, heavy metals and pharmaceuticals. Discov Environ. https://doi.org/10.1007/s44274-025-00217-7

Cui Y, Ning Z, Li M, Qin X, Yue X, Chen X et al (2025) Microbial network-driven remediation of saline-alkali soils by salt-tolerant plants. Front Microbiol. https://doi.org/10.3389/fmicb.2025.1565399

Efremenko E, Stepanov N, Senko O, Aslanli A, Maslova O, Lyagin I (2024) Using fungi in artificial microbial consortia to solve bioremediation problems. Microorganisms 12(3):470. https://doi.org/10.3390/microorganisms12030470

Faridy N, Torabi E et al (2024) Efficacy of novel bacterial consortia in degrading fipronil and thiobencarb in paddy soil: a survey for community structure and metabolic pathways. Front Microbiol 15. https://doi.org/10.3389/fmicb.2024.1366951

Gao L, Okoye CO, Wang C, Lou F, Jiang J (2024a) Enhanced remediation of phenanthrene and naphthalene by corn-bacterial consortium in contaminated soil. Plants 13(20):2839. https://doi.org/10.3390/plants13202839

Garcia-Ruiz R, Hernández M, Torres E (2017) Microscopy and morphometric analysis of biofilm architecture. Microb Ecol Rev 9(1):23–35

Harrison D, Smith J (2016) Measuring microbial respiration in soil microcosms. Methods Ecol 5(8):456–462

Hosseini S, Sharifi R, Habibi A (2025) Efficient bioremediation of crude oil contaminated soil by a consortium of in-situ biosurfactant producing hydrocarbon-degraders. Sci Rep 15:19852

Iqbal S, Begum F, Nguchu BA, Uzabakiriho PC, Shaw P (2025) The invisible architects: microbial communities and their transformative role in soil health and global climate changes. Environ Microbiome 20:36. https://doi.org/10.1186/s40793-025-00694-6

Jaiswal DK, Ade S, Belwal R, Pereira R, Verma JP (2025) Editorial: microbial co-cultures: a new era of synthetic biology and metabolic engineering, volume II. Front Microbiol. https://doi.org/10.3389/fmicb.2025.xxxxx

Lee C, Patel S (2018) Culturing unculturable microbes: bridging the gap between molecular and traditional microbiology. Appl Microbial Tech 22(4):308–320

Lethielleux-Juge C (2025) Review: roles of mycorrhizal symbioses and associated soil microbiomes in ecological restoration. Front Microbiol 16:1456041. https://doi.org/10.3389/fmicb.2025.1456041

Li Y et al (2025a) Eco-engineered remediation: microbial and rhizosphere-based strategies for heavy-metal detoxification. Environ Res 249:119417. https://doi.org/10.3389/fmicb.2023.1229828

Li Y et al (2025b) Eco-engineered remediation: microbial and rhizosphere-based strategies for heavy-metal detoxification. Environ Res 249:119417. https://doi.org/10.1016/j.envres.2025.119417

Ma S, Chen Q, Zheng Y, Ren T, He R, Cheng L et al (2025) A tale for two roles: root-secreted methyl ferulate inhibits Phytophthora nicotianae and enriches rhizosphere Bacillus. Microbiome 13:33. https://doi.org/10.1186/s40168-024-02008-3

MacLean BM, Lopes FAC, Haensel C, Santoro AE (2024) Combined bacterial–fungal inoculants enhance biofilms on hydrophobic plastics in aquatic environments. Environ Microbiome 19:77. https://doi.org/10.1186/s40793-024-00511-0

Martinez L, Johansen A, Steward F (2015) Enzyme assays as functional indicators of microbial activity in soil ecosystems. Soil Biol Biochem 80:123–130

Mitra D, Rodríguez-Seijo A, Adhikari P, Villalobos S d S, Panneerselvam P, Del Gallo M, Pellegrini M (2024) Unearthing and harnessing the power of the soil microbiome and mycorrhizas to enhance plant nutrient utilization under climate stress. Front Soil Sci 4:1485795. https://doi.org/10.3389/fsoil.2024.1485795

Nguyen HT, Lee J, Choi K (2022) Tracking hydrocarbon biodegradation kinetics in contaminated sediments using isotopic labeling. Environ Sci Technol 56(10):1234–1242

Ningombam L, Mana T, Pradhan S, Apum G, Singh YD (2025) Fungal bioremediation in environmental pollution and recent strategies. Discover Environment 3:100. https://doi.org/10.1007/s44274-025-00267-x

Qattan SYA (2025a) Harnessing bacterial consortia for effective bioremediation: targeted removal of heavy metals, hydrocarbons, and persistent pollutants. Environ Sci Eur 37:85. https://doi.org/10.1186/s12302-025-01103-y

Qattan IA (2025b) Biodegradation of conventional plastics by bacterial communities: a systematic review. Environ Sci Eur 37:37. https://doi.org/10.1016/j.scitotenv.2023.163908

Qurbani K et al (2025) Synergistic enhancement of heavy metal tolerance and reduction by indigenous bacterial consortia of *Pseudomonas putida* and *Pasteurella aerogenes*. Sci Rep 15:24663

Rezaei E, Moghimi SM (2024) Modeling the fungal highway to understand and improve bioremediation. Biotechnol Bioeng. https://doi.org/10.1016/j.ecoenv.2024.116543

Ruan Z, Chen K, Cao W, Meng L, Yang B, Xu M et al (2024) Engineering natural microbiomes toward enhanced bioremediation by microbiome modeling. Nat Commun 15:4694. https://doi.org/10.1038/s41467-024-49098-z

Shukla S, Didwania N, Choudhary R (2025) Arbuscular mycorrhizal fungi: a pathway to sustainable soil health, carbon sequestration, and greenhouse gas mitigation. J Saudi Soc Agric Sci 24:22. https://doi.org/10.1007/s44447-025-00023-w

Smith R, Jones M (2020) Quantitative PCR for monitoring microbial populations in ecological studies. Ecol Mol Methods 14(3):200–212

Thompson BL, Williams D, Garcia O (2019) Metabarcoding approaches for assessing microbial biodiversity in aquatic systems. Front Microbial Ecol 10:567

Walker P, Chen L, Roberts M (2018) Adaptive management in bioremediation: integrating monitoring and intervention. J Environ Manag 210:120–130

Wang Y, Liu H, Zhang D (2024a) Root exudates facilitate the regulation of soil microbial community function. Front Plant Sci 15:1461893. https://doi.org/10.3389/fpls.2024.1461893

Wang Z, Zhang X, Lu Y, Wei J, Qi S, Wu B, Cheng S (2024b) Sustainable remediation of soil and water utilizing arbuscular mycorrhizal fungi: a review. Microorganisms 12(7):1255. https://doi.org/10.3390/microorganisms12071255

Wibowo AN, Anggraeni E, Wulandari AP, Suyanto A, Sudrajat R, Rifa'i M (2024) Fungal and bacterial technologies to utilize waste materials and compounds for bioremediation: a review. Front Mater 11:1348171. https://doi.org/10.3389/fmats.2024.1416445

Xie X, Li S, Jiang X, Chai Y, Liu G (2024) Bridging the gap: biofilm-mediated establishment of Bacillus velezensis on Trichoderma guizhouense mycelia. Biofilm 7:100239. https://doi.org/10.1016/j.bioflm.2024.100239

Zhang J, Fan B, Zhao L, Zhao C, Yang F (2024) Biochar promotes compost humification by regulating bacterial and fungal communities. Front Microbiol 15:1470930. https://doi.org/10.3389/fmicb.2024.1470930

Zhao X, Liu Q, Li J (2025) Rhizosphere response and resistance to fertilization. Commun Earth Environ 6:2567. https://doi.org/10.1038/s43247-025-02567-9

Zheng H et al (2025) Rhizosphere effects on cadmium fractionation in contaminated soils. Sci Rep 15:14241. https://doi.org/10.1038/s41598-025-14241-3

Bottlenecks in Mycoremediation: Solutions and Prospects for This Fungal Bioremediation Process

8

Rahul Kumar Mishra, Arpita Roy, Subhasree Ray, Elvis Fosso-Kankeu, and Soumya Pandit

Abstract

Mycoremediation, using fungi for environmental detoxification of xenobiotics, is a versatile and multidisciplinary biotechnology with potential to deal with contamination problems globally. Although this approach offers several advantages over classical remediation methods (cost efficiency, environmental compatibility, and substrate versatility), the use of mycoremediation is hindered by several bottlenecks that prevent its large-scale application. This review seeks to address key limitations of efficient mycoremediation, which range from fungal strain selection and optimization through environmental parameters issues, scalability of the technology, regulatory obstacles, and technical limitations. Based on the existing research results and cases, this article analyzes some of the key barriers that need to be addressed and makes recommendations for breaking these bottlenecks. The critique encompasses both scientific platform issues and implementation hurdles, which suggest future research directions to enhance the efficiency and cost-effectiveness of mycoremediation approaches.

Keywords

Mycoremediation · Xenobiotics · Fungi · Contamination · Bottlenecks · Environment

R. K. Mishra · S. Ray · S. Pandit (✉)
Department of Life Sciences, Sharda School of Bioscience and Technology, Sharda University, Greater Noida, Uttar Pradesh, India

A. Roy
Department of Biotechnology, Sharda School of Bioscience and Technology, Sharda University, Greater Noida, UP, India

E. Fosso-Kankeu
Department of Metallurgy, Faculty of Engineering and the Built Environment, University of Johannesburg, Johannesburg, South Africa

A. Roy et al. (eds.), *Mycoremediation of Xenobiotics*, Emerging Paradigms in Pharmaceutical Research, https://doi.org/10.1007/978-981-95-5743-1_8

8.1 Introduction

Xenobiotics, such as persistent organic pollutants (POPs), heavy metals, pharmaceuticals, and industrial chemicals, could lead to ecological and human threats to environmental contamination. Costly and energy-consuming traditional practices, like excavation, burning, and chemical treatment, often lead to environmental disturbance (Briffa et al. 2020; Vaksmaa et al. 2023). To alleviate these disadvantages, bioremediation is being pursued as an environmentally safe technology, and this approach draws the greatest attention due to its high advantages and applicability.

Mycoremediation is the use of fungi to play a role in the degradation, addition, or transformation of pollutants. They have many attributes that make them suitable candidates for bioremediation, such as the ability to produce a wide variety of extracellular enzymes and tolerate toxic conditions, with possibly some large hyphal network allowing contact with and uptake of the contaminant.

Fungal metabolism is versatile for breaking down recalcitrant, complex organic compounds that are resistant to bacterial decomposition. Their bioaccumulation ability makes them applicable in heavy metal remediation (Dinakarkumar et al. 2024). A schematic diagram representing the challenges of mycoremediation is shown in Fig. 8.1.

The area of mycoremediation has expanded significantly within the last 20 years, and a number of laboratory case studies have shown successful decontamination with a variety of types of contaminants. White-rot fungi, including types such as the *Phanerochaete chrysosporium*, *Trametes versicolor*, and *Pleurotus ostreatus*, are used to degrade lignin-type structures of many persistent organic pollutants (Bautista-Zamudio et al. 2023). In the same line, a number of fungal species are

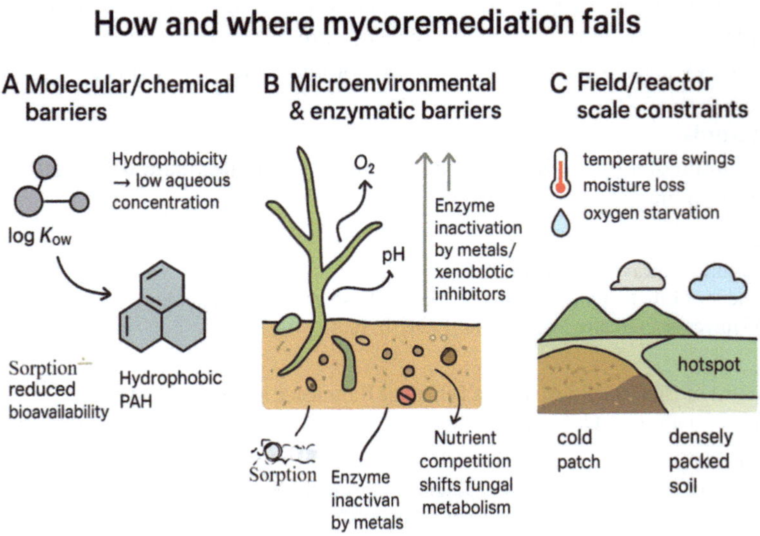

Fig. 8.1 This figure represents the challenges faced in mycoremediation

effective for the immobilization of heavy metals by biosorption, bioaccumulation, and precipitation processes.

However, this successful application of these results from the laboratory to the field scale is nontrivial. There are several bottlenecks restricting the application of mycoremediation from being widespread, including basic limitations from a scientific perspective as well as engineering and regulatory perspectives. The identification and alleviation of such bottlenecks are essential for further exploitation of mycoremediation as an alternative form of environmental remediation.

8.2 Selection and Optimization of Fungal Strains

Among critical bottlenecks of mycoremediation, the isolation and development of fungal strains suitable for different types of contamination is one of the most relevant. The success of mycoremediation relies on the matching between fungal abilities and contaminant features and environmental settings; however, this procedure is still mostly empirical and time-consuming (Akpasi et al. 2023). There are approximately 1.5 million known species of fungi, and similar numbers of as-yet-undiscovered strains exist, creating an immense variety of consultant strains to meet drug development needs.

The existing strategies for selecting strains in general depend on screening methods that do not necessarily correlate with the performance under field conditions. Laboratory screenings are always performed under controlled conditions, which are largely different from polluted places in reality. Parameters, such as the nutrient status, pH dynamics, temperature fluctuation, and co-contaminant composition, can significantly affect fungal performance but are rarely, if at all, included in screening procedures (Akpasi et al. 2023). As a result, the strains that show great potential for remediation in laboratory conditions may not be efficient enough when implemented in field treatments. The diversity of fungal strains and their ability to degrade a wide spectrum of contaminants are summarized in Table 8.1.

Then, a second important aspect relates to the genetic stability of the selected fungal strains. Many species of fungus have defined large genetic plastic pathways, since under environmental conditions, mutations and adaptation can be rapidly generated (Sauters and Rokas 2025). This type of flexibility can be an asset to survive in a polluted site, but it comes at the expense of sieveability, which in such cases may be lowered or even nonexistent. However, prolonged correct behavior (i.e., non-declining attributes) across full remediation needs a constant check and, if necessary, re-inoculation with virgin fungal matter.

One way to solve this problem is the genetic engineering of strains, but that involves other problems. Traditional breeding methodologies are hampered by the sexual reproductive lifestyle of most fungi, although novel genetic modification techniques have not found acceptance with both regulators and consumers (Salazar-Cerezo et al. 2023). Moreover, degradation pathways are generally polygenic because of the complex metabolization of any pollutant and hence multigene regulation that can become technically difficult to manipulate genetically.

Table 8.1 Representative fungal strains used in mycoremediation, their target pollutants, and associated mechanisms of action

Fungal strain	Target contaminant(s)	Mechanism of action	References
Phanerochaete chrysosporium	Polycyclic aromatic hydrocarbons (PAHs), dyes, pesticides	Produces lignin peroxidase and manganese peroxidase for oxidative degradation	Tang et al. (2005)
Pleurotus ostreatus	Textile dyes, hydrocarbons, phenols	Laccase-mediated degradation and biosorption	Arroyo-Becerra et al. (2016)
Aspergillus niger	Heavy metals (Pb, Cd, Cr), dyes	Biosorption through cell wall binding; enzyme activity	Ahmad et al. (2006)
Trametes versicolor	Pharmaceuticals, dyes, endocrine-disrupting compounds	Laccase production for oxidative breakdown	Mir-Tutusaus et al. (2014)
Fusarium solani	Pesticides, petroleum hydrocarbons	Enzymatic degradation and adaptation to hydrocarbons	Okal et al. (2023)
Ganoderma lucidum	Phenolic compounds, textile effluents	Laccase and peroxidase activity; biosorption	Alshiekheid et al. (2023)
Mucor hiemalis	Synthetic dyes, azo compounds	Adsorption on fungal biomass; partial enzymatic action	Lu et al. (2016)

The procedure stresses the "need to include a standardized screening process in strategies of strain selection." These are as follows: realistic experimental conditions, multi-stressor additivity, as well as long-term impact of stress. Lastly, having these well-characterized whole collections published should help eliminate unnecessary "reinvention of the wheel" by making strain selection less challenging and repetitive work that has popped up in various labs, and even agencies.

8.3 Environmental Factors and Site-Specific Limitations

The potential for the mycoremediation of polluted sites is determined by environmental conditions, which are frequently not homogeneous and hard to manage. Development and synthesis of metabolites might be influenced by temperature, moisture, pH-oxygen conditions, and the availability of nutrient sources (Mishra and Srivastava 2020). It is therefore important to know the optimal value of these parameters for maximum remediation potential when considering field applications in light of strict environmental regulations.

Temperature sensitivities may be a particularly important challenge for the adaptability of mycoremediation to varied locales throughout the year. Furthermore, most fungi used in clean-up activities show an optimum temperature range that is unlikely to correspond to the temperature at which contaminated soil is found. As seasonal fluctuations in temperature may occasionally result in limited or no adiabatic development, thereby increasing treatment times and possibly relocating pollutants (Akpasi et al. 2023). Applications in cold climates are particularly

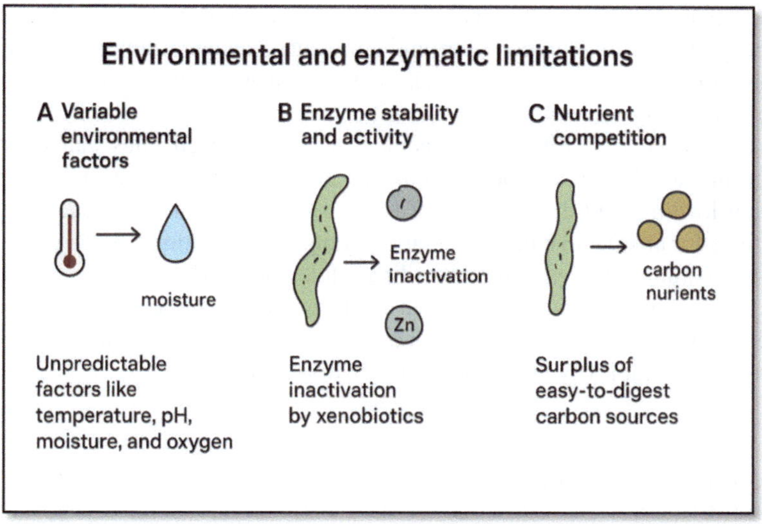

Fig. 8.2 This figure represents Environmental and enzymatic limitations

challenging, and the low temperature can strongly inhibit the metabolism and growth rate of fungi (Fig. 8.2).

Moisture controls also constitute one of the few environmental variables to exhibit consistent effects on the success of mycoremediation. Fungi need a certain amount of moisture to grow and metabolize, but "too much" will create anaerobic conditions that will limit the aerobic degradation. On the other hand, low water restricts the colonization and activity of fungi. Obtaining optimum soil moisture content in the field situation will involve a proper understanding of soil properties, drainage characteristics, and local rainfall data. Effective moisture management solutions required via engineering on a site-by-site basis add complication and cost to remediation (Akpasi et al. 2023).

Soil pH strongly affects the physiology of fungi and the bioavailability of a contaminant, yet adjustment of pH is difficult and costly in practice. A large number of contaminated sites are characterized by such extreme pH conditions as a result of geogenic or industrial influences. Although some fungi are known to be either acidophilic or alkalinophilic, the best results in remediation generally occur at a pH within a narrower range. The soil buffering system may make pH modification hard to adjust and maintain for an extended time ("Redistribution of soil water by a saprotrophic fungus enhances carbon mineralization | Request PDF," 2020).

The availability and balance of the nutrients serve as another challenge for optimization of mycoremediation. Fungi rely on an adequate supply of carbon and nitrogen for growth and enzyme production; however, in contaminated substrates, nutrient imbalances or limitations are often apparent. The addition of external nutrients not only can increase fungal activity but also potentially lead to the growth of other competing microorganisms or changes in site geochemistry that are not optimal. An optimal nutrient amendment approach would need a well-defined site description and very careful monitoring of all operations during a remediation. The efficiency of mycoremediation is strongly influenced by environmental parameters, such as temperature, pH, nutrient levels, and oxygen availability, which are outlined in Table 8.2.

The presence of inhibitory molecules in contaminated sites can drastically restrict fungal colonization and performance. In particular, heavy metals, high contaminant loading, and the presence of toxic co-contaminants can generate hostile conditions for an effective implementation of mycoremediation (Thapa et al. 2024). Although certain fungi exhibit extraordinary tolerance towards adverse conditions, limits for inhibition have been shown to differ between species and pollutants. Pretreatment or dilution may be needed in heavily contaminated locations, which increases the complexity and cost of remediation efforts.

Table 8.2 Environmental factors influencing the efficiency of mycoremediation and their impacts on fungal performance

Factor	Optimal range	Negative impact if deviated	Remarks/mitigation strategy	References
Temperature	25–30 °C (mesophilic fungi)	Low temperature slows metabolism; high temperature denatures enzymes	Selection of thermotolerant/ psychrotolerant strains	Akpasi et al. (2023)
pH	4.5–6.5	Extreme acidic/ alkaline pH reduces enzyme activity	Buffering agents; pH adjustment	Fernández-López et al. (2023)
Moisture	60–80% of water-holding capacity	Low moisture inhibits growth; excess leads to anaerobic conditions	Controlled irrigation and aeration	Geris et al. (2024)
Nutrients (C/N/P)	Balanced availability (C:N ≈ 20–30:1)	Nutrient deficiency slows fungal growth; excess nitrogen represses ligninolytic enzymes	Supplementation with organic wastes/ low-cost substrates	Akhtar and Mannan (2020)
Oxygen Availability	Aerobic conditions preferred	Oxygen limitation suppresses ligninolytic enzyme production	Aeration strategies or bioventing	Akpasi et al. (2023)
Competition with other microbes	Moderate coexistence possible	Overgrowth by bacteria or native fungi may reduce efficiency	Bioaugmentation, use of consortia, or pretreatment	Efremenko et al. (2024)
Contaminant concentration	Moderate levels (< toxic threshold)	Very high levels can inhibit fungal metabolism or be lethal	Adaptation or use of tolerant fungal strains	Akhtar and Mannan (2020)

8.4 Scalability and Engineering Challenges

One of the key limitations in mycoremediation adoption has been the reluctance to move from an experimental/phase testing laboratory setup to a field-scaled phytoremediation site. The scaling of mycoremediation processes is accompanied by a series of technical problems associated with inoculum production, distribution, monitoring, and control of the process. Engineering needs in large-scale mycoremediation are frequently quite different than those in small-scale studies and can require specially developed equipment and methods.

Inoculum supply and delivery are challenging logistical issues for large mycoremediation initiatives. Adequate volumetric production of viable fungal material for field applications involves technical facilities and knowledge, which are not available in all geographical areas. The storage and delivery of fungal inocula with acceptable viability are complicated transport issues that contribute to the expense. Various inoculum formulations (liquid suspension, solid substrate, and immobilized preparations), each with specific values as well as constraints for dispersing in the field (Akhtar and Mannan 2020).

The realization of a homogeneous spread of a fungus over contaminated areas is a major technical obstacle. In contrast to chemical treatments that can be homogenized and spread relatively easily, living fungal inocula require careful manipulation and application for successful establishment. The application timing, incorporation depth, and protection from environmental stresses must be strictly controlled to ensure a successful outcome over large treatment areas. The specialization of application equipment and methods is an area that is still being developed (Hyde et al. 2019). A diagram for major bottlenecks in mycoremediation of xenobiotics, such as enzyme instability, toxic by-products, environmental variability, scale-up challenges, and competition with native microbiota, is shown in Fig. 8.3.

Process control and monitoring. In reality, field-level bioremediation with mycotechnology would be difficult due to the biological systematics complexity and varying environments. There is a process difficulty in identifying suitable engineering methods to be followed: conventional ones (based on the strict control of process parameters) are not easily applied to living biological systems, wherein behavior changes as a response to stimuli. Integrated monitoring programs should integrate biological, chemical, and physical records to describe the efficacy of remediation efforts and system performance (Roy et al. 2023).

Additionally, the more contaminated sites are heterogeneous, and thus, it is difficult to scale up (depending on the zones of these sites, they may require proper treatments or the use of a different fungal strain). Specifics of contamination patterns, soil properties, and environmental conditions are very uneven across the sites; hence, management strategies need to be extremely dynamic to cope with this spatial variability. One of the most critical engineering challenges to more sustainable design innovation is the development of treatment units that are flexible and adaptable to specific site conditions (Padhye et al. 2023).

Economic considerations are an important factor in the potential for mycoremediation at large, as it will only be economical on a large scale if it costs less than

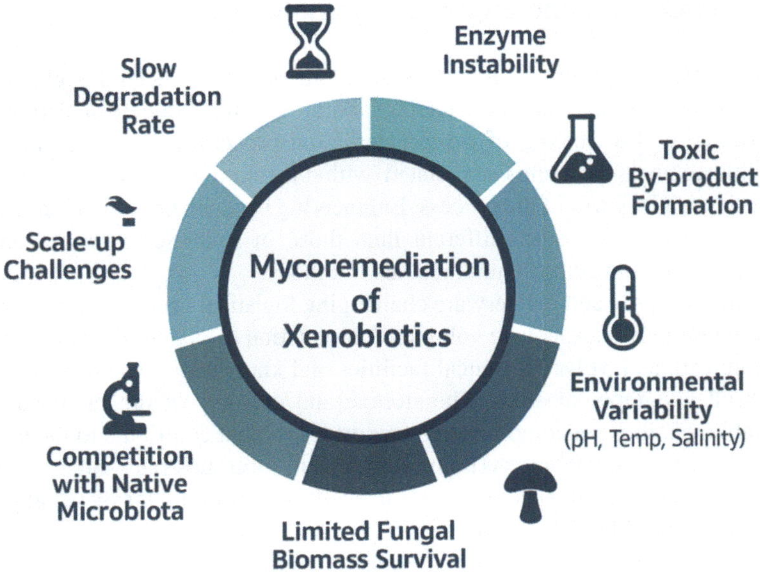

Fig. 8.3 This figure represents major bottlenecks in mycoremediation of xenobiotics

most other remediation processes. The cost of mycoremediation economics relies on inoculum production, its application and monitoring cost, and treatment duration. Scale efficiency is about minimizing costs at the components of the system, but it also translates to innovations in business models or finance.

8.5 Contaminant-Specific Limitations

Mycoremediation is challenged by several classes of environmental pollutants, and understanding these contaminant-specific limitations is required to design an efficient and effective treatment system. Because of this, many different types of xenobiotics can be found in the environment, and their chemical nature, bioavailability, toxicity, and degradation pathways require specific approaches to treatment. Fungi have been shown to be quite successful at treating some pollutants in controlled laboratory conditions (Padhye et al. 2023). Sadly, in-field measurements are less optimistic due to various matrix effects, interactions with co-contaminants, and environmental conditions (Kumar et al. 2025).

POPs are a highly recalcitrant class of compounds difficult to remediate by mycoremediation since many possess both chemical persistence and recalcitrance to biological degradation. Though white-rot fungi possess non-specific oxidative enzyme systems that have been shown to degrade a wide range of POPs, the efficiency and completeness of degradation are frequently less than ideal. Substances like PCBs, dioxins, and some pesticides have been only partially converted into other substances, instead of being fully mineralized, with still unidentified metabolites (Thapa et al. 2024).

The bioavailability of pollutants plays a critical role in the effectiveness of mycoremediation but is usually poorly understood and challenging to predict. Highly sorbed or sequestered in micropores, contaminants may also exist as separate phases and can be inaccessible to fungal enzymes even though they are theoretically degradable. Aging of contaminated soils can continue to decrease bioavailability with time, rendering older contamination more difficult to remediate than new spills. Other methods for increasing water accessibility to DEHP may be needed (e.g., addition of surfactants or physical processing) and could introduce additional considerations (Ganguly et al. 2024).

Heavy metals pose a special problem to mycoremediation, as they cannot be "processed" into a less harmful form and are toxic regardless of the size or properties of their particles. Despite the many options for metal tolerance and accumulation mechanisms in fungi, the efficiency of it heavily depends on the availability of metals, both in terms of speciation and soil chemistry, and environmental factors. Metal remobilization is quite likely under different conditions, constituting long-term risks that need to be analyzed and managed cautiously (Briffa et al. 2020).

Mixed contaminations, on the other hand, which are often encountered in industrial sites, contribute supermultiply to the sum such that the combined effect of all contaminants exceeds that resulting from pooling up their separate actions. These multiple interactions between classes of pollutants could, in unpredictable ways, modify bioavailability, toxicity, and biodegradation paths. Several alternatives could account for their observation: some effectors can stimulate the activity of an enzyme, compete with the fungal system for a preferred substrate, and divert carbon flow into a different path, whereas others would have an inhibitory effect on the fungus, causing, in concert, the response observed. Interventions for therapy can be developed once one knows the types of interactions contaminants have on fungal physiology (Ganguly et al. 2024).

Or, if not, it would be because emerging contaminants, such as pharmaceuticals, personal care products, and engineered nanomaterials, have all been classified in this group (Carvalho et al. 2018a), which might represent considerable difficulty for mycoremediation to even happen at all due to the lack of information on their biodegradation and effects on fungi. The overzealous entry of new chemicals into the biosphere is faster than the cure trajectory and demands quick screening and flexible remediation.

8.6 Regulatory and Standardization Barriers

Regulatory influence on mycoremediation has been difficult because standards have not emphasized key properties of biodegradation, instead focusing on traditional clean-up technologies. Without a shared set of protocols, performance guidelines, and rules, there is little for technology suppliers and their end users to hang on to so that they can begin development or commercialization.

Mycology technology has to go through biological approval and verification mechanisms, so when there is a bioassay (bioindicator for the toxicity of an environment), you spend tons of cash to have this prove very well documented that you get them in stereo. The asymmetric biological nature of mycoremediation systems has prevented the current options from being easily applied to treatment with relatively safe organisms that present no safety issues in handling and application, low biosafety risk of contamination, potential reduction or lack of increase in environmental pollution over traditional (chemical and physical) treatments, and genetic stability factors. We are not aware of any current regulation that could require a tiered risk assessment and include exposure and hazard to non-target organisms, ecosystem functions, or human health (Negi and Das 2023).

Introducing nonindigenous fungi with the intention of employing them in bioremediation further addresses particular unanswered questions related to bioinvasions and ecological disruption. Most fungi used in mycoremediation are known species that have been found in nature and have safety records; still, their use in nonintended ecosystems or at high concentrations needs to attract more attention. It will (or should) be a consideration to develop appropriate containment approaches and prevention measures that reduce the risk of dissemination for regulatory approval (Medaura et al. 2021).

The standardization of mycoremediation techniques and standardized performance criteria is still in its embryonic stages, making it difficult to compare technologies as well as quality control/regulatory perspectives. Unlike the traditional practices of mycoremediation, for which there are established engineering standards and performance specifications, "mycoremediation" is not standardized with respect to its use against various classes of contaminants. The institution of industrial standards would facilitate the technology transfer, quality control, and regulatory bodies in their evaluations (Kafle et al. 2022).

8.7 Technical and Methodological Limitations

There are technical limitations that limit the development of technologies applied to mycoremediation research and practice. Analytical techniques available to monitor fungal activity, contaminant degradation, and system performance are not generally sensitive or specific enough, convenient, or practical for effective process control and optimization. The need for more advanced analytical techniques and monitoring concepts is a priority in the development of efficient mycoremediation.

Measurement of biomass and viability remains arguably the two most challenging monitoring and control factors in mycoremediation. Fungal biomass levels of wild-type and mutant *Aspergillus* were determined by traditional methods, such as ergosterol content analysis or microscopic counting, which might not yield representative or real-time estimates of actively growing fungal populations within complex environmental matrices. Progress in the development of rapid, dependable approaches for assessing fungal activity and vitality could significantly enhance process-monitoring capabilities to facilitate better control of remediation systems (Pasanen et al. 1999).

Revealing contamination fate and transformation pathways is still a challenge, attributable to the complexity of fungal metabolism and the possibility of multiple degradation pathways. Knowledge of the fate of pollutants during mycoremediation is important to evaluate treatment efficiency and to predict potential risk for the formation of final metabolites. "Omics"-based, quantitative mass spectrometry and isotope tracer-based approaches are now starting to expand the toolkit of methods for pathway elucidation, but these methods are generally expensive and technically challenging (Hajnajafi and Iqbal 2025).

Application of molecular biology techniques to mycoremediation, supplemented with screening, monitoring, and optimization processes, is a promising approach; however, they also face technical and economic barriers. Genomic and transcriptomic approaches can provide insight into fungal physiology and metabolism, but making this knowledge applicable is not straightforward. The frequency and distribution of TBFVs throughout the endemic range are largely unknown in Namibia but could be addressed by molecular assays that are fieldable and with simplified analytical methods (McGuinness et al. 2024).

It has to be acknowledged that models and predictability of mycoremediation systems are inevitably constrained by the complexity of biological processes and environmental conditions. If there existed a mathematical model that could adequately describe how these remediation systems function under different conditions, then the overall system design and optimization might be improved. Modeling such processes is a complex task: biological (metabolic) as well as chemical and physical processes on length scales between the somewhat macroscopic and down to the micro are involved technically.

Quality control (QC) is, in fact, a software program that concerns service and good quality management among other studies providers and fields, which may certainly be excellent. The development of uniform protocols for strain maintenance, experimental methodology, and result reporting could also facilitate better quality disseminatable research on mycoremediation and, thus, likewise, the ease with which the technology propagates between labs or is applied to landscapes can become less problematic (Sharmile et al. 2025).

8.8 Economic and Commercial Feasibility Challenges

The economics of the mycoremediation process will greatly affect the acceptance and potential commercial use of this technology. Even though mycoremediation could offer the advantage of being cost-efficient in comparison to traditional remediation in some cases, for these organic devices, several economic handicaps must be overcome for them to become more economically viable than other commercially available units.

On the other hand, a solid economic model and cost-benefit analysis are required to mobilize investments and demonstrate the final beneficiaries' value. As such, the development and implementation of mycoremediation systems is generally a capital-intensive exercise in terms of project upstart costs, particularly for large-scale operations that require sophisticated equipment and associated infrastructure, including trained personnel (Akpasi et al. 2023). Also, the plants for the production of GMOs and for applying inoculants, as well as monitoring systems, are an expensive investment that must be distributed across many applications to become cost-effective. There are no highly developed supply-chain networks for custom-designed vehicles, nor large specialized service providers, and they can also add to the cost of the vehicle and place limits on the market (Eckerstorfer et al. 2025).

Operation costs for a mycoremediation system include those associated with monitoring, maintaining, supplementing nutrients to, and managing the process. These values might be less costly than those of traditional remediation techniques, but they are still subject to considerable variations due to the site-specific conditions, contaminant properties, and the length of treatment. The biological basis of mycoremediation means that knowledge of treatment times can be uncertain, making it difficult to estimate costs and plan projects. The longer treatment times required for mycoremediation, when compared to those of traditional petroleum use, can be a factor in economic attractiveness that influences the degree of acceptance, especially if fast site closure is desired (Barrech 2018). Although natural attenuation and biotreatment systems would achieve greater reductions in contaminants with the passage of time—albeit at less cost—long-term carrying costs and the opportunity costs for walk-away options can largely offset any savings. An important consideration for commercially relevant strategies is to increase treatment rates while ensuring cost-effectiveness.

However, several challenges in the market development of mycoremediation technologies, namely, technology awareness, risk perception, and regulatory uncertainty, were identified. Many potential end users neither know that these biological remediation technologies exist nor comprehend the comparative risk of using such technologies over traditional waste treatment methods. To achieve trust in your technology by the markets, there need to be actual cases and certified performance, and you need to train the influential people—regulators, consultants, and site owners (Thapa et al. 2024).

Challenges for mycoremediation commercialization: The fragmented state of the environmental remediation sector is both a bane and a boon. An illustration of this aspect is that, in many cases, different types of contaminants will lead to several competitive domains of the mycoremediation sector while simultaneously also necessitating the adaptation of several technological types and market strategies. For smaller-scale projects, this means economies of scale are limited and new business models evolve around the processes to make them economically viable.

8.9 Case Histories: Field Application

This highlights the importance of studying practical mycoremediation applications to reveal real, as opposed to potential, limitations (or challenges) on site. These examples constitute milestones for the promises and pitfalls confronting the "scaling-up" of laboratory-to-field applications of mycoremediation and provide important lessons for the design of new technologies and technology transfer strategies.

An early but nonfield application of mycoremediation is to treat toxic contamination from wood treatment facilities using *Phanerochaete chrysosporium* (Tang et al. 2005). Although fungal PAH remediation may be environmentally feasible, compared to bacteria, few studies have attempted to assess at-field fungal remediation potential; such a study at a Pacific Northwest, creosote-contaminated site demonstrated fungal potential to remediate these compounds, though significant hurdles related to successful establishment and control of fungi in situ were experienced (Baker et al. 2009). Despite successful PAH biodegradation in the lab, their field studies indicated that seasonal temperature changes, moisture fluctuations, and competition with native microorganisms could produce variable results. This indicated the importance of optimizing the environment and the strong inoculation capabilities needed under hard field conditions (Czaplicki et al. 2020).

A Canadian mycoremediation system, targeted to address petroleum hydrocarbons over macroscopic scales, provided data on the limitations of scaling mycoremediation systems. *Pleurotus ostreatus* was used in this project, and it is sometimes used as a potential agent for bioremediation of larger areas of diesel fuel-contaminated soil. However, while initial results were promising in terms of contaminant degradation, the high variability in soil properties and contaminant loading made it challenging to maintain bioactivity throughout the treatment zone in practice. This research defended the use of adaptive management for situations where variation is site-dependent and highlighted the need for detailed characterization of a site prior to treatment (Anderson and Juday 2016).

These have responded to new challenges from emergent pollution and complex mixture effects in pharmaceuticals and mycoremediation applications. Field-scale pilot studies of antibiotic-tainted soil close to a pharmaceutical manufacturing plant are characterized in recent reports as containing the potential for fungal degradation to functional application to pharmaceutical treatment, yet mixed contaminations present a challenge. The optimization of treatment and follow-up was especially complicated because of the number of pharmacological components with various chemical characteristics and degradation mechanisms. The emphasis on flexible treatment modalities and an extensive analytic battery addresses the challenge of emerging contaminants implementation (Khan 2025).

Nonetheless, urban mycoremediation data specifically have exposed the challenges of accommodating biological treatment systems in the developed world, at high population densities, in well-regulated environments. We chose a PAH remediation project from a former gas station urban site with the belief that the societal, regulatory, and serial cost has the potential to be an implementation barrier.

Stakeholder engagement and treatment processes that do not disrupt the surrounding operation were also highlighted as part of the project (Khan 2025).

Together, these insights from both case studies provide key messages indicating that Mycoremediation has a high potential to treat a wide variety of pesticide-contaminated environments; however, site-specific parameters (e.g., process design and adaptive management) mean that care should be taken when extrapolating the findings of this study to other field sites. They bring valuable lessons for the future of mycoremediation and illustrate the research and development still required for these to serve as effective remediation technologies.

8.10 Emerging Solutions and Technological Advances

Unfortunately, mycoremediation technology has not fulfilled its potential as a robust treatment alternative due to a number of historically stagnant pressures; however, recent advances in biotechnology, engineering, and analytics are beginning to overcome these barriers. These new approaches are multidisciplinary and have great promises for addressing existing challenges while further broadening the use of fungal-mediated remediation technologies.

Novel genetic and synthetic biology-based approaches are also emerging as tools for the engineering of improved fungal strains with augmented remediation capabilities. New technologies of genetic engineering, like CRISPR-Cas systems and tools of synthetic biology, enable modification of metabolism in fungal strains that are promising for contaminant degradation, enhancing contaminant degradation pathways, stress resistance, and enzyme production. Various studies demonstrated that fungi can be given new capacity for degrading specific xenobiotics, including engineered fungi with enhanced heavy metal accumulation capability and laccase-overexpressing fungi with enhanced secretion of ligninolytic enzymes (Dinakarkumar et al. 2024).

Overcoming problems with fungal delivery, establishment, and control under field conditions has been facilitated by the development of immobilization and encapsulation tools. These advances in materials science have provided new carrier systems, which not only protect the fungi during transport and application but also ensure optimal nutrient power/supply and a microenvironment for growth (Briffa et al. 2020). Alternatives are emerging. Likely to limit fungal viability and efficacy under harsh conditions (Ferreyra-Suarez et al. 2024), such as biofilm-supported immobilization systems, polymer encapsulation techniques, and prepared, designed support matrices.

Emergence of new types of "omics" technologies (genomics, transcriptomics, proteomics, metabolomics) is fundamentally altering our view of fungal physiology/metabolism as they perform practical applications relevant to remediation. Such strategies produce a molecular perspective on the degradation of contaminants and on the responses to stress that can provide clues for both informed strain selection and process improvement. In the future. Such multi-omics data will be integrated with advanced bioinformatics tools to both discover novel degradation pathways and prediction models correlating various fungal characteristics (Ball et al. 2020).

Novel possibilities for an improved application of mycoremediation based on the use of mycoremediation enabling nanotechnologies are being offered by nanotechnologies able to enhance bioavailability and thus the efficiency of fungal interaction with pollutants and thus improvement in their performance. An array of nanoparticles can be designed to preserve nutrients for fungal cells while providing an environment where the protein is mixed with the nanoparticles to stabilize the enzyme. In addition to reduced application costs, nanoencapsulation of fungal inocula and nutrients for efficient delivery of inocula and nutrients at the nano level is emerging as a tool to improve the establishment success (Gholami-Shabani et al. 2018).

In unison, evolved supervisory and commanding systems, through actuators, automated systems, and AI, are recognizing problems of process supervision and enhancement. On the other hand, obtaining parameters related to fungal activity or contaminant concentration allows for adaptive process management and optimization. Machine learning techniques are now being explored to predict optimal process conditions and to prevent potential problems beforehand before system-wide performance is impacted.

Promising hybrid strategies, which integrate mycoremediation with other biotic, abiotic, or physicochemical treatments, are emerging as potential solutions for complex pollution situations that could go beyond the tolerance of separate technologies. Flow schemes, sequential treatments, bioaugmentation protocols, and further combined treatment training that harness the best of each method and temper its drawbacks are also promising. This would be especially useful toward the broad-spectrum and challenging analytes.

8.11 Future Research Directions and Opportunities

Specialized basic and applied research is required to develop mycoremediation. Recommendations include clarifying the ecology and physiology of fungi in contaminated sites, improving predictive models of biophysical and biochemical processes, and developing fast assessment tools for monitoring (high-priority areas). The efficiency of these processes can benefit from the bioengineering of fungal strains, the application of synthetic biology, and the optimization of large-scale growth systems. Synergies with carbon sequestration, circular economy principles, and sustainable agriculture offer potential value propositions that may transcend pollution control.

Similarly important are social, regulatory, and market factors. Ensuring public acceptance, cost-effectiveness, and procedural standardization will be fundamental to broader adoption. Collaboration across disciplines—integrating biology, engineering, data science, and social sciences—can bridge scalability hurdles and expedite translation from bench to field. Funding pilot projects and standard screening approaches will confirm technologies and minimize bottlenecks. New funding models and global collaborations can further enhance commercialization and worldwide application.

8.12 Conclusions

Mycoremediation holds great potential as a green environmental cleanup technology, but major scientific, technical, and regulatory challenges must be overcome. Overcoming these will take long-term, concerted effort by researchers, industry, and policymakers. With focused R&D, standardization, and creative business models, mycoremediation can go from being a specialty solution to becoming a mainstream, globally relevant method of environmental remediation.

In addition, the intersection of biogenetics, nanotechnology, and artificial intelligence provides novel opportunities to increase efficiency and scalability. By utilizing these cross-disciplinary innovations in combination with good environmental management practices, mycoremediation can provide consistent, responsive, and affordable solutions. With increased investment and international cooperation, it could become a pillar of green technologies as a means of countering environmental pollution.

References

Ahmad I, Ansari MI, Aqil F (2006) Biosorption of Ni, Cr and Cd by metal tolerant Aspergillus niger and Penicillium sp. using single and multimetal solution. Indian J Exp Biol 44:73–76

Akhtar N, Mannan MA (2020) Mycoremediation: expunging environmental pollutants. Biotechnol Rep (Amst) 26:e00452. https://doi.org/10.1016/j.btre.2020.e00452

Akpasi SO, Anekwe IMS, Tetteh EK, Amune UO, Shoyiga HO, Mahlangu TP, Kiambi SL (2023) Mycoremediation as a potentially promising technology: current status and prospects—a review. Appl Sci 13:4978. https://doi.org/10.3390/app13084978

Alshiekheid MA, Umar A, Ameen F, Alyahya SA, Dufossé L (2023) Biodegradation of chromium by laccase action of Ganoderma multipileum. J King Saud Univ Sci 35:102948. https://doi.org/10.1016/j.jksus.2023.102948

Anderson C, Juday G (2016) Mycoremediation of petroleum: a literature review. JESE 5. https://doi.org/10.17265/2162-5298/2016.08.002

Arroyo-Becerra A, Garrido Bazán V, Téllez-Téllez M, Herrera-Estrella A, Godinez G, Galicia S, Villalobos MA, Bibbins M (2016) Effect of textile dyes on activity and differential regulation of laccase genes from Pleurotus ostreatus grown in submerged fermentation. ABM Express 6:93

Ball B, Langille M, Geddes-McAlister J (2020) Fun(gi)omics: advanced and diverse technologies to explore emerging fungal pathogens and define mechanisms of antifungal resistance. MBio 11:e01020–20. https://doi.org/10.1128/mBio.01020-20

Barrech D (2018) A review on mycoremediation—the fungal bioremediation. PAB 7. https://doi.org/10.19045/bspab.2018.70042

Bautista-Zamudio PA, Flórez-Restrepo MA, López-Legarda X, Monroy-Giraldo LC, Segura-Sánchez F (2023) Biodegradation of plastics by white-rot fungi: a review. Sci Total Environ 901:165950. https://doi.org/10.1016/j.scitotenv.2023.165950

Briffa J, Sinagra E, Blundell R (2020) Heavy metal pollution in the environment and their toxicological effects on humans. Heliyon 6:e04691. https://doi.org/10.1016/j.heliyon.2020.e04691

Czaplicki LM, Redfern LK, Cooper EM, Ferguson PL, Vilgalys R, Gunsch CK (2020) Investigating the Mycobiome of the Holcomb Creosote Superfund Site. Chemosphere 252:126208. https://doi.org/10.1016/j.chemosphere.2020.126208

Dinakarkumar Y, Ramakrishnan G, Gujjula KR, Vasu V, Balamurugan P, Murali G (2024) Fungal bioremediation: an overview of the mechanisms, applications and future perspectives. Environ Chem Ecotoxicol 6:293–302. https://doi.org/10.1016/j.enceco.2024.07.002

Eckerstorfer MF, Dolezel M, Miklau M, Greiter A, Heissenberger A, Kastenhofer K, Schulz F, Hagen K, Otto M, Engelhard M (2025) Environmental applications of GM microorganisms: Tiny Critters posing huge challenges for risk assessment and governance. Int J Mol Sci 26:3174. https://doi.org/10.3390/ijms26073174

Efremenko E, Stepanov N, Senko O, Aslanli A, Maslova O, Lyagin I (2024) Using fungi in artificial microbial Consortia to solve bioremediation problems. Microorganisms 12:470. https://doi.org/10.3390/microorganisms12030470

Fernández-López MG, Batista-García RA, Aréchiga-Carvajal ET (2023) Alkaliphilic/Alkalitolerant fungi: molecular, biochemical, and biotechnological aspects. J Fungi (Basel) 9:652. https://doi.org/10.3390/jof9060652

Ferreyra-Suarez D, García-Depraect O, Castro-Muñoz R (2024) A review on fungal-based biopesticides and biofertilizers production. Ecotoxicol Environ Saf 283:116945. https://doi.org/10.1016/j.ecoenv.2024.116945

Ganguly N, Roy A, Banerjee S, Pandit S (2024) Contaminated Soils: remedies, hurdles, and perspectives, pp 251–268. https://doi.org/10.1007/978-3-031-58331-5_13

Geris R, Malta M, Soares LA, de Souza Neta LC, Pereira NS, Soares M, da Reis VS, de Pereira MG (2024) A review about the mycoremediation of soil impacted by war-like activities: challenges and gaps. J Fungi 10:94. https://doi.org/10.3390/jof10020094

Gholami-Shabani M, Gholami-Shabani Z, Shams-Ghahfarokhi M, Razzaghi-Abyaneh M (2018) Application of nanotechnology in mycoremediation: current status and future prospects. In: Prasad R, Kumar V, Kumar M, Wang S (eds) Fungal nanobionics: principles and applications. Springer, Singapore, pp 89–116. https://doi.org/10.1007/978-981-10-8666-3_4

Hajnajafi K, Iqbal MA (2025) Mass-spectrometry based metabolomics: an overview of workflows, strategies, data analysis and applications. Proteome Sci 23:5. https://doi.org/10.1186/s12953-025-00241-8

Hyde KD, Xu J, Rapior S, Jeewon R, Lumyong S, Niego AGT, Abeywickrama PD, Aluthmuhandiram JVS, Brahamanage RS, Brooks S, Chaiyasen A, Chethana KWT, Chomnunti P, Chepkirui C, Chuankid B, de Silva NI, Doilom M, Faulds C, Gentekaki E, Gopalan V, Kakumyan P, Harishchandra D, Hemachandran H, Hongsanan S, Karunarathna A, Karunarathna SC, Khan S, Kumla J, Jayawardena RS, Liu J-K, Liu N, Luangharn T, Macabeo APG, Marasinghe DS, Meeks D, Mortimer PE, Mueller P, Nadir S, Nataraja KN, Nontachaiyapoom S, O'Brien M, Penkhrue W, Phukhamsakda C, Ramanan US, Rathnayaka AR, Sadaba RB, Sandargo B, Samarakoon BC, Tennakoon DS, Siva R, Sriprom W, Suryanarayanan TS, Sujarit K, Suwannarach N, Suwunwong T, Thongbai B, Thongklang N, Wei D, Wijesinghe SN, Winiski J, Yan J, Yasanthika E, Stadler M (2019) The amazing potential of fungi: 50 ways we can exploit fungi industrially. Fungal Divers 97:1–136. https://doi.org/10.1007/s13225-019-00430-9

Kafle A, Timilsina A, Gautam A, Adhikari K, Bhattarai A, Aryal N (2022) Phytoremediation: mechanisms, plant selection and enhancement by natural and synthetic agents. Environ Adv 8:100203. https://doi.org/10.1016/j.envadv.2022.100203

Khan MF (2025) Fungi for sustainable pharmaceutical remediation: enzymatic innovations, challenges, and applications—a review. Processes 13:1034. https://doi.org/10.3390/pr13041034

Kumar V, Singh V, Pandit S (2025) Advanced omics approach and sustainable strategies for heavy metal microbial remediation in contaminated environments. Bioresour Technol Rep 29:102040. https://doi.org/10.1016/j.biteb.2025.102040

Lu 陆涛 T, Zhang Q-L, Yao S-J (2016) Application of biosorption and biodegradation functions of fungi in wastewater and sludge treatment, pp 65–90. https://doi.org/10.1007/978-3-319-42852-9_4

McGuinness AJ, Stinson LF, Snelson M, Loughman A, Stringer A, Hannan AJ, Cowan CSM, Jama HA, Caparros-Martin JA, West ML, Wardill HR (2024) From hype to hope: considerations in conducting robust microbiome science. Brain Behav Immun 115:120–130. https://doi.org/10.1016/j.bbi.2023.09.022

Medaura MC, Guivernau M, Moreno-Ventas X, Prenafeta-Boldú FX, Viñas M (2021) Bioaugmentation of native fungi, an efficient strategy for the bioremediation of an aged industrially polluted soil with heavy hydrocarbons. Front Microbiol 12:626436. https://doi.org/10.3389/fmicb.2021.626436

Michael-Igolima U, Abbey SJ, Ifelebuegu AO (2022) A systematic review on the effectiveness of remediation methods for oil contaminated soils. Environ Adv 9:100319. https://doi.org/10.1016/j.envadv.2022.100319

Mir-Tutusaus JA, Masís-Mora M, Corcellas C, Eljarrat E, Barceló D, Sarrà M, Caminal G, Vicent T, Rodríguez-Rodríguez CE (2014) Degradation of selected agrochemicals by the white rot fungus Trametes versicolor. Sci Total Environ 500–501:235–242. https://doi.org/10.1016/j.scitotenv.2014.08.116

Mishra M, Srivastava D (2020) Mycoremediation: a step towards sustainability. IJPE 6:298–305. https://doi.org/10.18811/ijpen.v6i04.09

Negi BB, Das C (2023) Mycoremediation of wastewater, challenges, and current status: a review. Bioresour Technol Rep 22:101409. https://doi.org/10.1016/j.biteb.2023.101409

Okal EJ, Heng G, Magige EA, Khan S, Wu S, Ge Z, Zhang T, Mortimer PE, Xu J (2023) Insights into the mechanisms involved in the fungal degradation of plastics. Ecotoxicol Environ Saf 262:115202. https://doi.org/10.1016/j.ecoenv.2023.115202

Padhye LP, Srivastava P, Jasemizad T, Bolan S, Hou D, Shaheen SM, Rinklebe J, O'Connor D, Lamb D, Wang H, Siddique KHM, Bolan N (2023) Contaminant containment for sustainable remediation of persistent contaminants in soil and groundwater. J Hazard Mater 455:131575. https://doi.org/10.1016/j.jhazmat.2023.131575

Pasanen A-L, Yli-Pietilä K, Pasanen P, Kalliokoski P, Tarhanen J (1999) Ergosterol content in various fungal species and biocontaminated building materials. Appl Environ Microbiol 65:138–142. https://doi.org/10.1128/aem.65.1.138-142.1999

Redistribution of soil water by a saprotrophic fungus enhances carbon mineralization | Request PDF, ResearchGate (2020). https://www.researchgate.net/publication/283687098_Redistribution_of_soil_water_by_a_saprotrophic_fungus_enhances_carbon_mineralization. Accessed September 18 2025.

Roy R, Samanta S, Pandit S, Naaz T, Banerjee S, Rawat J, Chaubey K, Saha R (2023) An overview of bacteria-mediated heavy metal bioremediation strategies. Appl Biochem Biotechnol 196:1–40. https://doi.org/10.1007/s12010-023-04614-7

Salazar-Cerezo S, de Vries RP, Garrigues S (2023) Strategies for the development of industrial fungal producing strains. J Fungi (Basel) 9:834. https://doi.org/10.3390/jof9080834

Sauters TJC, Rokas A (2025) Patterns and mechanisms of fungal genome plasticity. Curr Biol 35:R527–R544. https://doi.org/10.1016/j.cub.2025.04.003

Sharmile N, Chowdhury RR, Desai S (2025) A comprehensive review of quality control and reliability research in micro–nano technoLOGY. Technologies 13:94. https://doi.org/10.3390/technologies13030094

Tang L, Zeng G-M, Wang H, Shen G-L, Huang D-L (2005) Amperometric detection of lignin-degrading peroxidase activities from Phanerochaete chrysosporium. Enzyme Microb Technol 36:960–966. https://doi.org/10.1016/j.enzmictec.2005.02.009

Thapa BS, Pandit S, Mishra RK, Joshi S, Idris AM, Tusher TR (2024) Emergence of per- and polyfluoroalkyl substances (PFAS) and advances in the remediation strategies. Sci Total Environ 916:170142. https://doi.org/10.1016/j.scitotenv.2024.170142

Vaksmaa A, Guerrero-Cruz S, Ghosh P, Zeghal E, Hernando-Morales V, Niemann H (2023) Role of fungi in bioremediation of emerging pollutants. Front Mar Sci 10. https://doi.org/10.3389/fmars.2023.1070905

Novel Approaches and Prospects of Mycoremediation

9

Rakhi Rajput and Soumya Pandit

Abstract

Mycoremediation, the biotechnological application of fungi for the removal, degradation, or immobilization of environmental pollutants, has emerged as a sustainable alternative to conventional remediation technologies. Fungi possess unique enzymatic systems, particularly ligninolytic enzymes such as laccases, manganese peroxidases, and lignin peroxidases, that enable the breakdown of structurally diverse contaminants, including hydrocarbons, pesticides, dyes, pharmaceuticals, mycotoxins, and heavy metals. Their filamentous growth and extensive mycelial networks enhance pollutant accessibility in soil micropores, while their metabolic versatility ensures survival under extreme environmental conditions. Compared with chemical and physical methods, mycoremediation is cost-effective, eco-friendly, and capable of providing dual benefits of contaminant removal and ecological restoration. Recent advancements include the use of agricultural residues as low-cost substrates to promote circular bioeconomy, genetic engineering of fungi to enhance degradation efficiency, and integration with bacterial, algal, or plant-based systems for synergistic remediation. Despite its potential, challenges, such as large-scale implementation, field-level consistency, and regulatory frameworks, remain. This review highlights the principles, mechanisms, applications, benefits, and limitations of mycoremediation while discussing recent innovations and future prospects. Overall, mycoremediation holds great promise as a green, versatile, and sustainable technology for addressing global environmental pollution challenges.

R. Rajput · S. Pandit (✉)
Department of Life Sciences, Sharda School of Bioscience and Technology, Sharda University, Greater Noida, Uttar Pradesh, India

A. Roy et al. (eds.), *Mycoremediation of Xenobiotics*, Emerging Paradigms in Pharmaceutical Research, https://doi.org/10.1007/978-981-95-5743-1_9

Keywords

Mycoremediation · Bioremediation · Fungi · Environmental pollution · Sustainable development · Biotechnology Introduction

9.1 Introduction

The need for research and innovation has increased considerably due to the growing world population. Environmental pollution has been one of the most serious problems in the twenty-first century, and it comes from man-made industrial activities, urbanization, and agricultural practices that have led to widespread pollution of soil, water, and air systems (Moghimi Dehkordi et al. 2024). Conventional remediation methods, although successful in some applications, are typically expensive and energy-intensive and could potentially cause or induce secondary pollution (Lee et al. 2024). To curb these drawbacks, Biological remediation methods that capitalize upon the intrinsic abilities of living organisms to either degrade and detoxify or isolate contaminants present in their environment are becoming more common among researchers (Azubuike et al. 2016).

Mycoremediation has emerged as one of the promising upgradations in bioremediation options employing the mycelium of fungi and offers an effective, economical, and eco-friendly alternative for cleaning up polluted habitats (Dinakarkumar et al. 2024). Mycologist Paul Stamets has coined the term “mycoremediation” to describe the use of fungi and their enzymatic byproducts in clean-up efforts. This strategy takes advantage of the distinctive physiological and biochemical characteristics associated with fungi, such as their vast hyphal network, arsenal of enzymes, and ability to grow in a variety of habitats (Dinakarkumar et al. 2024).

The importance of mycoremediation is not only technical, but it also follows the current tendency for sustainable development and the search for “green” technologies worldwide. Given the increasingly stringent environmental legislation and heightened public awareness of environmental issues, it is of great significance to develop effective remediation technologies with minimum environmental impact, as well as economic success (Akpasi et al. 2023). Mycoremediation fulfills both of these requirements by employing the organisms and processes that Mother Nature has already provided, which results in the least possible impact on the environment from remediation efforts (Akhtar and Mannan 2020).

This review attempts to cover all the aspects of mycoremediation in detail, including basic principles, recent technological advancements, and future perspectives. The various applications of mycoremediation on different environmental matrices and different types of contaminants will be discussed in the paper, considering also challenges and limitations that should be addressed for large-scale implementation. In addition, emerging trends and new strategies are also described in this review that are contributing to the mycoremediation of the future, such as how advanced biotechnological tools can be combined and hybrid-based remediation systems.

9.2 Fundamental Principles of Mycoremediation

The potential for mycoremediation is based on the distinctive biological and metabolic features of fungi. Fungi have a number of unique properties, which make them excellent candidates for the development of environmental applications compared to other microorganisms, as given in Table 9.1 (Akpasi et al. 2023). Understanding these basics is necessary for assessing the capabilities and limitations of mycoremediation technologies.

Mycelial structure and colonization potential: Fungi are eukaryotic organisms that grow in a filamentous form consisting of networks of hyphae (Akhtar and Mannan 2020). These hyphal structures, collectively referred to as mycelium, can extend several meters in soil within a few weeks, depending on nutrient availability. For example, *Pleurotus ostreatus* can colonize contaminated soil patches at rates of 3–5 mm/day under optimal conditions (Ibrahim et al. 2024). The high surface

Table 9.1 Selected fungi used in mycoremediation and their pollutant removal efficiencies

Fungal species	Target pollutant(s)	Removal efficiency/ uptake capacity	Mechanism involved	Reference
Phanerochaete chrysosporium	Polycyclic aromatic hydrocarbons (PAHs)	70–90% degradation in 30 days	LiP, MnP, laccase secretion	Wang et al. (2018)
Trametes versicolor	Azo dyes	~95% degradation within 10–12 days	Laccase, peroxidase	Singh et al. (2025)
Pleurotus ostreatus	Soil hydrocarbons, pesticides	60–80% removal within 4–6 weeks	Extracellular enzymes, mycelial colonization	Ibrahim et al. (2024)
Aspergillus niger	Petroleum hydrocarbons	40–60% degradation in 21 days	Hydrocarbon metabolism, biosurfactant release	Ene et al. (2014)
Rhizopus arrhizus	Cadmium (Cd^{2+}), lead (Pb^{2+})	Cd^{2+}: 85–95% removal (100 mg/L); Pb^{2+}: 140 mg/g biomass	Biosorption via cell wall polysaccharides	Dusengemungu et al. (2020)
Penicillium chrysogenum	Chromium (Cr(VI))	Up to 320 mg/g biomass uptake	Bioaccumulation, reduction to Cr(III)	Pazouki et al. (2007)
Aspergillus fumigatus	Lead (Pb^{2+}), copper (Cu^{2+})	Pb^{2+} uptake ~140 mg/g dry biomass	Biosorption, ion exchange	Dusengemungu et al. (2020)
Aspergillus spp., *Penicillium* spp.	Acidic mine drainage, industrial effluents	Active at pH 3–9; 70% activity at 37 °C, pH 4.5	Acid- and temperature-tolerant enzymes	Zhang et al. (2025)

area-to-volume ratio of hyphae facilitates rapid nutrient interception and metabolite exchange. Importantly, the mycelial structure enables fungi to penetrate soil micropores (<2 μm diameter), which conventional remediation methods often cannot access (Sun et al. 2023).

Enzymatic arsenal: The enzymatic repertoire of fungi is central to their remediation potential. White-rot fungi such as *Phanerochaete chrysosporium* produce extracellular lignin peroxidase (LiP), manganese peroxidase (MnP), and laccase, which are capable of degrading 70–90% of polycyclic aromatic hydrocarbons (PAHs) such as benzo[a]pyrene and anthracene within 30 days of treatment (Wang et al. 2018). Fungal enzymes are often non-specific; laccases, for instance, oxidize a wide range of phenolic and non-phenolic xenobiotics, making them effective even at mixed-contaminant sites. In controlled trials, *Trametes versicolor* degraded up to 95% of azo dyes within 10–12 days, compared to <30% removal by bacterial consortia under similar conditions (Singh et al. 2025).

Metabolic flexibility: Fungi can act as saprophytes, symbionts, or pathogens, switching their metabolic lifestyles according to environmental conditions. Some species can metabolize pollutants as carbon or energy sources. For example, *Aspergillus niger* and *Fusarium solani* were shown to utilize crude oil hydrocarbons, achieving 40–60% degradation of alkanes and aromatics within 21 days in contaminated soils (Ene et al. 2014). This metabolic plasticity allows fungi to thrive where high pollutant concentrations suppress bacterial activity.

As shown in Fig. 9.1, biosorption and bioaccumulation of heavy metals: Another remediation mechanism is biosorption, mediated by fungal cell wall components (chitin, glucans, mannoproteins), which carry functional groups (–OH, –COOH, –NH_2) that bind metal ions. Quantitatively, *Aspergillus fumigatus* biomass demonstrated lead (Pb^{2+}) uptake of 140 mg/g dry weight, while *Rhizopus arrhizus* achieved cadmium (Cd^{2+}) removal efficiencies of 85–95% at concentrations up to 100 mg/L

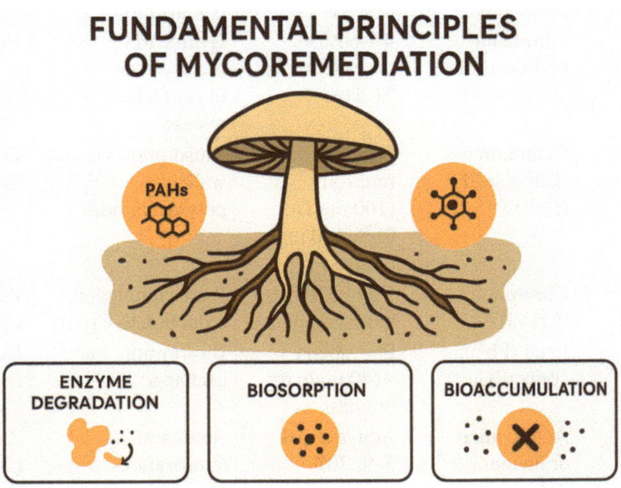

Fig. 9.1 Fundamental principles of mycoremediation

(Dusengemungu et al. 2020). Similarly, *Penicillium chrysogenum* was reported to bioaccumulate Cr(VI) up to 320 mg/g biomass under laboratory conditions (Pazouki et al. 2007).

Environmental adaptability: Unlike many bacteria that function optimally only at neutral pH, fungi are more versatile. *Aspergillus* spp. and *Penicillium* spp. remain metabolically active across pH ranges 3–9, making them suitable for acidic mine drainage or alkaline industrial effluents. Additionally, many fungi tolerate temperature fluctuations (15–35 °C) and low oxygen availability, broadening their applicability. For instance, *Phanerochaete chrysosporium* retained >70% enzymatic activity at 37 °C and pH 4.5, conditions often encountered in contaminated soils (Zhang et al. 2025).

9.3 Transformation and Degradation of Contaminants

The efficiency of mycoremediation is largely governed by biochemical and physiological processes that regulate the interaction between fungi and environmental pollutants. These processes include enzymatic degradation, biosorption, bioaccumulation, biotransformation, and co-metabolism, which collectively enhance the remediation potential of fungal systems (Negi and Das 2023). As given in Table 9.2, among these, enzymatic degradation represents the primary mode of action for the breakdown of organic pollutants. Fungi are prolific secretors of extracellular enzymes, such as lignin peroxidase (LiP), manganese peroxidase (MnP), and laccase (Lac), which can oxidize structurally diverse aromatic compounds (Wang et al. 2018). For instance, *Phanerochaete chrysosporium* has been shown to degrade nearly 90% of benzo[a]pyrene, a persistent polycyclic aromatic hydrocarbon (PAH), within 30 days through LiP and MnP activity, while *Trametes versicolor* laccase decolorized 95% of Reactive Blue 19 dye within eight days (Zhang et al. 2023). This broad oxidative capability, mediated by radical-driven reactions, contrasts with the substrate-specific enzymes of most bacteria and allows fungi to attack recalcitrant contaminants, such as PAHs, PCBs, pesticides, and synthetic dyes with efficiencies often ranging between 70% and 95% under optimized conditions (Tesfaye et al. 2025). In addition to enzymatic degradation, biosorption is another important mechanism that involves the passive binding of pollutants to fungal biomass via cell wall functional groups, such as amino, hydroxyl, carboxyl, and phosphate groups (Tesfaye et al. 2025). *Aspergillus niger* biomass, for example, was reported to remove 98% of Pb^{2+} from aqueous solutions at concentrations of 100 mg/L within 24 h, while *Rhizopus arrhizus* exhibited cadmium uptake capacities of up to 95% under similar conditions. Pretreatment strategies, such as alkali modification, can further enhance fungal biosorption, with *Aspergillus terreus* showing a 40% increase in Cr(VI) binding after chemical treatment (Bai 2001). Complementing biosorption, bioaccumulation involves active uptake of pollutants by metabolically active fungal cells. This is achieved through energy-dependent transport systems and sequestration in intracellular compartments (Kalita and Baruah 2023). *Penicillium chrysogenum* has been shown to accumulate Cr(VI) up to 320 mg/g dry biomass, reducing it

Table 9.2 Fungal species involved in the transformation and degradation of contaminants

Fungal species	Pollutant/contaminant	Degradation or uptake efficiency	Mechanism involved	Reference
Phanerochaete chrysosporium	Benzo[a]pyrene (PAH)	~90% degradation in 30 days	LiP & MnP enzymatic oxidation	Zhang et al. (2023)
Trametes versicolor	Reactive Blue 19 (azo dye)	95% decolorization in 8 days	Laccase-mediated oxidation	Tesfaye et al. (2025)
Pleurotus ostreatus	Endosulfan (pesticide)	90% degradation in 15 days (with corn stover)	Co-metabolism + ligninolytic enzymes	Chakraborty et al. (2025)
Aspergillus niger	Lead (Pb^{2+}); DDT	98% Pb^{2+} removal (100 mg/L); ~60% DDT transformed in 21 days	Biosorption (Pb^{2+}); biotransformation (DDT → DDE/DDD)	Yu et al. (2021)
Rhizopus arrhizus	Cadmium (Cd^{2+})	85–95% removal (100 mg/L)	Biosorption via cell wall functional groups	Bai (2001)
Penicillium chrysogenum	Chromium (Cr(VI))	Up to 320 mg/g dry biomass uptake	Bioaccumulation + reduction to Cr(III)	Fierro et al. (2022)
Aspergillus flavus	Copper (Cu^{2+})	~70% uptake at 50 mg/L (living cells)	Active bioaccumulation	Kalita and Baruah (2023)
Cunninghamella elegans	Pentachlorophenol (PCP)	Significant detoxification via hydroxylation/methylation	Biotransformation (phase I & II metabolism)	Yu et al. (2021)
Trametes hirsuta	17α-ethinylestradiol (synthetic estrogen)	>80% reduction in estrogenic activity	Hydroxylation + aromatic ring cleavage	Suzuki et al. (2003)
Phanerochaete chrysosporium	PCBs (polychlorinated biphenyls)	50–70% degradation with glucose supplementation	Co-metabolism + ligninolytic enzyme activity	Šrédlová et al. (2021)

intracellularly to the less toxic Cr(III), while *Aspergillus flavus* bioaccumulated nearly 70% of Cu^{2+} at concentrations of 50 mg/L, compared to less than 30% in heat-killed controls (Fierro et al. 2022). These findings highlight the ability of fungi not only to adsorb pollutants but also to actively concentrate and detoxify them from dilute environmental sources (Li et al. 2024).

As given in Fig. 9.2, fungal biotransformation processes further enhance remediation capacity by converting toxic pollutants into less harmful or less mobile metabolites without complete mineralization (Li et al. 2024). For instance, *Aspergillus niger* was shown to transform nearly 60% of DDT into less bioavailable metabolites, such as DDE and DDD, within 21 days, while *Cunninghamella elegans* metabolized pentachlorophenol through hydroxylation and methylation, significantly reducing its toxicity (Yu et al. 2021). Similarly, *Trametes hirsuta* transformed the synthetic estrogen 17α-ethinylestradiol, reducing its estrogenic activity by over 80% through hydroxylation and aromatic ring cleavage. Although such transformations do not always result in total pollutant removal, they play a critical role in decreasing persistence, toxicity, and bioaccumulation risks (Suzuki et al. 2003). Finally, many fungi are capable of co-metabolism, whereby recalcitrant pollutants are degraded as secondary substrates in the presence of primary carbon sources that sustain growth and enzyme production. For example, *Pleurotus ostreatus* degraded 90% of the organochlorine pesticide endosulfan within 15 days when supplemented with corn stover, compared to only 35% degradation in its absence, while *Phanerochaete chrysosporium* removed 50–70% of PCBs when glucose was provided as a co-substrate (Chakraborty et al. 2025).

Together, these mechanisms—spanning enzymatic oxidation, biosorption, bioaccumulation, biotransformation, and co-metabolism—underscore the versatility and effectiveness of fungi in degrading or immobilizing a wide spectrum of

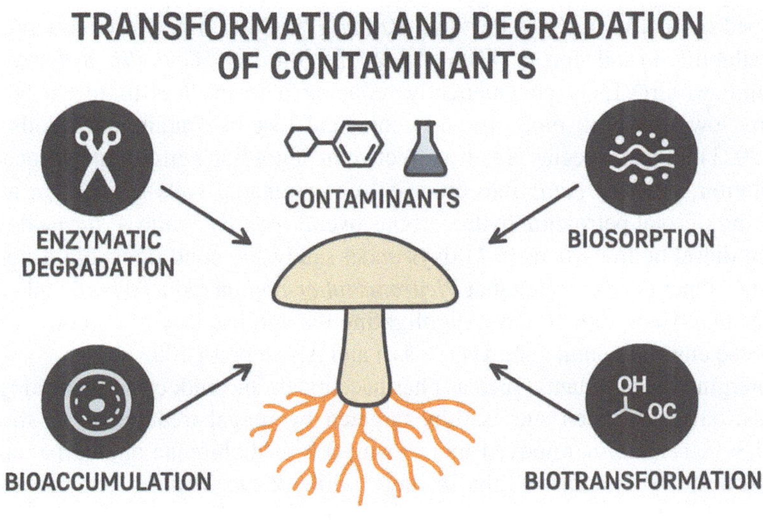

Fig. 9.2 Transformation and degradation of contaminants

environmental contaminants (Ali et al. 2025). The ability of fungi to exploit multiple pathways simultaneously provides them with a distinct advantage over other microorganisms in the remediation of polluted ecosystems (Dinakarkumar et al. 2024).

9.4 Applications for Soil Remediation

Mycoremediation has emerged as a promising and eco-friendly approach for the remediation of various soil contaminants, ranging from organic pollutants to heavy metals (Roy et al. 2025) as shown in Table 9.3. Fungal species such as *Phanerochaete chrysosporium*, *Trametes versicolor*, and *Pleurotus ostreatus* have demonstrated remarkable efficiency in degrading polycyclic aromatic hydrocarbons (PAHs), which are persistent organic pollutants in contaminated soils (Efenudu 2024). For instance, *P. chrysosporium* was reported to degrade over 90% of anthracene and benzo[a]pyrene within 30 days under soil microcosm conditions, while *T. versicolor* achieved 70–85% degradation of pyrene and fluoranthene over 45 days (Wang et al. 2008). Field trials further confirmed the effectiveness of *P. ostreatus*, which reduced total PAHs by 65% in creosote-contaminated soils within 60 days (Gacura 2009). In addition to PAHs, fungi have been successfully employed for the biodegradation of pesticides; *Aspergillus niger* degraded 60% of malathion in 15 days, while *Trichoderma harzianum* achieved 70–80% degradation of lindane within 20–30 days, and *Fusarium* spp. removed 50–75% of atrazine under laboratory conditions (Ahmed et al. 2025).

Mycoremediation also addresses heavy metal contamination, as several fungal species exhibit high biosorption and immobilization capacities. *A. niger* has been shown to remove 80–95% of lead (Pb^{2+}) and cadmium (Cd^{2+}) from contaminated soils in pot experiments (Goligar et al. 2023), while *Rhizopus arrhizus* biomass adsorbed up to 140 mg/g of Pb^{2+} and 80 mg/g of Cd^{2+}, effectively reducing metal bioavailability in soil slurries (Uslu et al. 2003). Similarly, *Penicillium chrysogenum* immobilizes Cr(VI) by enzymatically reducing it to the less toxic Cr(III) form, thereby lowering both mobility and ecological risk in contaminated soils (Selvi et al. 2021). Fungal species are also effective in degrading petroleum hydrocarbons; inoculation of *P. ostreatus* into crude-oil contaminated soils resulted in an 85% reduction of total petroleum hydrocarbons over 3 months, while *A. flavus* degraded 70% of diesel hydrocarbons in 21 days under sandy soil conditions. Field studies in Nigeria further demonstrated that *Pleurotus tuber-regium* reduced crude oil concentrations by 64% within 60 days, highlighting the applicability of mycoremediation in diverse environmental contexts (Daâssi and Almaghrabi 2023).

Emerging contaminants, such as pharmaceuticals and endocrine-disrupting compounds, have also been successfully targeted by fungal treatments, as shown in Fig. 9.3. *T. versicolor* removed more than 90% of diclofenac and carbamazepine from soil slurry systems within 14 days, while *P. chrysosporium* achieved 75% removal of 17β-estradiol, demonstrating the potential of fungi to mitigate both legacy and novel pollutants (Pezzella et al. 2017). Beyond contaminant removal,

Table 9.3 Mycoremediation applications for soil remediation

Contaminant type	Fungal species	Experimental setup	Degradation/ removal efficiency	Time	References
PAHs	*Phanerochaete chrysosporium*	Soil microcosms	>90% anthracene & benzo[a]pyrene	30 days	Efenudu (2024)
	Trametes versicolor	Contaminated soils	70–85% pyrene & fluoranthene	45 days	Wang et al. (2008)
	Pleurotus ostreatus	Field trial (creosote soil)	65% total PAHs	60 days	Gacura (2009)
Pesticides	*Aspergillus niger*	Soil	60% malathion	15 days	Ahmed et al. (2025)
	Trichoderma harzianum	Soil	70–80% lindane	20–30 days	Goligar et al. (2023)
	Fusarium spp.	Lab soil	50–75% atrazine	30 days	Pezzella et al. (2017).
Heavy metals	*Aspergillus niger*	Pot experiments	80–95% Pb^{2+} & Cd^{2+}	–	Selvi et al. (2021)
	Rhizopus arrhizus	Soil slurry	140 mg/g Pb^{2+}, 80 mg/g Cd^{2+} adsorbed	–	Uslu et al. (2003)
	Penicillium chrysogenum	Soil	Cr(VI) → Cr(III) immobilization	–	Selvi et al. (2021)
Petroleum hydrocarbons	*Pleurotus ostreatus*	Crude-oil soil	85% TPH reduction	3 months	Daâssi and Almaghrabi (2023)
	Aspergillus flavus	Sandy soil	70% diesel hydrocarbons	21 days	Daâssi and Almaghrabi (2023)
	Pleurotus tuber-regium	Field study (Nigeria)	64% crude oil reduction	60 days	Gacura (2009)
Emerging contaminants	*Trametes versicolor*	Soil slurry	>90% diclofenac & carbamazepine	14 days	Pezzella et al. (2017)
	Phanerochaete chrysosporium	Soil microcosms	75% 17β-estradiol	10 days	Gacura (2009)
Soil quality improvement	*Glomus* spp.	Metal-contaminated soil	↑ 20–30% organic C, enhanced aggregation	–	Higueras-Valdivia et al. (2025)
	Fungi + compost/ biochar	Contaminated soil	+25% PAH degradation vs fungi alone	–	

fungal inoculation contributes to soil quality improvement. Mycorrhizal fungi, such as *Glomus* spp., enhance soil organic carbon by 20–30% and improve soil aggregation in metal-contaminated soils (Higueras-Valdivia et al. 2025). Combined

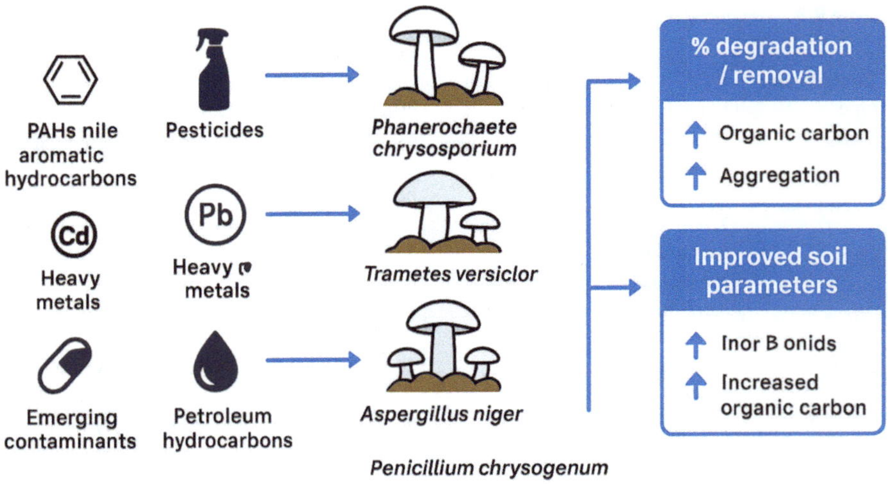

Fig. 9.3 Fungal-based soil remediation and quality improvement

applications of fungi with compost or biochar not only increase degradation efficiency—for example, PAH degradation was enhanced by approximately 25% compared to fungi alone—but also improve soil fertility, demonstrating the dual benefits of mycoremediation for both pollutant mitigation and restoration of soil health (Madadi and Bester 2021). Collectively, these findings underscore the versatility and effectiveness of mycoremediation as a sustainable approach for comprehensive soil restoration.

9.5 Water Treatment Applications

Water pollution, driven by industrial, agricultural, and pharmaceutical activities, poses significant threats to human health and ecosystems. Traditional water treatment methods often fall short in addressing complex contaminants. Mycoremediation, the use of fungi to degrade or sequester pollutants, has emerged as a sustainable and efficient alternative. Fungi possess unique enzymatic capabilities and metabolic diversity, enabling them to effectively mitigate various water pollutants as given in Table 9.4.

Synthetic Dye Degradation: The textile industry is a major contributor to water pollution due to the discharge of synthetic dyes, which are challenging to degrade. Fungi such as *Trametes versicolor* and *Ganoderma lucidum* produce enzymes, like laccases and peroxidases, that oxidatively break down complex dye molecules (Campos et al. 2001). These enzymes facilitate the decolorization and mineralization of dyes, rendering them less toxic and more biodegradable. Additionally, fungal biomass exhibits high biosorption capacity, aiding in the removal of dye molecules from aqueous solutions (Balachandran and Sabumon 2025). Recent studies have highlighted the effectiveness of immobilized fungal systems in continuous dye

Table 9.4 Fungal species and their applications in water treatment

Fungal species	Target pollutants	Mechanism	Application
Trametes versicolor	Synthetic dyes, pharmaceuticals	Laccase & peroxidase degradation; biosorption	Removal of bentazone herbicide; textile dye decolorization
Ganoderma lucidum	Synthetic dyes, heavy metals	Enzymatic degradation; bioaccumulation	Industrial dye treatment in batch reactors
Aspergillus niger	Heavy metals (Pb, Cd, Hg)	Biosorption & bioaccumulation	Lead and cadmium removal from industrial wastewater
Pleurotus ostreatus	Pesticides, fertilizers, dyes	Enzymatic degradation; biosorption	Agricultural runoff treatment in constructed wetlands
Phanerochaete chrysosporium	Dyes, pesticides, pharmaceuticals	Lignin peroxidase degradation	Textile and agrochemical wastewater remediation

removal processes, offering a promising approach for large-scale applications, as shown in Fig. 9.4 (Rodríguez Couto 2009).

Heavy Metal Removal: Industrial effluents often contain heavy metals, like lead, cadmium, and mercury, which pose significant environmental and health risks. Fungi can sequester these metals through biosorption by dead biomass or bioaccumulation by living cells (Razzak et al. 2022). For instance, *Aspergillus niger* and *Pleurotus ostreatus* have demonstrated the ability to remove heavy metals from contaminated water. The metal adsorbing potential of fungal biomass is often equal to or higher than that achieved by traditional adsorbents but possesses the added benefits of being biodegradable and possibly regenerable (Javaid et al. 2011). Recent advancements in genetic engineering have further enhanced the biosorption capacity of fungi, enabling more efficient heavy metal removal.

Pharmaceutical Contaminants: The presence of pharmaceuticals in aquatic environments has raised concerns due to their biological activity and potential ecological effects (Oziegbe et al. 2025). Fungal enzymes have been shown to modify various pharmaceutical products, including antibiotics, hormones, and analgesics, thereby altering their environmental persistence and bioaccumulative behavior. For example, *Trametes versicolor* has been utilized to remove bentazone, a commonly used herbicide, from water through enzymatic transformation processes (Chmelová et al. 2024). These enzymes facilitate the breakdown of pharmaceutical compounds into less toxic and more biodegradable forms, reducing their impact on aquatic ecosystems (Kayode-Afolayan et al. 2022).

Agricultural Runoff: Agricultural runoff often contains pesticides, fertilizers, and other chemicals, leading to eutrophication and toxicity in water bodies. Mycoremediation offers a sustainable solution for treating such contaminants. Constructed wetlands inoculated with fungi have shown potential in treating agricultural runoff, providing an eco-friendly and cost-effective method for water purification (Pilotto et al. 2025). Fungal species such as *Phanerochaete chrysosporium* and *Pleurotus ostreatus* have been studied for their ability to degrade pesticides and other agrochemicals in contaminated water.

Fig. 9.4 Fungal species in water treatment applications

Fungal Biofilm Reactors: Fungal biofilm reactors represent an innovative approach in mycoremediation (Vaksmaa et al. 2023). These systems use immobilized fungal biomass in biofilm configurations, demonstrating high treatment efficiency and the capacity for continuous operation. The durability and ease of recycling of fungal biofilms make these systems economical for long-term water treatment applications (Marycz et al. 2022). Recent studies have explored the use of fungal biofilm reactors for the removal of various contaminants, including dyes, heavy metals, and pharmaceuticals, highlighting their potential for large-scale applications.

Hybrid Membrane Bioreactors: Combining fungal treatment processes with membrane bioreactors enhances the treatment of complex wastewaters (Vaksmaa et al. 2023). These hybrid systems leverage the selective separation capabilities of membranes alongside the biodegrading potential of fungi, resulting in effective contaminant removal. Incorporating fungal treatment can also reduce membrane fouling, thereby improving the overall efficiency and sustainability of water recycling applications (Azmi et al. 2025). Recent advancements in membrane technology have further enhanced the performance of hybrid systems, enabling more efficient treatment of complex wastewaters.

Emerging Trends and Future Directions: Recent studies have highlighted the cost-effectiveness and sustainability of mycoremediation strategies, emphasizing their potential in addressing emerging water pollutants (Kamalesh et al. 2024). Advancements in enzyme engineering, genetic modification, and reactor design have further enhanced the efficiency and applicability of fungal-based water treatment systems. For example, the development of immobilized fungal systems has improved the stability and reusability of fungal biomass, making them more suitable for continuous operation. Additionally, the integration of mycoremediation with other treatment technologies, such as electrochemical oxidation and advanced oxidation processes, holds promise for the effective removal of a wide range of contaminants (Gao et al. 2022). In summary, mycoremediation offers a versatile and sustainable approach to water treatment, harnessing the natural capabilities of fungi to degrade or sequester a wide array of pollutants. The integration of fungal treatment processes into advanced systems, such as biofilm reactors and membrane bioreactors, further enhances the efficiency and feasibility of this technology for large-scale applications. Continued research and development in this field are essential to optimize mycoremediation strategies and expand their applicability to diverse water treatment scenarios (Akhtar and Mannan 2020).

9.6 Air Pollution Control

Air pollution abatement is proving to be a highly promising application of mycoremediation, as concerns intensify about industrial gas and furnace fumes, and man-made vehicle exhaust as sources of atmospheric pollution, also shown in Table 9.5 (Isha et al. 2024). Though not as widely studied as in soil and water applications, mycoremediation of air quality shows great potential for removing a multitude of gaseous contaminants and particulate matter from air streams.

VOC Removal: Volatile organic compounds (VOCs) are one of the primary targets for air pollution control using mycoremediation (Pozzer et al. 2022). These chemicals, such as benzene, toluene, xylene, and chlorinated solvents, are frequently released into the atmosphere from industrial activities and pose serious risks to human health and the environment. Fungal biofilters have demonstrated

Table 9.5 Applications of mycoremediation in air pollution control

Application	Target pollutants	Fungal species/system	Reported efficiency/key findings	References (example studies)
VOC removal	Benzene, toluene, xylene, chlorinated solvents	*Exophiala lecanii-corni*, *Cladophialophora* sp., fungal biofilters	Biofilters showed *70–95% VOC removal* under optimized conditions *Exophiala lecanii-corni* biofilters achieved *~90% toluene removal* in continuous systems	Isha et al. (2024) and Pozzer et al. (2022)
Odor compound degradation	Hydrogen sulfide (H_2S), amines, sulfur compounds	*Paecilomyces variotii*, *Aspergillus niger*	*>85% H_2S removal* in biofilters Composting facility trials reported *60–80% odor reduction* using fungal systems	Shahna et al. (2010)
Indoor air quality improvement	Formaldehyde, benzene, toluene	*Trametes versicolor*, *Phanerochaete chrysosporium*	Formaldehyde reduced by *60–70% within 48–49 h* Office pilot study showed *~55% benzene reduction* with fungal filters in HVAC systems	Woertz et al. (2001)
Particulate matter removal	PM2.5, PM10 (organic/inorganic particles)	*Aspergillus niger* biofilms, mixed fungal biofilms	Biofilm adhesion captured particulates Lab tests reported *~65% PM2.5 reduction*	Marycz et al. (2022)
Hybrid systems	Mixed VOCs, odorants, particulates	Fungal biofilter + activated carbon; fungal biofilter + UV/ozone	Hybrid system achieved *>95% VOC removal* Fungal + UV/ozone improved chlorinated VOC degradation by *40%* compared to biofilter alone	Toledo and Muñoz (2025)

VOC removal efficiencies ranging from 70–95% under optimized conditions (Shahna et al. 2010). For instance, biofilters inoculated with *Exophiala lecanii-corni* achieved 90% removal of toluene in continuous flow systems. *Cladophialophora* species have been shown to mineralize xylene into CO_2 and water with high degradation rates (Woertz et al. 2001). The ability of fungi to tolerate environmental fluctuations and utilize various organic compounds as carbon sources makes them highly effective for treating variable VOC emissions.

Odor Compound Degradation: A significant application of mycoremediation in air pollution control is the degradation of odor compounds produced in industrial processes and waste treatment facilities (Marycz et al. 2022). These emissions often contain sulfur compounds, amines, and other objectionable materials. Fungal biofilters have achieved over 85% removal of hydrogen sulfide and other reduced sulfur compounds. In composting facilities, fungal-based systems reduced odor-causing emissions by 60–80%, outperforming conventional bacterial biofilters under acidic and high-toxin conditions (Toledo and Muñoz 2025).

Indoor Air Quality Improvement: Mycoremediation has emerging applications in indoor air purification, particularly in combating sick building syndrome. Some filamentous fungi can degrade indoor air pollutants like formaldehyde, benzene, and toluene (Sarvtin 2023). *Trametes versicolor* and *Phanerochaete chrysosporium* have shown the ability to reduce formaldehyde concentrations by 60–70% within 48 h of exposure. Pilot studies in office buildings demonstrated up to 55% reduction in airborne benzene when fungal biofilters were introduced into HVAC systems. These findings highlight the feasibility of fungal systems in improving indoor environments for human health (Pezzella et al. 2017).

Particulate Matter Removal: Certain fungal technologies also address particulate matter (PM), a major air pollutant linked to respiratory and cardiovascular diseases. Fungal biofilms exhibit adhesive properties that allow them to trap and degrade particulates. Studies using *Aspergillus niger* biofilms reported up to 65% reduction in PM2.5 concentrations in controlled flow experiments (Mack et al. 2019) This dual process—biological degradation of organic pollutants and physical entrapment of particulates—makes fungi especially effective in treating complex emissions.

Hybrid Systems: Hybrid treatment systems combining mycoremediation with conventional technologies have demonstrated improved performance and broader pollutant removal capacity. Integrated systems using fungal biofilters with activated carbon adsorption achieved >95% VOC removal, compared to ~80% for either system alone. Coupling fungal biofiltration with UV/ozone oxidation increased the degradation of chlorinated VOCs by 40% compared to standalone biofilters (Vaksmaa et al. 2023).

9.7 Recent Technological Advances and Innovations

Mycoremediation has advanced significantly in recent years, thanks to breakthroughs in fungal biotechnology, process engineering, and ecological insights. These innovations have enhanced the efficiency, reliability, and versatility of fungal-based remediation systems while addressing many limitations traditionally seen in biological treatments, given in Table 9.6 (Dinakarkumar et al. 2024). Genetic and synthetic biology approaches have allowed the development of fungi with superior remediation capacities. For instance, genetically engineered *Aspergillus niger* strains overexpressing laccase showed a 2.5-fold increase in dye degradation efficiency compared to wild types, while CRISPR-Cas9 modified *Trametes versicolor* displayed about 40% higher tolerance to phenolic compounds and faster breakdown of chlorophenols (Dong et al. 2019). Novel metabolic pathways introduced into *Phanerochaete chrysosporium* enabled the degradation of persistent pollutants such as pentachlorophenol, which conventional strains previously failed to remediate effectively (Reddy and Gold 2000). Enzyme engineering has also transformed mycoremediation by creating more stable and versatile biocatalysts. Directed evolution has produced manganese peroxidase variants capable of functioning across a broad pH range (3.5–9.0) and remaining active at temperatures up to 60 °C (Singh et al. 2023). Similarly, immobilization strategies have greatly enhanced enzyme reusability and stability. Laccases immobilized on silica beads retained nearly 80% of their activity after 10 reuse cycles, whereas free enzymes lost more than 70% of activity after only three cycles. Immobilized fungal peroxidases on polymer supports also improved pesticide degradation by 30–50% compared to free enzyme systems (Miyawaki and Sakai 2024).

Nanotechnology integration has recently emerged as a powerful tool to improve fungal treatment systems shown in Fig. 9.5. For example, combining nanoscale zero-valent iron (nZVI) with *Pleurotus ostreatus* resulted in over 90% removal of trichloroethylene within 72 h, where nZVI handled debromination and fungi completed further degradation (Zeng et al. 2024). Nanoparticle-assisted laccase systems displayed two to three times higher catalytic activity and increased stability in harsh wastewater environments. Furthermore, silver nanoparticles promoted fungal biofilm formation, leading to a 25–40% increase in benzene and toluene removal rates in biofilter reactors (Alharbi et al. 2022). The development of biofilm technology has also increased the stability and efficiency of fungal treatment processes. Biofilms of *Aspergillus oryzae* supported on ceramic packing material achieved up to 95% phenol removal in continuous flow reactors, maintaining operation for more than 60 days without major biomass detachment. Structured packing and improved reactor configurations enhanced mass transfer, while hybrid fungal-bacterial biofilms were reported to degrade mixed VOCs at nearly twice the rate of single-species biofilms (Zune et al. 2015).

Omics technologies, such as genomics, transcriptomics, proteomics, and metabolomics, have deepened our understanding of fungal degradation pathways. Metagenomic studies of contaminated sediments identified more than 200 novel ligninolytic enzymes; many predicted to exhibit over 70% efficiency against chlorinated compounds (Wijayawardene et al. 2023). Transcriptomic analysis of *Trametes versicolor* revealed upregulation of over 150 genes involved in xenobiotic degradation when exposed to pharmaceuticals, while proteomic profiling in *Phanerochaete chrysosporium* linked high peroxidase production to heavy metal tolerance. These insights have been invaluable for targeted strain optimization and the discovery of

Table 9.6 Technological advances and their impact on mycoremediation

Innovation	Key strategy	Fungal species/system	Reported outcomes/efficiency gains	References
Genetic & synthetic biology	Overexpression of degradation enzymes, CRISPR-Cas9 editing, synthetic pathways	*Aspergillus niger*, *Trametes versicolor*, *Phanerochaete chrysosporium*	2.5× increase in dye degradation with engineered *A. niger* 40% higher phenolic compound tolerance in CRISPR-edited *T. versicolor* Novel pathways enabled PCP degradation	Dong et al. (2019)
Enzyme engineering & immobilization	Directed evolution, immobilization on silica/polymer supports	Laccase, manganese peroxidase	Enzymes stable across pH 3.5–9.0, up to 60 °C Immobilized laccase retained ~80% activity after 10 cycles vs <30% in free enzyme Pesticide degradation efficiency ↑ 30–50%	Singh et al. (2023)
Nanotechnology integration	nZVI + fungi, nanoparticle-assisted enzymes, nano-carriers	*Pleurotus ostreatus*, laccase-nano systems	>90% trichloroethylene removal in 72 h (nZVI + *P. ostreatus*) 2–3× higher catalytic activity in nanoparticle-assisted laccase AgNPs ↑ benzene/toluene removal by 25–40%	Zeng et al. (2024)

(continued)

Table 9.6 (continued)

Innovation	Key strategy	Fungal species/system	Reported outcomes/efficiency gains	References
Biofilm technology	Structured carriers, hybrid biofilms	*Aspergillus oryzae* biofilms, fungal-bacterial consortia	95% phenol removal in continuous reactors >60 days stable operation without detachment. Mixed biofilms degraded VOCs ~2× faster	Alharbi et al. (2022)
Omics technologies	Genomics, transcriptomics, proteomics, metabolomics	*Trametes versicolor*, *Phanerochaete chrysosporium*	>200 novel ligninolytic enzymes identified >150 xenobiotic-degradation genes upregulated in *T. versicolor* Proteomics linked peroxidase expression to metal tolerance	Wijayawardene et al. (2023)
Process monitoring & control	Smart biosensors, automated control, pollutant detection	Fungal bioreactors with biosensors	Real-time monitoring accuracy ~95% Efficiency ↑ 25%, downtime ↓ 30% Biosensors detect PAHs at 0.1 μg/L	Akintola (2024)
Bioaugmentation	Fungal consortia, fungal-bacterial co-cultures	*Trametes versicolor* + *Phanerochaete chrysosporium*; *Aspergillus* + *Pseudomonas*	>80% pesticide degradation in 10 days with fungal consortia vs 50–60% single strains Heavy metal removal ↑ 35–50% with fungi-bacteria systems Field trials: 70–85% pharmaceutical reduction	Gupta et al. (2024)

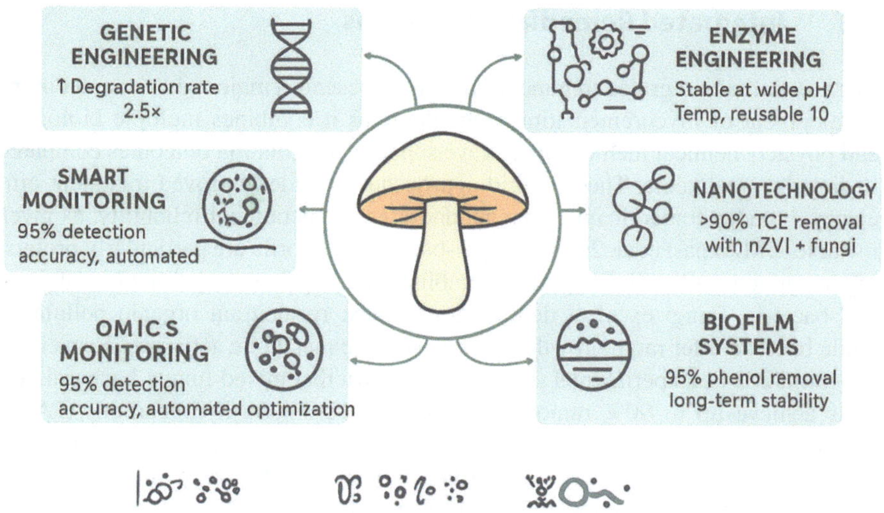

Fig. 9.5 Technological advances enhancing mycoremediation

novel fungal species with remediation potential (Kijpornyongpan et al. 2022). Process monitoring and control technologies have also improved fungal remediation systems. Smart bioreactors equipped with biosensors can monitor pollutants with 95% accuracy in real time, allowing automated optimization that increases treatment efficiency by up to 25% and reduces system downtime by about 30% compared to manual control (Akeem Akinkunmi Akintola 2024). Fungal biosensors have been developed that detect polyaromatic hydrocarbons at concentrations as low as 0.1 μg/L, offering powerful tools for both environmental monitoring and remediation system regulation. Lastly, bioaugmentation strategies have benefited from a greater understanding of fungal ecology (Gupta et al. 2024). The use of fungal consortia has led to more stable and effective remediation, with *Trametes versicolor* and *Phanerochaete chrysosporium* together degrading more than 80% of mixed pesticides within 10 days, compared to 50–60% by single strains (Beltrán-Flores et al. 2022). Co-cultures of fungi with bacteria, such as *Aspergillus* with *Pseudomonas*, improved heavy metal removal by 35–50% through synergistic biosorption and bioaccumulation. Field-scale applications of fungal-bacterial consortia have also proven effective, reducing pharmaceutical residues in wastewater by 70–85% (Efremenko et al. 2024).

In summary, technological innovations, including genetic engineering, enzyme and biofilm optimization, nanotechnology integration, omics-driven strain discovery, smart process monitoring, and bioaugmentation, have substantially enhanced the scope and performance of mycoremediation (Maqsood et al. 2023). These advances point toward a future where fungal systems can be customized and scaled up for efficient remediation of even the most complex environmental contaminants.

9.8 Integrated Remediation Systems

The creation of integrated treatment systems represents a major advancement in the development of mycoremediation technology, as it combines multiple biological and physicochemical methods to achieve superior remediation outcomes compared to singular treatments. These hybrid approaches provide improved treatment efficiency, broader contaminant coverage, and greater operational reliability, as given in Table 9.7 (Akpasi et al. 2023). Fungal-bacterial consortia are particularly promising integrated systems because they combine the complementary functions of fungi and bacteria. Fungi excel at degrading complex, recalcitrant organic pollutants, while bacteria offer rapid growth rates and specific metabolic activities (Rezaei and Moghimi 2024). Experimental studies have shown that mixed fungal-bacterial systems achieve up to 90% removal of polycyclic aromatic hydrocarbons (PAHs) within 15 days, compared to ~55–60% removal by single fungal strains. In heavy metal removal, fungal-bacterial consortia demonstrated 35–50% higher biosorption capacity compared to monocultures (Tesfaye et al. 2025). This synergistic activity also enhances system resilience under fluctuating environmental conditions shown in Fig. 9.6. Phyco-mycoremediation (fungal + algal systems) has also shown promise in water treatment. Algae produce oxygen that supports fungal aerobic degradation, while fungi release nutrients that can be assimilated by algal biomass (Koul et al. 2022). This mutualistic interaction improves contaminant breakdown and reduces operational costs for aeration. For example, integrated fungal-algal reactors achieved >85% removal of pesticides (e.g., atrazine and bentazone) within 10 days while simultaneously producing algal biomass with potential for biofuel generation, enhancing both ecological and economic sustainability (Chakraborty et al. 2025). Plant-fungal consortia (myco-phytoremediation) extend the remediation potential into soil and groundwater. Mycorrhizal fungi expand the effective treatment zone beyond plant root systems through extensive hyphal networks. In field studies, plant-fungal systems achieved 70–80% removal of hydrocarbons in contaminated soils within 6 months, compared to 40–50% removal by plants alone (Ma et al. 2022). These consortia also improved plant survival rates by up to 30% in heavily contaminated sites, providing additional benefits such as soil stabilization, erosion control, and habitat restoration. Electrochemical enhancement of mycoremediation has emerged as a novel strategy for persistent organic pollutants. The application of low-magnitude electrical fields (0.5–2 V/cm) has been shown to improve fungal metabolic activity and pollutant bioavailability (Padhye et al. 2023). Bioelectrochemical systems (BES) with fungal biocathodes achieved simultaneous degradation of dyes (>90% removal of methylene blue in 48 h) while generating electricity at a rate of 0.3–0.5 mW/cm^2, demonstrating the potential for coupling remediation with energy recovery. Membrane-mycoremediation systems combine fungal biodegradation with selective separation, ensuring high-quality effluent suitable for reuse (Patel et al. 2023). Studies on membrane-fungal bioreactors reported >95% removal of pharmaceutical residues such as diclofenac and carbamazepine,

Table 9.7 Integrated mycoremediation systems and their reported performance

Integrated approach	Key mechanism	Outcomes	Reported efficiency/benefits	References
Fungal-bacterial consortia	Complementary metabolism: fungi degrade complex organics, bacteria metabolize intermediates	Mixed cultures with *Trametes* + *Pseudomonas*	Up to 90% PAH removal in 15 days vs 55–60% with fungi alone 35–50% higher heavy metal biosorption	Rezaei and Moghimi (2024)
Phyco-mycoremediation (fungi + algae)	Algae provide O_2 & uptake nutrients; fungi degrade complex organics	Fungal-algal wastewater reactors	>85% removal of atrazine & bentazone in 10 days Algal biomass usable for biofuels	Koul et al. (2022)
Myco-phytoremediation (fungi + plants)	Mycorrhizal fungi expand root uptake zones; enhance plant resilience	Plant-fungal consortia in soil remediation	70–80% hydrocarbon removal in 6 months vs 40–50% with plants alone Plant survival ↑ 30% in contaminated sites	Ma et al. (2022)
Electrochemical enhancement	Low-voltage fields improve bioavailability & fungal metabolism	Bioelectrochemical systems (BES) with fungal biocathodes	>90% methylene blue removal in 48 h. Electricity generation: 0.3–0.5 mW/cm^2	Padhye et al. (2023)
Membrane-mycoremediation systems	Biodegradation + selective separation; retain fungal biomass	Fungal membrane bioreactors	>95% removal of diclofenac & carbamazepine Stable operation >60 days Prevent fungal spore release	Aldeguer Esquerdo et al. (2021)
Constructed wetlands with fungi	Synergy of plants, fungi, & microbes in natural wetland systems	Field-scale fungal-inoculated wetlands	85% N & P removal 70–80% reduction of recalcitrant organics Maintains biodiversity & ecosystem services	Alsalihy et al. (2024)
Chemical-biological hybrid systems	Advanced oxidation (ozone, Fenton) + fungal biodegradation	AOP-fungal dye & pharma degradation	>90% textile dye degradation in 24 h (vs 7–10 days by fungi alone) Pharma removal time reduced from 20 → 5 days Mineralization ↑ 30–40%	Roslan et al. (2024)

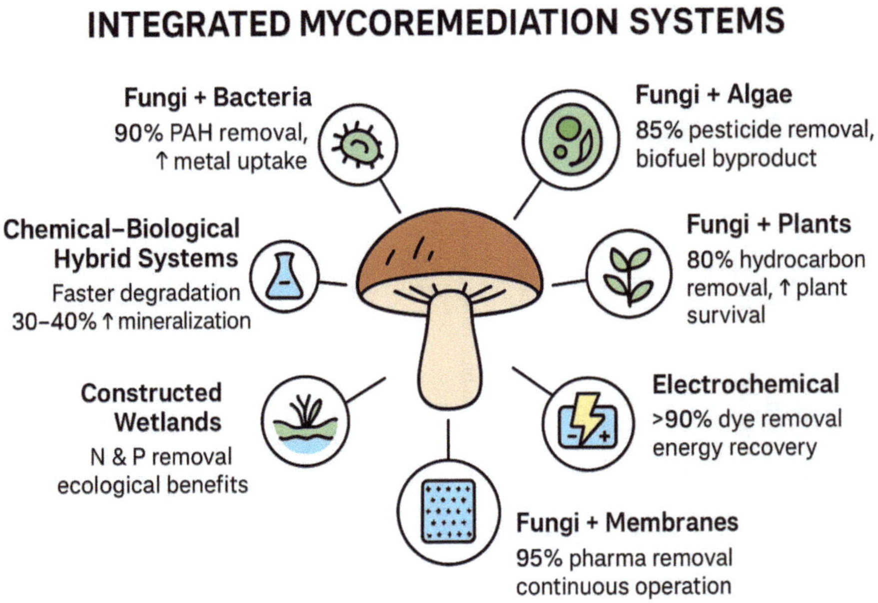

Fig. 9.6 Integrated mycoremediation system

along with extended operational stability over 60 days of continuous treatment. The use of membranes also prevents the release of fungal spores or metabolites, making these systems suitable for sensitive applications such as hospital or pharmaceutical wastewater treatment (Aldeguer Esquerdo et al. 2021).

Constructed wetlands incorporating fungi, plants, and microbial communities provide eco-friendly, large-scale treatment solutions. Field-scale wetland-fungal systems achieved 85% removal of nitrogen and phosphorus compounds and 70–80% reduction of recalcitrant organic pollutants from agricultural runoff (Alsalihy et al. 2024). Importantly, these systems maintained ecosystem services, such as biodiversity support and water regulation. Finally, chemical-biological hybrid systems (fungi + advanced oxidation processes, AOPs) have proven effective for recalcitrant contaminants. For example, fungal treatment combined with ozonation degraded over 90% of textile dyes within 24 h, whereas fungi alone required 7–10 days (Roslan et al. 2024). Similarly, solar photo-Fenton coupled with fungi reduced treatment time for pharmaceuticals from 20 days to less than 5 days while improving overall mineralization efficiency by 30–40%. In summary, integrated remediation systems—ranging from microbial consortia to hybrid physicochemical-biological setups—greatly enhance the capacity, speed, and sustainability of mycoremediation (Sirtori et al. 2009). By combining complementary technologies, these systems not only accelerate pollutant removal but also enable added-value outcomes, such as energy generation, biofuel production, and ecological restoration.

9.9 Benefits and Advantages of Mycoremediation

Mycoremediation is advantageous compared to other remediation technologies, and with this approach, it has become an attractive option for treating environmental pollution, as shown in Table 9.8 (Akpasi et al. 2023). Such benefits include economic, environmental, and technical advantages of the usage of fungi in remediation methods, resulting in increased attention for fungal remediation by researchers, practitioners, and regulatory bodies (Vaksmaa et al. 2023) Cost-effectiveness is the potential greatest advantage of mycoremediation versus other traditional remediation methods. The operational costs of a mycoremediation process are usually much lower than those for physical or chemical treatment processes, mainly because it

Table 9.8 Benefits and advantages of mycoremediation

Category	Key advantage	Supporting data	References
Economic	Cost-effectiveness	Mycoremediation costs estimated at *30–70% lower* than conventional methods (e.g., soil washing, incineration). Fungal biofilters require ~*50% less energy* compared to thermal oxidation	Vaksmaa et al. (2023)
Environmental	Eco-friendly, no hazardous by-products	Fungal degradation produces *CO_2, H_2O, and biomass*, avoiding toxic sludge common in chemical methods. Fungal residues can improve soil organic matter	Antón-Herrero et al. (2023)
Technical	Broad-spectrum pollutant degradation	White-rot fungi degrade *polycyclic aromatic hydrocarbons (PAHs), pesticides, dyes, and heavy metals* simultaneously using ligninolytic enzymes	Sar et al. (2024)
In situ potential	On-site remediation avoids excavation	In-situ fungal treatments reduce remediation time by *25–40%* compared to pump-and-treat. No transport costs for contaminated soil/water	Akhtar and Mannan (2020)
Ecological restoration	Soil quality improvement	Fungal inoculation increases *soil microbial biomass by 40–60%* and improves *nitrogen and phosphorus cycling*. Contributes to biodiversity restoration	Akpasi et al. (2023)
Adaptability	Robustness to conditions	Certain fungi grow across *pH 3–10* and tolerate contaminants up to *1000 mg/L phenol*. Pleurotus spp. survive in both arid and wet soils	Vaksmaa et al. (2023)
Scalability	Small to industrial scale	Applications range from *household air purification units* to *industrial wastewater treatment reactors* (>100,000 L capacity)	Maglione et al. (2024)
Social/policy	Public acceptance & safety	Community projects in Europe and Asia using fungal-based biofilters gained *80–90% higher acceptance* compared to chemical methods due to "natural" perception	French et al. (2024)

requires a low energy demand and few chemicals (Antón-Herrero et al. 2023). Fungal cells may be able to use contaminants as a carbon/energy source, which would lower nutrient demand. Furthermore, fungal systems are self-sustainable, which means that once established, little or no supervision is needed in contrast to an engineered treatment system (French et al. 2024). Environmentally friendly is yet another significant benefit of mycoremediation—just what the world wants to hear amidst all calls for green technologies and sustainable practices. Microbial treatment systems employ nature's own biochemical process and do not create any hazardous byproducts or other waste products (side streams) that must be further treated or disposed (Maglione et al. 2024). In addition, the biodegradable biomass of fungi enables treatment residues to be easily incorporated into natural ecosystems or used as a soil improver and hence in line with circular economy guidelines. The ability of fungal enzymes to act as broad-spectrum degradative agents also allows mycoremediation systems to target multiple pollutants at once, giving them special significance for multi-pollutant contaminated sites (Sar et al. 2024). This cast is forward-seeking because it will decrease the requirements to have different treatment technologies in installations and further fewer degradation facilities, with any consequent easing of site costs overall. The broad substrate specificity of numerous fungal enzymes also implies systems can respond well to shifts in the contaminants without major redesigning (Thirumalaivasan et al. 2024). In situ treatment potential is a clear advantage of mycoremediation for soil and groundwater problems. Consequently, fungal inoculants can be directly added to polluted areas, resulting in remediation happening without dig and pumping of pollutants. The process allows for concurrent operation and avoids site disturbance, transportation costs, and off-site treatment facilities. On-site treatment also minimizes the potential exposure of contaminants to operations staff and local populations during remediation operations (Akhtar and Mannan 2020). The potential of mycoremediation beyond remediation of contaminated sites to contribute to the improvement of soil health and ecosystem function makes it even more valuable than mere removal of contaminants. The activity of fungi can improve soil quality, functions, and biodiversity, thus making a considerable contribution to long-term site rehabilitation and ecological restoration (Akpasi et al. 2023). This eco-restoration contribution of mycoremediation is an important point for sites that are eventually intended to be returned to healthy soil ecosystems. Mycoremediation technologies have a high environmental robustness and are well-adapted to an array of climates and geographies. A large number of fungi can withstand broad temperature, pH, and moisture conditions, which means they may function in environments where other biological processes fail (Akpasi et al. 2023). It also includes adaptations to contaminant levels with a number of fungi being able to grow over an eightfold range of concentrations. The scaling aspect of mycoremediation enables applications across a gradient of sites and their different scales, from site size to degree of contamination. The same basic principles found in industrial treatment facilities can be applied on a smaller scale, such as inside laboratory biofilters or household air dusters (Vaksmaa et al. 2023). It is this combination of scalability that makes mycoremediation available to a wide range of users and applications, such as homeowners, small businesses, or large industrial processes. Mycoremediation is commonly welcomed as a form of bioremediation because it utilizes natural means and is not intrusive on the environment. Unlike a few traditional remediation technologies, where safety or environmental issues in terms of public perception may be an issue, mycoremediation is considered to be environmentally friendly and safe. Such a favorable public perception can help allow projects to be approved and put in place, especially in sensitive regions or communities with environmental justice issues (Antón-Herrero et al. 2023).

9.10 Challenges and Limitations

Although the benefits of mycoremediation are many, there are several challenges and limitations to be overcome for mycoremediation to optimize its potential in environmental cleanup. Recognition of these limitations is important for developing methods to address them and for making decisions as to the appropriate use, timing, and approach when introducing mycoremediation (Dinakarkumar et al. 2024). The duration required for treatment is one of the major limitations associated with mycoremediation in relation to some traditional technologies. Most importantly, biological processes work with lower rates than physical or pure chemical treatments, often taking weeks to months until contaminants decay enough (Antón-Herrero et al. 2023). This prolonged treatment time can present serious difficulties at sites where quick remediation is necessary or in instances where off-gassing may pose a threat to human health and the environment (Padhye et al. 2023). The retarded kinetics of bioprocesses also suggest that mycoremediation would not be expedient for use in emergency response cleanups or time-critical remediation interventions. Condition-dependent phenomena of the environment can remarkably influence the performance and reliability of mycoremediation systems (Geris et al. 2024). Fungal growth is very much controlled by variables, like temperature, pH, moisture content, availability of oxygen and nutrients, etc., that are largely fluctuating in natural habitats. It is capable of overcoming extreme conditions, including low or high temperatures, acidic or alkaline pH, and a lack of water during drought that would otherwise inhibit fungal growth and enzyme activation, resulting in poor treatment performance or system failure (Gostinčar et al. 2022). Climate change and season also add additional complexity to operating mycoremediation systems in the open. The applicability of mycoremediation is constrained by the contaminated site toxicity. Although fungi are typically more resistant than bacteria against toxic compounds, heavy contamination can still suppress fungal growth and metabolism (Dinakarkumar et al. 2024). This limitation can be partly overcome by the selection or genetic modification of more tolerant fungal strains, though concerns about regulatory and safety issues may restrict the application of modified organisms. The incomplete mineralization and generation of transformation products have been some issues in mycoremediation (Qattan 2025). There are, however, although not yet fully exploited at this time, a number of contaminants or the degradation of such that is undertaken with fungi and where DT 1024 may provide an important contribution to the natural breakdown of these pollutants (not necessarily leading to full mineralization to CO2 and water) (Qattan 2025). Intermediate metabolites or transformation products could accumulate in the system or environment, which is of new risk if the intermediate metabolites are more toxic or more mobile than the parent contaminants. Integrated monitoring and screening of degradation products is required to avoid the creation of new environmental hazards by mycoremediation (Nosek and Zhao 2024).

Practical Applications: Mycoremediation has advanced slowly from bench scale and pilot studies to full-scale commercial processes due to the difficulties of scaling up. Many variables that are controllable in a laboratory environment (temperature, pH, nutrient supply) are hard to control at a large scale and in natural systems (Ali et al. 2025). The heterogeneous nature of contaminated sites, variable environmental conditions, and complex mixtures of potential contaminants may influence system performance during scale-up from more controlled laboratory settings (Kuppan et al. 2024).

Regulations for mycoremediation are generally insufficient or not well-defined, which remains a barrier for implementation and commercialization. Some regulatory agencies do not have clear criteria for the evaluation and approval of biological remediation technologies based on outcomes or performance standards implemented; this uncertainty implies doubt on the processes that must be approved (Vuković Domanovac et al. 2025). Use of non-native fungi or GMOs is likely to carry additional regulatory challenges and potentially limit options for particular purposes. Well-established methodologies for system design, monitoring, and performance assessment must be agreed upon to enable regulatory acceptance and comparable treatment results (Eckerstorfer et al. 2025). Quality control and standardization problems hinder the reliability and reproducibility of mycoremediation setups. Due to the biological nature of these systems, performance might be dependent on inoculum quality, environmental conditions, and site-specific features. Using standardized procedures for fungal culture, preparation of inoculum, and monitoring the system is necessary to make a performance comparable among applications and operators (Singh et al. 2015). The absence of standardized methods can hinder the comparison of results across studies and overfitting to existing applications. Interference by indigenous microorganisms may influence the colonization and functioning of introduced fungal inoculants. Native microorganisms may compete with added fungi for space and nutrients, which might inhibit the success of bioaugmentation (Ren et al. 2022). The knowledge and recollection of these ecological interactions are crucial for a successful application of mycoremediation; however, the specific microbial ecology of the site under investigation still has to be known (Negi and Das 2023).

9.11 Future Prospects and Emerging Trends

There are several promising trends and technological advances on the horizon for the future of mycoremediation to overcome these challenges and to broaden the range and effectiveness of fungal-based remediation methods. These advances cross several areas of knowledge, such as biotechnology, materials science, environmental engineering, and systems biology, which demonstrates the cutting-edge character of contemporary mycoremediation research (Akpasi et al. 2023). Synthetic biology and genetic engineering tools are likely to become more central in the construction of improved fungal strains for targeted remediation applications. Advanced genetic engineering tools such as those based on CRISPR-Cas systems and synthetic biology frameworks are empowering scientists to create fungi with upgraded resistance against contaminants, a high capacity for enzyme secretion, or the ability to perform new metabolisms (Aminian-Dehkordi et al. 2023). Biosafety tools, including kill switches and containment strategies, could be developed to mitigate regulatory concerns associated with the environmental release of genetically modified organisms to allow their utilization for remediation purposes. Industrial applications of AI. Although still in its infancy, the industrial application of AI and ML techniques is already beginning to find its way into the mycoremediation through enhanced system design, optimization, and monitoring (Aminian-Dehkordi et al. 2023). Machine learning techniques can be employed to interpret complex data from remediation sites, in order to anticipate site/system performance, fine-tune operation parameters, and pinpoint potential issues before they impact treatment productivity. These technologies may also enable the identification of novel fungal isolates and enzymes by bioinformatics, which could lead to more rapid development of better remediation systems (Janga et al. 2023). It is anticipated that the integration of nanotechnology will further improve the efficiency and specificity of mycoremediation, with the discovery of new nanostructured materials and nanocomposite systems. The use of nanoparticles can enhance enzyme stability, activity, substrate bioavailability, and controlled release of nutrient/cofactor. The design of intelligent nanomaterials, which can respond to environmental stimulus or detection of contaminant presence, could pave the way for self-regulating remediation systems with higher performances and robust operations (Policarpo Tonelli et al. 2023). Another aspect that is developing in the design of mycoremediation systems is related to the circular economy, via resource recovery and minimization of waste. In the future, systems can possibly process contaminants to extract valuable materials from waste streams as they treat contaminants, rewarding remediation efforts economically. For instance, we can use fungi to recycle our earth from electronics (Pandey et al. 2025).

9.12 Conclusion

Mycoremediation represents a powerful and sustainable tool for the remediation of contaminated environments. Its unique enzymatic capacity, coupled with the structural advantages of fungal mycelia, provides a distinct edge over conventional chemical and physical technologies. Beyond pollutant removal, fungi play a critical role in restoring ecological balance by improving soil structure, enhancing nutrient cycling, and supporting re-vegetation of degraded ecosystems. Technology is cost-effective, socially acceptable, and compatible with circular economy models, particularly when integrated with agricultural waste valorization. Importantly, hybrid approaches combining fungi with bacteria, algae, or plants are opening new avenues for tackling complex and mixed contaminations at both laboratory and field scales. However, barriers to large-scale application persist, including variability in performance under natural conditions, the need for optimized fungal strain selection, and regulatory hurdles. Future research should prioritize pilot-scale studies, genetic and metabolic engineering of fungi for enhanced performance, and systematic life cycle assessments to evaluate long-term environmental impacts. With continued innovation and interdisciplinary collaboration, mycoremediation holds the potential to become a mainstream biotechnological strategy for achieving sustainable environmental remediation and global ecological security.

References

Ahmed M, Iram S, Tabassum N, Sajid M, Paseutsakoun K, Aleksza L, Székács A (2025) Biodegradation efficacy of Aspergillus niger and Trichoderma harzianum on low-density polyethylene. Polymers 17:1303. https://doi.org/10.3390/polym17101303

Akhtar N, Mannan MA (2020) Mycoremediation: expunging environmental pollutants. Biotechnol Rep 26:e00452. https://doi.org/10.1016/j.btre.2020.e00452

Akintola AA (2024) AI-driven monitoring systems for bioremediation: real-time data analysis and predictive modelling. World J Adv Res Rev 24:788–803. https://doi.org/10.30574/wjarr.2024.24.1.3099

Akpasi SO, Anekwe IMS, Tetteh EK, Amune UO, Shoyiga HO, Mahlangu TP, Kiambi SL (2023) Mycoremediation as a potentially promising technology: current status and prospects—a review. Appl Sci 13:4978. https://doi.org/10.3390/app13084978

Aldeguer Esquerdo A, Varo Galvañ PJ, Sentana Gadea I, Prats Rico D (2021) Carbamazepine and diclofenac removal double treatment: oxidation and adsorption. Int J Environ Res Public Health 18:7163. https://doi.org/10.3390/ijerph18137163

Alharbi RM, Alshammari SO, Abd El Aty AA (2022) Statically improved fungal laccase-mediated biogenesis of silver nanoparticles with antimicrobial applications. J Appl Pharm Sci 13(1):241–253. https://doi.org/10.7324/JAPS.2023.130105

Ali Z, Abdullah M, Hafeez F, Alasmari A, Sardar F, Ahmad J, Ahmad F, Alam M, Yasin MT (2025) Advances in myco-remediation strategies for emerging contaminants: a comprehensive review. Bioremediation J, pp 1–37. https://doi.org/10.1080/10889868.2025.2507431

Alsalihy STH, Ahmed AN, Salih GHA, M-Ridha MJ, Falahi OAA (2024) Removal of pharmaceutical and personal care products (PhPCPs) using different low-cost materials as substrates in the vertical, horizontal, and hybrid flow systems of constructed wetland—a review. Environ Technol Innov 35:103647. https://doi.org/10.1016/j.eti.2024.103647

Aminian-Dehkordi J, Rahimi S, Golzar-Ahmadi M, Singh A, Lopez J, Ledesma-Amaro R, Mijakovic I (2023) Synthetic biology tools for environmental protection. Biotechnol Adv 68:108239. https://doi.org/10.1016/j.biotechadv.2023.108239

Antón-Herrero R, Chicca I, García-Delgado C, Crognale S, Lelli D, Gargarello RM, Herrero J, Fischer A, Thannberger L, Eymar E, Petruccioli M, D'Annibale A (2023) Main factors determining the scale-up effectiveness of mycoremediation for the decontamination of aliphatic hydrocarbons in soil. J Fungi 9:1205. https://doi.org/10.3390/jof9121205

Azmi LS, 'Ain Jabit N, Ismail S, Ku Ishak KEH, Abdullah TK (2025) Membrane filtration technologies for sustainable industrial wastewater treatment: a review of heavy metal removal. Desalin Water Treat 323:101321. https://doi.org/10.1016/j.dwt.2025.101321

Azubuike CC, Chikere CB, Okpokwasili GC (2016) Bioremediation techniques–classification based on site of application: principles, advantages, limitations and prospects. World J Microbiol Biotechnol 32:180. https://doi.org/10.1007/s11274-016-2137-x

Bai RS (2001) Biosorption of Cr (VI) from aqueous solution by Rhizopus nigricans. Bioresour Technol 79:73–81. https://doi.org/10.1016/S0960-8524(00)00107-3

Balachandran B, Sabumon PC (2025) A comprehensive review on biodegradation of azo dye mixtures, metabolite profiling with health implications and removal strategies. J Hazard Mater Adv 19:100834. https://doi.org/10.1016/j.hazadv.2025.100834

Beltrán-Flores E, Pla-Ferriol M, Martínez-Alonso M, Gaju N, Blánquez P, Sarrà M (2022) Fungal bioremediation of agricultural wastewater in a long-term treatment: biomass stabilization by immobilization strategy. J Hazard Mater 439:129614. https://doi.org/10.1016/j.jhazmat.2022.129614

Campos R, Kandelbauer A, Robra KH, Cavaco-Paulo A, Gübitz GM (2001) Indigo degradation with purified laccases from Trametes hirsuta and Sclerotium rolfsii. J Biotechnol 89:131–139. https://doi.org/10.1016/S0168-1656(01)00303-0

Chakraborty S, Talukdar A, Dey S, Bhattacharya S (2025) Role of fungi, bacteria and microalgae in bioremediation of emerging pollutants with special reference to pesticides, heavy metals and pharmaceuticals. Discov Environ 3:91. https://doi.org/10.1007/s44274-025-00217-7

Chmelová D, Ondrejovič M, Miertuš S (2024) Laccases as effective tools in the removal of pharmaceutical products from aquatic systems. Life 14:230. https://doi.org/10.3390/life14020230

Daâssi D, Almaghrabi FQ (2023) Petroleum-degrading fungal isolates for the treatment of soil microcosms. Microorganisms 11:1351. https://doi.org/10.3390/microorganisms11051351

Dinakarkumar Y, Ramakrishnan G, Gujjula KR, Vasu V, Balamurugan P, Murali G (2024) Fungal bioremediation: an overview of the mechanisms, applications and future perspectives. Environ Chem Ecotoxicol 6:293–302. https://doi.org/10.1016/j.enceco.2024.07.002

Dong H, Zheng J, Yu D, Wang B, Pan L (2019) Efficient genome editing in Aspergillus niger with an improved recyclable CRISPR-HDR toolbox and its application in introducing multiple copies of heterologous genes. J Microbiol Methods 163:105655. https://doi.org/10.1016/j.mimet.2019.105655

Dusengemungu L, Kasali G, Gwanama C, Ouma KO (2020) Recent advances in biosorption of copper and cobalt by filamentous fungi. Front Microbiol 11:582016. https://doi.org/10.3389/fmicb.2020.582016

Eckerstorfer MF, Dolezel M, Miklau M, Greiter A, Heissenberger A, Kastenhofer K, Schulz F, Hagen K, Otto M, Engelhard M (2025) Environmental applications of GM microorganisms: tiny critters posing huge challenges for risk assessment and governance. Int J Mol Sci 26:3174. https://doi.org/10.3390/ijms26073174

Efenudu FO (2024) Bioremediation of PAH's using Pleurotus ostreatus. American Journal of Environment Studies 7:1–15. https://doi.org/10.47672/ajes.1721

Efremenko E, Stepanov N, Senko O, Aslanli A, Maslova O, Lyagin I (2024) Using fungi in artificial microbial consortia to solve bioremediation problems. Microorganisms 12:470. https://doi.org/10.3390/microorganisms12030470

Ene IV, Brunke S, Brown AJP, Hube B (2014) Metabolism in fungal pathogenesis. Cold Spring Harb Perspect Med 4:a019695–a019695. https://doi.org/10.1101/cshperspect.a019695

Fierro F, Vaca I, Castillo NI, García-Rico RO, Chávez R (2022) Penicillium chrysogenum, a vintage model with a cutting-edge profile in biotechnology. Microorganisms 10:573. https://doi.org/10.3390/microorganisms10030573
French V, Kleiman M, Bhalla K, Moreira R, Jalaee A, Kronstad JW, Foster EJ (2024) Fungal-derived adsorption membrane to capture potentially toxic elements. Chem Eng J 488:151028. https://doi.org/10.1016/j.cej.2024.151028
Gacura M (2009) Effect of Pleurotus ostreatus on bioremediation of PAH contaminated river sediment. https://doi.org/10.13140/RG.2.2.21295.94888
Gao Y, Shah K, Kwok I, Wang M, Rome LH, Mahendra S (2022) Immobilized fungal enzymes: innovations and potential applications in biodegradation and biosynthesis. Biotechnol Adv 57:107936. https://doi.org/10.1016/j.biotechadv.2022.107936
Geris R, Malta M, Soares LA, De Souza Neta LC, Pereira NS, Soares M, Reis VDS, Pereira MDG (2024) A review about the mycoremediation of soil impacted by war-like activities: challenges and gaps. J Fungi 10:94. https://doi.org/10.3390/jof10020094
Goligar N, Saadatmand S, Khavarinejad RA (2023) Mycoremediation of lead and cadmium by lignocellulosic enzymes of Pleurotus eryngii. AMB Express 13:127. https://doi.org/10.1186/s13568-023-01626-8
Gostinčar C, Zalar P, Gunde-Cimerman N (2022) No need for speed: slow development of fungi in extreme environments. Fungal Biol Rev 39:1–14. https://doi.org/10.1016/j.fbr.2021.11.002
Gupta N, Koley A, Banerjee S, Ghosh A, Hoque RR, Balachandran S (2024) Nanomaterial-mediated strategies for enhancing bioremediation of polycyclic aromatic hydrocarbons: a systematic review. Hybrid Adv 7:100315. https://doi.org/10.1016/j.hybadv.2024.100315
Higueras-Valdivia M, Silva-Castro GA, Paniagua-López M, Romero-Freire A, García-Romera I (2025) Mitigation of heavy metal soil contamination: a novel strategy with mycorrhizal fungi and biotransformed olive residue. J Hazard Mater Adv 17:100570. https://doi.org/10.1016/j.hazadv.2024.100570
Ibrahim AE, Salem HA, Abdelhalim A (2024) Environmental implications of three Pleurotus strain growths for water remediation in the perspective of climate change in New Egyptian Delta. Environ Sci Pollut Res 31:22588–22603. https://doi.org/10.1007/s11356-024-32412-z
Isha, Ali S, Khalid A, Naseer IA, Raza H, Chang Y-C (2024) Bioremediation of smog: current trends and future perspectives. Processes 12:2266. https://doi.org/10.3390/pr12102266
Janga JK, Reddy KR, Raviteja KVNS (2023) Integrating artificial intelligence, machine learning, and deep learning approaches into remediation of contaminated sites: a review. Chemosphere 345:140476. https://doi.org/10.1016/j.chemosphere.2023.140476
Javaid A, Bajwa R, Shafique U, Anwar J (2011) Removal of heavy metals by adsorption on Pleurotus ostreatus. Biomass Bioener 35:1675–1682. https://doi.org/10.1016/j.biombioe.2010.12.035
Kalita N, Baruah PP (2023) Cyanobacteria as a potent platform for heavy metals biosorption: uptake, responses and removal mechanisms. J Hazard Mater Adv 11:100349. https://doi.org/10.1016/j.hazadv.2023.100349
Kamalesh R, Karishma S, Saravanan A, Yaashikaa PR (2024) Emerging breakthroughs in membrane filtration techniques and their application in agricultural wastewater treatment: reusability aspects. Sustain Chem Environ 8:100183. https://doi.org/10.1016/j.scenv.2024.100183
Kayode-Afolayan SD, Ahuekwe EF, Nwinyi OC (2022) Impacts of pharmaceutical effluents on aquatic ecosystems. Sci Afr 17:e01288. https://doi.org/10.1016/j.sciaf.2022.e01288
Kijpornyongpan T, Schwartz A, Yaguchi A, Salvachúa D (2022) Systems biology-guided understanding of white-rot fungi for biotechnological applications: a review. iScience 25:104640. https://doi.org/10.1016/j.isci.2022.104640
Koul B, Sharma K, Shah MP (2022) Phycoremediation: a sustainable alternative in wastewater treatment (WWT) regime. Environ Technol Innov 25:102040. https://doi.org/10.1016/j.eti.2021.102040
Kuppan N, Padman M, Mahadeva M, Srinivasan S, Devarajan R (2024) A comprehensive review of sustainable bioremediation techniques: eco friendly solutions for waste and pollution management. Waste Manag Bull 2:154–171. https://doi.org/10.1016/j.wmb.2024.07.005

Lee H, Sam K, Coulon F, De Gisi S, Notarnicola M, Labianca C (2024) Recent developments and prospects of sustainable remediation treatments for major contaminants in soil: a review. Sci Total Environ 912:168769. https://doi.org/10.1016/j.scitotenv.2023.168769

Li Q, Zhang M, Wei B, Lan W, Wang Q, Chen C, Zhao H, Liu D, Gadd GM (2024) Fungal biomineralization of toxic metals accelerates organic pollutant removal. Curr Biol 34:2077–2084.e3. https://doi.org/10.1016/j.cub.2024.04.005

Ma Y, Ankit, Tiwari J, Bauddh K (2022) Plant-mycorrhizal fungi interactions in phytoremediation of geogenic contaminated soils. Front Microbiol 13:843415. https://doi.org/10.3389/fmicb.2022.843415

Mack SM, Madl AK, Pinkerton KE (2019) Respiratory health effects of exposure to ambient particulate matter and bioaerosols. In: Terjung R (ed) Comprehensive physiology. Wiley, pp 1–20. https://doi.org/10.1002/cphy.c180040

Madadi R, Bester K (2021) Fungi and biochar applications in bioremediation of organic micropollutants from aquatic media. Mar Pollut Bull 166:112247. https://doi.org/10.1016/j.marpolbul.2021.112247

Maglione G, Zinno P, Tropea A, Mussagy CU, Dufossé L, Giuffrida D, Mondello A (2024) Microbes' role in environmental pollution and remediation: a bioeconomy focus approach. AIMS Microbiol 10:723–755. https://doi.org/10.3934/microbiol.2024033

Maqsood Q, Sumrin A, Waseem R, Hussain M, Imtiaz M, Hussain N (2023) Bioengineered microbial strains for detoxification of toxic environmental pollutants. Environ Res 227:115665. https://doi.org/10.1016/j.envres.2023.115665

Marycz M, Brillowska-Dąbrowska A, Muñoz R, Gębicki J (2022) A state of the art review on the use of fungi in biofiltration to remove volatile hydrophobic pollutants. Rev Environ Sci Biotechnol 21:225–246. https://doi.org/10.1007/s11157-021-09608-7

Miyawaki A, Sakai S (2024) Immobilization of laccases on mechanically ground silk fibroin nanofibers for enhanced stability. Int J Biol Macromol 282:136745. https://doi.org/10.1016/j.ijbiomac.2024.136745

Moghimi Dehkordi M, Pournuroz Nodeh Z, Soleimani Dehkordi K, Salmanvandi H, Rasouli Khorjestan R, Ghaffarzadeh M (2024) Soil, air, and water pollution from mining and industrial activities: sources of pollution, environmental impacts, and prevention and control methods. Results Eng 23:102729. https://doi.org/10.1016/j.rineng.2024.102729

Negi BB, Das C (2023) Mycoremediation of wastewater, challenges, and current status: a review. Bioresour Technol Rep 22:101409. https://doi.org/10.1016/j.biteb.2023.101409

Nosek K, Zhao D (2024) Transformation products of diclofenac: formation, occurrence, and toxicity implication in the aquatic environment. Water Res 266:122388. https://doi.org/10.1016/j.watres.2024.122388

Oziegbe O, Oziegbe EJ, Ojo-Omoniyi O (2025) Bioremediation of heavy metals in aquatic environment: a review. Clean Chem Eng 11:100193. https://doi.org/10.1016/j.clce.2025.100193

Padhye LP, Srivastava P, Jasemizad T, Bolan S, Hou D, Shaheen SM, Rinklebe J, O'Connor D, Lamb D, Wang H, Siddique KHM, Bolan N (2023) Contaminant containment for sustainable remediation of persistent contaminants in soil and groundwater. J Hazard Mater 455:131575. https://doi.org/10.1016/j.jhazmat.2023.131575

Pandey VK, Rustagi S, Singh G (2025) The role of mycoremediation in the circular economy. In: Abdel Latef AAH, Zayed EM, Omar AA (eds) Sustainable remediation for pollution and climate resilience. Springer Nature Singapore, Singapore, pp 421–441. https://doi.org/10.1007/978-981-96-5674-5_16

Patel D, Johnson J, Madamwar D, Desai C (2023) Bioelectrochemical systems for the treatment of textile dye wastewaters. In: Current developments in bioengineering and biotechnology. Elsevier, pp 649–678. https://doi.org/10.1016/B978-0-323-91235-8.00017-6

Pazouki M, Keyanpour-Rad M, Shafie S, Shahhoseini S (2007) Efficiency of Penicillium chrysogenum PTCC 5037 in reducing low concentration of chromium hexavalent in a chromium electroplating plant wastewater. Bioresour Technol 98:2116–2122. https://doi.org/10.1016/j.biortech.2006.08.005

Pezzella C, Macellaro G, Sannia G, Raganati F, Olivieri G, Marzocchella A, Schlosser D, Piscitelli A (2017) Exploitation of Trametes versicolor for bioremediation of endocrine disrupting chemicals in bioreactors. PLoS One 12:e0178758. https://doi.org/10.1371/journal.pone.0178758

Pilotto F, Golovko O, Hultberg M (2025) Water quality and pesticide contamination in agricultural streams: assessing white-rot fungi and biochar treatments as remediation strategies. J Environ Manag 394:127282. https://doi.org/10.1016/j.jenvman.2025.127282

Policarpo Tonelli FM, Roy A, Ananda Murthy HC (eds) (2023) Green nanoremediation: sustainable management of environmental pollution. Springer, Cham. https://doi.org/10.1007/978-3-031-30558-0

Pozzer AC, Gómez PA, Weiss J (2022) Volatile organic compounds in aquatic ecosystems—detection, origin, significance and applications. Sci Total Environ 838:156155. https://doi.org/10.1016/j.scitotenv.2022.156155

Qattan SYA (2025) Harnessing bacterial consortia for effective bioremediation: targeted removal of heavy metals, hydrocarbons, and persistent pollutants. Environ Sci Eur 37:85. https://doi.org/10.1186/s12302-025-01103-y

Razzak SA, Faruque MO, Alsheikh Z, Alsheikhmohamad L, Alkuroud D, Alfayez A, Hossain SMZ, Hossain MM (2022) A comprehensive review on conventional and biological-driven heavy metals removal from industrial wastewater. Environ Adv 7:100168. https://doi.org/10.1016/j.envadv.2022.100168

Reddy GVB, Gold MH (2000) Degradation of pentachlorophenol by Phanerochaete chrysosporium: intermediates and reactions involved. Microbiology 146:405–413. https://doi.org/10.1099/00221287-146-2-405

Ren W, Guo Y, Han X, Sun Y, Li Q, Wu B, Xia T, Shen K, Wu P, He Y (2022) Indigenous microorganisms offset Arbuscular Mycorrhizal fungi-induced plant growth and nutrient acquisition through negatively modulating the genes of phosphorus transport and nitrogen assimilation. Front Plant Sci 13:880181. https://doi.org/10.3389/fpls.2022.880181

Rezaei Z, Moghimi H (2024) Fungal-bacterial consortia: a promising strategy for the removal of petroleum hydrocarbons. Ecotoxicol Environ Saf 280:116543. https://doi.org/10.1016/j.ecoenv.2024.116543

Rodríguez Couto S (2009) Dye removal by immobilised fungi. Biotechnol Adv 27:227–235. https://doi.org/10.1016/j.biotechadv.2008.12.001

Roslan NN, Lau HLH, Suhaimi NAA, Shahri NNM, Verinda SB, Nur M, Lim J-W, Usman A (2024) Recent advances in advanced oxidation processes for degrading pharmaceuticals in wastewater—a review. Catalysts 14:189. https://doi.org/10.3390/catal14030189

Roy A, Gogoi N, Haider FU, Farooq M (2025) Mycoremediation for sustainable remediation of environmental pollutants. Biocatal Agric Biotechnol 64:103526. https://doi.org/10.1016/j.bcab.2025.103526

Sar T, Marchlewicz A, Harirchi S, Mantzouridou FT, Hosoglu MI, Akbas MY, Hellwig C, Taherzadeh MJ (2024) Resource recovery and treatment of wastewaters using filamentous fungi. Sci Total Environ 951:175752. https://doi.org/10.1016/j.scitotenv.2024.175752

Sarvtin MT (2023) A review of the role of indoor fungi in sick building syndrome. Int J Med Lab. https://doi.org/10.18502/ijml.v10i3.13744

Selvi A, AlSalhi MS, Devanesan S, Maruthamuthu MK, Mani P, Rajasekar A (2021) Characterization of biospheric bacterial community on reduction and removal of chromium from tannery contaminated soil using an integrated approach of bio-enhanced electrokinetic remediation. J Environ Chem Eng 9:106602. https://doi.org/10.1016/j.jece.2021.106602

Shahna FG, Golbabaei F, Hamedi J, Mahjub H, Darabi HR, Shahtaheri SJ (2010) Treatment of benzene, toluene and xylene contaminated air in a bioactive foam emulsion reactor. Chin J Chem Eng 18:113–121. https://doi.org/10.1016/S1004-9541(08)60331-6

Singh M, Srivastava PK, Verma PC, Kharwar RN, Singh N, Tripathi RD (2015) Soil fungi for mycoremediation of arsenic pollution in agriculture soils. J Appl Microbiol 119:1278–1290. https://doi.org/10.1111/jam.12948

Singh AK, Iqbal HMN, Cardullo N, Muccilli V, Fernández-Lucas J, Schmidt JE, Jesionowski T, Bilal M (2023) Structural insights, biocatalytic characteristics, and application prospects of lignin-modifying enzymes for sustainable biotechnology. Int J Biol Macromol 242:124968. https://doi.org/10.1016/j.ijbiomac.2023.124968

Singh AK, Abellanas-Perez P, De Andrades D, Cornet I, Fernandez-Lafuente R, Bilal M (2025) Laccase-based biocatalytic systems application in sustainable degradation of pharmaceutically active contaminants. J Hazard Mater 485:136803. https://doi.org/10.1016/j.jhazmat.2024.136803

Sirtori C, Zapata A, Oller I, Gernjak W, Agüera A, Malato S (2009) Decontamination industrial pharmaceutical wastewater by combining solar photo-Fenton and biological treatment. Water Res 43:661–668. https://doi.org/10.1016/j.watres.2008.11.013

Šrédlová K, Šírová K, Stella T, Cajthaml T (2021) Degradation products of polychlorinated biphenyls and their in vitro transformation by Ligninolytic fungi. Toxics 9:81. https://doi.org/10.3390/toxics9040081

Sun K, Jiang H-J, Pan Y-T, Lu F, Zhu Q, Ma C-Y, Zhang A-Y, Zhou J-Y, Zhang W, Dai C-C (2023) Hyphosphere microorganisms facilitate hyphal spreading and root colonization of plant symbiotic fungus in ammonium-enriched soil. ISME J 17:1626–1638. https://doi.org/10.1038/s41396-023-01476-z

Suzuki K, Hirai H, Murata H, Nishida T (2003) Removal of estrogenic activities of 17β-estradiol and ethinylestradiol by ligninolytic enzymes from white rot fungi. Water Res 37:1972–1975. https://doi.org/10.1016/S0043-1354(02)00533-X

Tesfaye EL, Bogale FM, Aragaw TA (2025) Biodegradation of polycyclic aromatic hydrocarbons: the role of ligninolytic enzymes and advances of biosensors for in-situ monitoring. Emerging Contam 11:100424. https://doi.org/10.1016/j.emcon.2024.100424

Thirumalaivasan N, Gnanasekaran L, Kumar S, Durvasulu R, Sundaram T, Rajendran S, Nangan S, Kanagaraj K (2024) Utilization of fungal and bacterial bioremediation techniques for the treatment of toxic waste and biowaste. Front Mater 11:1416445. https://doi.org/10.3389/fmats.2024.1416445

Toledo M, Muñoz R (2025) Odour prevention strategies in wastewater treatment and composting plants: a review. J Environ Manag 375:124402. https://doi.org/10.1016/j.jenvman.2025.124402

Uslu G, Dursun AY, Ekiz Hİ, Aksu Z (2003) The effect of cd (II), Pb (II) and cu (II) ions on the growth and bioaccumulation properties of Rhizopus arrhizus. Process Biochem 39:105–110. https://doi.org/10.1016/S0032-9592(03)00032-3

Vaksmaa A, Guerrero-Cruz S, Ghosh P, Zeghal E, Hernando-Morales V, Niemann H (2023) Role of fungi in bioremediation of emerging pollutants. Front Mar Sci 10:1070905. https://doi.org/10.3389/fmars.2023.1070905

Vuković Domanovac M, Volf M, Šabić Runjavec M, Terzić I (2025) Sustainable and advanced strategies for bioremediation of highly contaminated wastewater. Processes 13:2250. https://doi.org/10.3390/pr13072250

Wang X, Gong Z, Li P, Zhang L, Hu X (2008) Degradation of Pyrene and benzo(a)pyrene in contaminated soil by immobilized fungi. Environ Eng Sci 25:677–684. https://doi.org/10.1089/ees.2007.0075

Wang X, Yao B, Su X (2018) Linking enzymatic oxidative degradation of lignin to organics detoxification. Int J Mol Sci 19:3373. https://doi.org/10.3390/ijms19113373

Wijayawardene N, Boonyuen N, Ranaweera C, De Zoysa H, Padmathilake R, Nifla F, Dai D-Q, Liu Y, Suwannarach N, Kumla J, Bamunuarachchige T, Chen H-H (2023) OMICS and other advanced technologies in mycological applications. J Fungi 9:688. https://doi.org/10.3390/jof9060688

Woertz JR, Kinney KA, McIntosh NDP, Szaniszlo PJ (2001) Removal of toluene in a vapor-phase bioreactor containing a strain of the dimorphic black yeast *Exophiala lecanii-corni*. Biotechnol Bioeng 75:550–558. https://doi.org/10.1002/bit.10066

Yu R, Liu J, Wang Y, Wang H, Zhang H (2021) Aspergillus niger as a secondary metabolite factory. Front Chem 9:701022. https://doi.org/10.3389/fchem.2021.701022

Zeng G, Ma Z, Zhang R, He Y, Fan X, Lei X, Xiao Y, Zhang M, Sun D (2024) The application of Nano zero-Valent iron in synergy with white rot fungi in environmental pollution control. Toxics 12:721. https://doi.org/10.3390/toxics12100721

Zhang W, Li Q, Wang J, Wang Z, Zhan H, Yu X, Zheng Y, Xiao T, Zhou L-W (2023) Biodegradation of benzo[a]pyrene by a white-rot fungus Phlebia acerina: surfactant-enhanced degradation and possible genes involved. J Fungi 9:978. https://doi.org/10.3390/jof9100978

Zhang R, Zhou L, Xie L, Lu L, Zhou H, Yang Y, Hu J (2025) Metabolite profiling and adaptation mechanisms of aspergillus cristatus under pH stress. Front Microbiol 16:1576132. https://doi.org/10.3389/fmicb.2025.1576132

Zune Q, Delepierre A, Gofflot S, Bauwens J, Twizere JC, Punt PJ, Francis F, Toye D, Bawin T, Delvigne F (2015) A fungal biofilm reactor based on metal structured packing improves the quality of a Gla::GFP fusion protein produced by aspergillus oryzae. Appl Microbiol Biotechnol 99:6241–6254. https://doi.org/10.1007/s00253-015-6608-z

Regulatory and Policy Considerations 10

Fernanda Maria Policarpo Tonelli, Flávia Cristina Policarpo Tonelli, Joice Cristina de Paula, Iasmin Fernandes Emediato, Luiz Felipe Elver Santana Soares, Vinicius Marx Silva Delgado, and Naony Sousa Costa Martins

Abstract

Environmental pollution is a concerning topic, since contaminants of diverse chemical nature can be present in air, water, and soil, threatening the balance of ecosystems and the survival of living beings. These pollutants, for being present in places in which they were not expected to be found (in living organisms or outside them), are considered xenobiotics and need to be removed in the most efficient and fast manner possible. However, pollution remediation strategies also need to be sustainable, avoiding worsening the contamination scenario that already exists. Among bio-based, eco-friendly remediation protocols are the ones that aim to promote mycoremediation: using fungi to restore contaminated areas. This promising approach, however, also presents limitations, especially when it comes to human and environmental health. This chapter is dedicated to exploring these limitations and regulatory and policy aspects associated with them.

Keywords

Environmental law · Human health · Xenobiotics · Mycoremediation · Safety

F. M. P. Tonelli (✉) · F. C. P. Tonelli · I. F. Emediato · L. F. E. S. Soares · V. M. S. Delgado
Federal University of São João del Rei, Centro-Oeste Campus, Divinópolis, MG, Brazil
e-mail: ftonelli@ufsj.edu.br

J. C. de Paula
Anhanguera College, Divinópolis, MG, Brazil

N. S. C. Martins
Federal University of Juiz de Fora, Law Department, Juiz de Fora, MG, Brazil

A. Roy et al. (eds.), *Mycoremediation of Xenobiotics*, Emerging Paradigms in Pharmaceutical Research, https://doi.org/10.1007/978-981-95-5743-1_10

10.1 Environmental Pollution and the Xenobiotics

Xenobiotics is a term used to refer to synthetic substances that are foreign to ecosystems or living organisms (e.g., dyes synthesized by industry) or to make reference to naturally occurring compounds that are present in some areas at concentrations that exceed their natural background levels (e.g., mycotoxins produced by fungi species as part of their defense mechanisms) (Soucek 2011).

These chemicals exhibit some characteristics that make them threatening to living beings, such as persistence, capacity to spread easily and to bioaccumulate, and toxicity. Xenobiotics can exhibit volatility, lipophilicity, and low biodegradability, which favors their accumulation in organisms' fat tissues and magnification along the food chain. Consequently, they are challenging to be removed from where they are present (Štefanac et al. 2021).

So, these compounds are considered concerning environmental pollutants; they can be micropollutants (present at small concentrations such as ng/L) or present at higher concentrations (Singh et al. 2023). They may enter ecosystems through diverse channels, such as agricultural runoff, industrial and household effluents, hospital discharges, atmospheric deposition, and solid waste leaching (Fig. 10.1). Xenobiotics can infiltrate water, soil, air, and biological systems, leading to undesirable effects (Štefanac et al. 2021; Miglani et al. 2022).

Pharmaceuticals, for example, can enter aquatic systems through wastewater or improper disposal. As a consequence of their presence, they can end up disrupting aquatic life, inducing hormonal imbalances, and causing genotoxicity. For example, if the xenobiotic is an antibiotic, it can also contribute to the selection of resistant microorganisms. Pesticides used in agriculture can also do damage when they reach soil, disrupting plant physiology, increasing cancer risks in exposed animals, and inhibiting their normal hormonal functions (Li et al. 2024). Microbial communities can also be altered in composition and function by xenobiotics; gene expression can suffer influences posing increased risks to human and ecosystem health (Zhang et al. 2022).

An example of threatening xenobiotics are polyfluoroalkyl substances; they are considered persistent organic pollutants that can bioaccumulate or be spread to different places, posing a risk to environmental and living beings' health. In Europe only, the estimation of annual healthcare costs associated with exposure to polyfluoroalkyl substances is elevated; this amount is estimated to be between €52 billion and €84 billion (Cordner et al. 2021).

To worsen the scenario of toxicity, many environmental contaminants are molecules resistant to environmental degradation and exhibiting the capacity to easily accumulate throughout ecosystems. Stable contaminants presenting mobility make conventional remediation approaches, such as incineration, most of the time insufficient (Brunn et al. 2023). So, special and more expansive procedures need to be used. It is estimated that the remediation of only one group of the polyfluoroalkyl substances, the perfluoroalkyl acids, would cost annually 20 to 7000 trillion dollars (Ling 2024). Expensive site testing, containment, and transport may be necessary to deal with these pollutants; conventional procedures (such as incineration) can

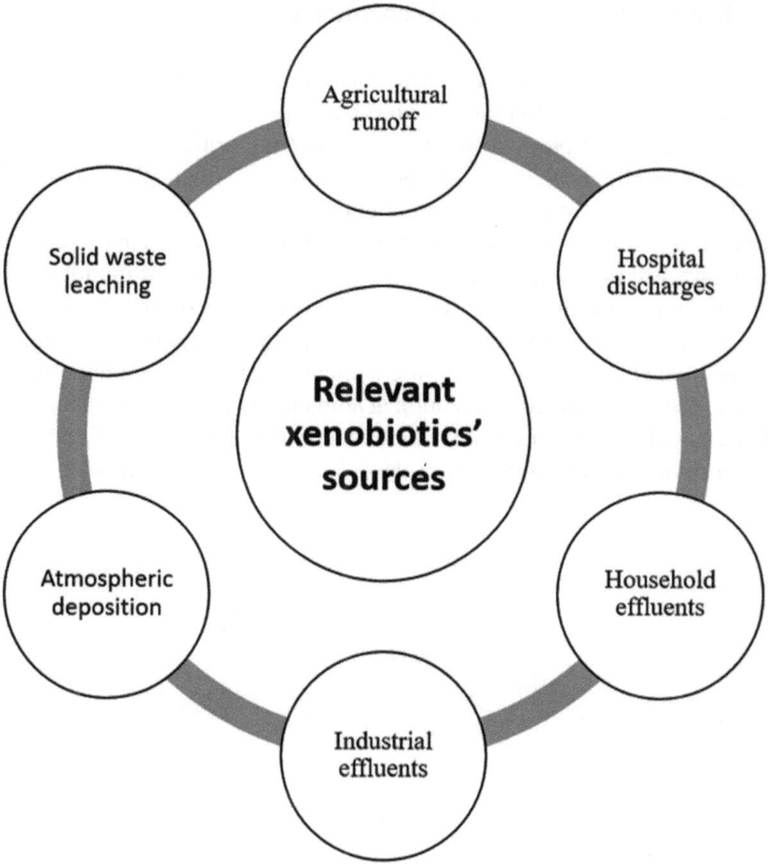

Fig. 10.1 Important sources of xenobiotics resulting in environmental pollution

also demand special parameters (such as extremely high temperatures) and, even in this manner, result in an inadequate method to the management of the contaminant (Cordner et al. 2021).

So, innovative and more sustainable methods to remediate environmental pollution associated with xenobiotics are necessary. Among these methods, the ones using fungi can be highlighted. These organisms can promote the mycoremediation of contaminants. Fungi can be used in strategies, such as biostimulation (which involves altering survival conditions to stimulate specific enzymes) and/or bioaugmentation (which involves introducing specialized species to native populations aiming to degrade specific pollutants) (Magan et al. 2010), for example.

Advanced omics tools (such as metagenomics, proteomics, and metabolomics) are very important to assist in the search for genes and enzymes responsible for xenobiotics' degradation. This type of information is crucial to allow the proposal of new eco-friendly management strategies and protocols based on metabolic pathways to efficiently, cost-effectively, and safely deal with these concerning environmental contaminants (Miglani et al. 2022).

10.2 Main Advantages of Mycoremediation

Mycoremediation presents a large array of advantages regarding other types of remediation strategies, including other bio-based protocols (Fig. 10.2). It involves using fungi's unique biology to offer broad-spectrum pollutant removal with low energy and low chemical inputs. Hyphal architecture (that allows organisms to physically access low-bioavailability pollutants in soils and porous matrices) and their oxidative enzymology favor the process of dealing with environmental contaminants in a sustainable manner by removing, transforming, or immobilizing pollutants. This type of remediation is versatile, bio-based and presents the potential for restoring contaminated water and soil, for example (Ningombam et al. 2025).

The mentioned low thermal and low chemical energy needs are interesting characteristics. They allow mycoremediation at ambient temperature and at near-neutral pH (Akpasi et al. 2023). Fungal enzymes, such as the laccase from *Trametes versicolor*, can retain good stability and a desirable activity at moderate temperatures

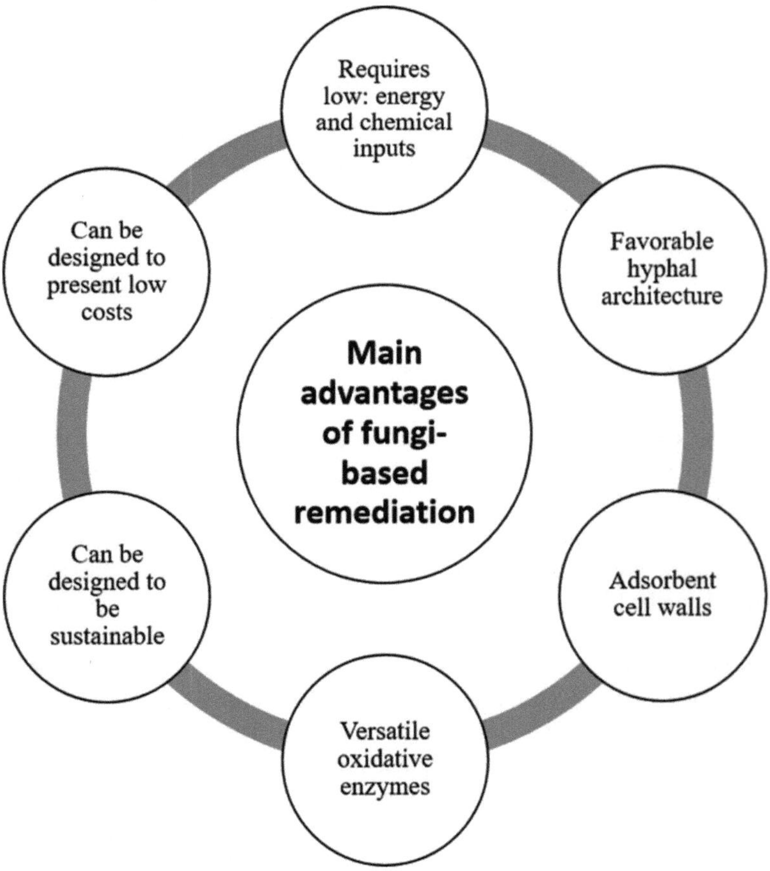

Fig. 10.2 Main advantages of mycoremediation

from approximately 25 °C to 30 °C and pH from 5.0 to 7.0. It makes the protein interesting to be used in *in situ* remediation applications without requiring external heating or pH adjustments, for example (Jin et al. 2016). Conventional treatments like incineration and advanced chemical oxidation commonly require higher temperatures and adjustments in the pH. Using fungi, however, it is possible, due to the mentioned aspects, to reduce costs, enhance efficiency and minimize negative impacts on the ecosystem. The favorable techno-economical scenario also favors process scale-up and control (Akhtar and Mannan 2020).

Among fungi species, the so-called ligninolytic "white-rot" ones (such as *Pleurotus* spp., *Phanerochaete chrysosporium*, and *Trametes versicolor*) stand out as they secrete oxidative enzymes, such as laccases, peroxidases (like manganese and lignin ones), and peroxygenases that are essential to mycoremediation (Torres-Farradá et al. 2024). Some of these enzymes are extracellular catalysts that present low substrate specificity. This feature allows them to work over a wide range of aromatic and halogenated compounds (Khan et al. 2023), like dyes, pesticides, polycyclic aromatic hydrocarbons (PAHs), pharmaceuticals, personal care products, and other emerging contaminants. This is an advantageous feature and makes fungi a good option when it comes to remediating environmental pollution.

Laccases are one of the aforementioned extracellular enzymes. They are oxidoreductases belonging to the family of multicopper oxidases (Shourie and Vijayalakshmi 2022). The protein has a conserved structure containing four copper atoms divided into three categories: T1, T2, and T3. T1 is coordinated by two histidines and a cysteine; it is responsible for oxidizing the substrate and it also confers the protein's characteristic blue color. At this site, cysteine is primarily responsible for structural stability, while histidine residues confer oxidation potency. Mutations on the histidine change its potency and, hence, may change the enzyme activity and the substrate range. For example, phenylalanine being present in the axial coordination position leads to a strong oxidation potency, which decreases when it is replaced by a leucine or methionine residue. However, those changes may also alter the thermodynamic stability and decrease the enzyme's catalytic rate. T2 and T3 form a trinuclear cluster, where oxygen is reduced to water (Strong and Claus 2011).

The reaction starts with the subtract being oxidized on the enzyme's active site by a copper ion. The reduced cation then donates the electron to an oxygen atom, reducing it to water in the trinuclear cluster, while the oxidized substrate dissociates and may further react non-enzymatically. The protein itself is already very promiscuous by nature, but the range of its substrates can be increased with mediators. Mediators are strong electron donors that can enter the enzyme's active site, be oxidized, and then diffuse away. Once it has donated an electron to the T1 copper, it leaves the active pocket as a strong oxidizing reagent that interacts with other molecules that are too big to fit into the protein's site (Chaurasia et al. 2013).

Another protein with mycoremediation properties that deserves to be highlighted is the manganese peroxidase. It has a heme structure as a prosthetic group and Mg^{2+} as ligand ions. The reaction starts with the hydrogen peroxide interacting with the heme group, forming an unstable compound named Mn porphyrin, which oxidizes the two Mg^{2+} to return to its natural heme state. Those two Mg^{2+} are released as

Mg^{3+}, which reacts nonspecifically with organic acids present in, for example, phenol pollutants and textile dyes (Saikia et al. 2023).

Fungi species such as *Phlebia acerina* (Zhang et al. 2023), *Septoriella* sp., and *Trichoderma* sp. (Crittenden et al. 2025) offered good efficiency, for example, in promoting the effective transformation of recalcitrant organics, such as PAHs. High removal efficiencies were observed, motivating the scientific community to dedicate more attention to this type of remediation protocol.

Fungal cell walls (rich in chitin, glucans, and multiple functional groups) can also facilitate the removal of some contaminants, such as heavy metals. These pollutants can also be bioaccumulated into fungal biomass, which consists in an economical and regenerable option for restoring metal-laden wastewater and mining-impacted sites, with the possibility of metal recovery from loaded biomass (Yang et al. 2024; Oziegbe et al. 2025).

It is important to mention that are also considered advantageous aspects associated with mycoremediation, its cost-efficiency, and the fact of being an environmentally friendly methodology applicable even to deal efficiently with emerging contaminants (Akpasi et al. 2023; Sośnicka et al. 2022). However relevant limitations associated with this type of strategy still need to receive further attention before large-scale field application.

10.3 Main Limitations of Mycoremediation

Mycoremediation has already proved to be very useful in the management of a large array of xenobiotics when performed at laboratory scale. However, it is still not commonly applied on a large scale in the field. That is due to the fact that there are important limitations associated with this type of remediation strategy that still need to receive attention before this large-scale use becomes feasible (Fig. 10.3).

The first one that can be highlighted is fungi's sensitivity to environmental conditions such as oxygen availability, pH, temperature, and moisture (Antón-Herrero et al. 2023). Although fungi can operate at mild temperature and pH, deviations from optimal ranges commonly result in reduced growth and also negatively impact enzyme turnover (non-immobilized catalysts, free enzymes, are more sensitive to these alterations). In environmental waters and industrial effluents, hardness and salinity (especially when it comes to the presence of anions such as chloride, sulfate, and phosphate) can diminish some enzymes' activity, such as laccase's activity (Mayr et al. 2024; Salem et al. 2024; Singh et al. 2025). Among the strategies that may allow the surpassing of these limitations are as follows: immobilization of fungal enzymes in carriers, controlling variables that interfere with mycoremediation's efficiency, engineering fungi or using organisms isolated from extreme environments, and adding stabilizers and/or mediators to the mycoremediation. Regarding immobilization, alginate and biochar, for example, can offer protection to fungi and their enzymes from fluctuations in pH and temperature, also favoring stability and reusability (Cheute et al. 2025). Regarding controlling the crucial variables, maintaining optimal pH, temperature, and moisture content, for example,

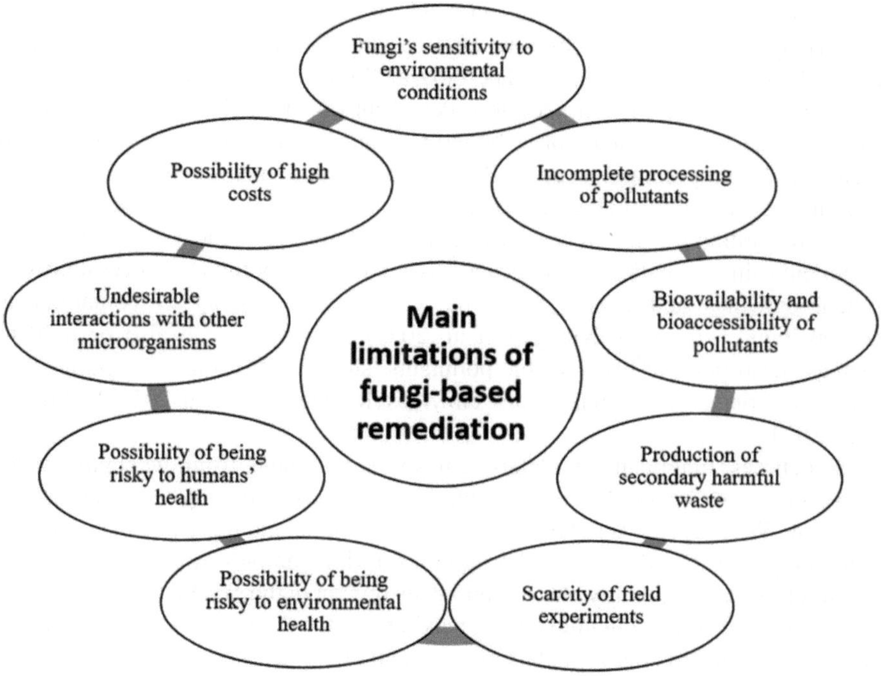

Fig. 10.3 Main limitations of mycoremediation

controlled bioreactors are an interesting option that can favor mycoremediation (Ningombam et al. 2025). Engineering fungi and/or isolating them from extreme environments allows an increase in robustness under fluctuating environmental conditions, once extremophilic or engineered fungi can maintain enzymatic activity, where mesophilic fungi may not (Mate and Alcalde 2015; Sahay 2022; Pham et al. 2024). Regarding mediators and stabilizers, syringaldehyde (a natural mediator) (Viana et al. 2025) and polyethylene glycol (a stabilizer) (Bou-Mitri and Kermasha 2018), for example, may favor mycoremediation by enhancing fungal enzymes' activity and stability.

The nutrient status also interferes with mycoremediation. White-rot fungi, for example, produce ligninolytic enzymes in larger amounts under nutrient-limited, lignocellulosic contexts. These conditions are commonly not guaranteed in polluted sites (Torres-Farradá et al. 2024). It is also possible that, in the field, competition between bacteria and other fungus species interferes with the pollutants' management. However, it is not always an undesirable scenario; mixed consortia may outperform single isolates when well-designed to do so (Thirumalaivasan et al. 2024).

It is also possible that fungi-based remediation results in the incomplete processing of pollutants. The rate of pollutant removal when compared with that of physical–chemical processes (such as advanced oxidation) can be lower, as it depends on enzyme production, the presence or absence of extracellular redox mediators, and the capacity of mass transfer to hyphae. Complete remediation is not guaranteed since transformation of pollutants can result in partially processed intermediates, which may retain, or in some cases increase, toxicity (Chen et al. 2024; Dinakarkumar et al. 2024; Navina et al. 2024). The bioavailability of contaminants in complex matrices may be a critical aspect, too. Persistent organic pollutants, for example, can be sequestered in micropores, bound to organic matter, or embedded in aged matrices, limiting fungal access. Bioavailability and bioaccessibility are rate-limiting aspects influencing mycoremediation's success (Antón-Herrero et al. 2023). Regarding the partial degradation, associating mycoremediation with a pre-treatment using ozonation or photocatalysis can increase pollutants' accessibility to fungal enzymes and contribute to the complete processing of contaminants (Kanakaraju et al. 2025; Xiangyu et al. 2025). Surfactants can also be used to increase desorption of hydrophobic pollutants (such as PAHs) from sorption sites, enhancing contact with hyphae and/or enzymes (Rathankumar et al. 2022; Mahmoud et al. 2024).

Sometimes, fungi can completely remove the pollutants from the environment. However, depending on the contaminant, it can end up producing secondary harmful waste. When it comes to heavy metals, for example, the inorganic materials are commonly remediated by fungi through biosorption/complexation to cell walls or precipitation. Consequently, the contaminants remain in the system (Xie 2024). It is also possible that the contaminant dose and/or chemical nature result in toxicity to fungi. Still analyzing heavy metals as pollutants, for example, even at small doses, they can inhibit fungal metabolism and consequently their growth and capacity to perform mycoremediation (Gao et al. 2024).

It is important to highlight that, prior to large-scale application, the mycoremediation protocols need to receive deep attention regarding a wide range of details. Until nowadays, field demonstrations of success are fewer than lab studies (AbuQamar et al. 2023; Wang et al. 2024). Challenges in scaling the process up from lab to field applications involve difficulties associated with guaranteeing stability of fungi enzymes, controlling/monitoring fungi themselves and their possible interactions with other microorganisms, trying to avoid high costs, and mitigating any risks to the environment and living beings' health. The scarcity of information on those aspects in field application and regarding the long-term monitoring needs remains an important barrier (Antón-Herrero et al. 2023).

From the mentioned aspects, when it comes to a large-scale application of mycoremediation, guaranteeing human and environmental health will be highlighted in the present chapter. The following topics are dedicated to exploring relevant regulatory and policy considerations on these topics.

10.4 Regulatory and Policy Considerations Regarding Human Health

Deploying living organisms or any type of biological product into the environment raises regulatory, biosafety, and especially human-health issues. Policymakers need to dedicate attention to them to allow mycoremediation to be scaled responsibly.

Environmental health policies are essential not only for environmental preservation but also for guaranteeing human health. In Brazil, for example, the right to health is addressed in the 1988 Brazilian Constitution and also in the 1990 Law N°. 8080 (which created the Unified Health System (SUS)). This mentioned law, in fact, not only established SUS in the country but also addressed the conditions for the promotion, protection, and organization of the health system in Brazil (which involves a balanced environment and the development of effective health policies (Brazil 1988, 1990)).

Regulation through public policies is essential for the health system to be effective for society. It involves, in addition to technical standardization, the state's capacity for oversight and adaptation. The health infrastructure guarantees the safety, effectiveness, and quality of products and services offered to the population, in addition to regulating the actions of public and private agents (WHO 2022).

So, it is not allowed to apply in the field a large-scale new remediation protocol using fungi without guaranteeing first the aspects associated with not negatively impacting, for example, human health. However, the regulation of this type of remediation is challenging. This is mainly due to the fact that the activity itself intersects many policy areas, including the following: environmental cleanup, product authorization, biosafety and public health provisions, and occupational safety.

Although mycoremediation is not the specific topic addressed by national and international instruments already existent, a large array of them provide explicit guidelines or requirements for deliberate environmental release of microorganisms or microbial products. As mycoremediation can result in introducing, in some areas, biological agents that could affect human health directly or indirectly (through food and water, for example) and the ecosystem in general, these guidelines or requirements should apply.

The Organization for Economic Cooperation and Development (OECD), for example, has long provided harmonized guidance for environmental safety evaluation of microbial agents used as biological control products. This guidance is frequently used as a template for regulators considering environmental releases of organisms and can also be useful when it comes to using fungi for the management of environmental pollution. Such guidance emphasizes aspects such as hazard identification, exposure pathways, and nontarget impacts (OECD 2012). The Environmental Protection Agency (EPA) of the United States provides guidance on microbial risk assessment, ecological risk assessment, and occupational protection; aspects like pathogenicity, antimicrobial resistance markers, persistence, and production of toxic metabolites, for example, are highlighted as topics that need to receive attention (EPA 2012).

Consequently, a responsible regulatory approach must dedicate attention to aspects associated with human health that can be affected by mycoremediation; for example, occupational exposure to fungi, spores, mycotoxins, and the risk of selecting antimicrobial resistant strains. When it comes to occupational exposure to fungal biomass and spores, it can be a problem for workers handling fungal cultures (Kontro et al. 2022). In this way, regulators need to apply occupational hygiene principles and require training, exposure controls, and right-to-know documentation. It is also possible that the residues from mycoremediation, if incorrectly disposed, generate aerosols carrying spores, cells, or particulate fungal-bound contaminants. In this scenario, not only site workers but also nearby residents, for example, can be affected. Therefore, exposure assessments must include realistic release scenarios and particle fate modeling (Liu et al. 2021; Farzin et al. 2023). It is important to highlight that some fungi can produce mycotoxins as secondary metabolites; these substances pose inhalation and/or ingestion risks to humans (Omotayo et al. 2019; Kępińska-Pacelik and Biel 2021). So, to avoid this type of negative impact on human health, regulatory dossiers must also require data on metabolite identity, toxicology, and persistence.

Regarding occupational exposure, it is also possible that workers suffer exposure to more toxic intermediates produced during mycoremediation. Biotransformation does not guarantee detoxification; intermediate metabolites formed during fungal degradation can be more toxic, more mobile, and/or also more bioaccumulative than initial compounds (Nascimento et al. 2025). Consequently, regulators need to be aware of this to require data on transformation products and their toxicological profiles.

When it comes to the selection of antimicrobial resistance genes, fungal strains carrying these genes may grow better than other strains not containing them and can also promote horizontal gene transfer (Morogovsky et al. 2022; Zhang et al. 2023; Habig et al. 2025). Regulatory assessments must not ignore these possibilities; although resistance to antifungal agents is not currently a more serious threat than drug-resistant bacteria, the diseases caused by fungi are estimated to affect 6.5 million people every year. Pathogens resistant to currently available antifungal substances are commonly involved in the approximately 3.8 million deaths annually caused by fungi-provoked diseases (van Rhijn et al. 2024).

So, as health is a fundamental human right, it must be a priority in regulations and social strategies. The development of public health policies is not limited to the formulation of legislation and government programs; it involves social participation in decision-making processes, resource allocation, and ensuring equity and accessibility. Globalization involves the intense circulation of goods and services, which poses regulatory challenges that require national and international cooperation. Consequently, there is a complexity associated with the establishment of health policies (Oliveira and Pepe 2024); however, regulating practices such as remediation is crucial to protect the population from potential harm to their health.

10.5 Regulatory and Policy Considerations Regarding the Environment

Environmental regulation is a crucial topic, as it aims to ensure sustainability, guaranteeing a balanced ecosystem. Especially in recent years, national, local, and international legislation has been elaborated to address problems related to pollution, climate change, deforestation, and biodiversity degradation. However, efforts in this sense are now new. In Brazil, for example, the 1988 Federal Constitution mentions the environment as a fundamental right, establishing the responsibility of public authorities and the community for its protection (Brazil 1988). The Law N° 6938 of 1981 established the National Environmental Policy in the country, introducing mechanisms for environmental licensing and impact assessment, and also presented aims to preserve, improve, and restore environmental quality (Brazil 1981).

Environmental management must focus not only on creating standards but also on implementing effective public policies that involve various social spheres. The goal, and at the same time, the challenge, is to create a development model that balances the productive and consumer sectors, consolidating a sustainable development chain. In this context, international collaboration and the acceptance of international commitments within countries are important to mitigate the negative effects anthropic actions have caused on the environment and to avoid future ones (da Silva 2019).

However, even to deal with these negative effects it is necessary to be aware of the risks the processes may cause. As previously mentioned, aiming for sustainable development, environmental pollution remediation measures are necessary in order to restore polluted areas. Various living organisms have been researched for application to offer this desirable remediation; however, mycoremediation has gained prominence due to advantageous features such as fungi's low-cost, versatility, and possibility to offer eco-friendly protocols for detoxifying a large number of pollutants (such as dyes, pharmaceutical ingredients, heavy metals, and pesticides) (Ghosh et al. 2023).

However, the lack of specific regulatory guidelines has been one of the main obstacles to the implementation of mycoremediation. These are ambiguous or even nonexistent provisions in environmental legislation, resulting in unpredictable compliance costs, lengthy permitting periods, and uncertain liability regimes. The lack of adequate regulation discourages researchers and investors from investing time and money in applying this type of bioremediation. Furthermore, when there is any aspect of regulation more specifically on mycoremediation, variability is noted between federal, state, and municipal levels, which creates confusion and complexity (Domanovac et al. 2025).

Another factor hindering the regulation of remediation techniques using fungi is the impossibility of establishing precise standards due to the variability of fungal mycelia and their responses to specific environments. Therefore, predicting remediation rates under heterogeneous field conditions is complex. Furthermore, international and regional regulations currently require microorganisms, especially genetically modified ones, to undergo specific risk assessment before being released into the field. Therefore, in many countries, the release of genetically modified fungi is strictly restricted or even prohibited, limiting advances in mycoremediation using engineered organisms (Ningombam et al. 2025).

One of the key allies of mycoremediation regulation is external policies for sustainability and circular economic models. An example of regulatory and policy support for mycoremediation is the European Union's LIFE MySOIL initiative. This project aims to treat oil-contaminated soils through fungal inoculation, achieving a reduction of over 91% in total petroleum hydrocarbons (TPHs), significantly reducing greenhouse gas emissions, energy use, costs, and toxicity compared to thermal remediation methods. Agricultural waste is also used as substrate to fungi, thus applying the principles of the circular economy (European Dissemination Agency 2023).

Another example worth noting was an initiative from Canada. Canada's Environmental Protection Act of 1999 regulates all new substances introduced into the environment, including microorganisms. The New Substances Program excludes risks to natural and genetically modified organisms prior to their introduction into the environment. This process involves transparent and structured public consultations and screening assessments for the approval of microbial bioremediation agents. Establishing and conducting this process favors the targeted implementation of microremediation (Canada 2025).

Therefore, regulatory and policy aspects associated with mycoremediation are still challenging. However, when it comes to environmental health, the principles of environmental law need to be observed while elaborating legislation. These principles include, for example, the Sustainable Development (United Nations 1987) and the Precautionary Principles (United Nations 1992). The first one received attention from the United Nations (UN) in "Our Common Future Report" from 1987, and the second one, in "Rio Declaration on Environment and Development." It is important to consider them when regulating mycoremediation to minimize risks: protecting human and environmental health, preventing future damage, distributing responsibilities fairly, and promoting sustainable development by aligning science and public policy (Rodrigues 2025).

10.6 Integration of Mycoremediation Into Sustainability Public Policies: Future Perspectives

Mycoremediation is a promising alternative in the management of xenobiotics and exhibits a strong potential for integration into public policies aimed at sustainability. It can favor, for example, a circular economy: agricultural and industrial residues can be employed as substrates for fungal growth; consequently, it would simultaneously promote environmental remediation and the reuse of residues that would be converted into by-products (European Dissemination Agency 2023; Ghosh et al. 2023).

This type of bioremediation is also strongly aligned with the Sustainable Development Goals (SDGs) proposed by the UN. Fungi-based remediation can directly contribute, for example, to SDG 3 (Good Health and Well-Being), SDG 6 (Clean Water and Sanitation), SDG 12 (Responsible Consumption and Production), and SDG 15 (Life on Land): highlighting its capacity to act as an interdisciplinary tool in both public health and environmental preservation (United Nations 2025).

However, to become a technology widely applied, it is expected that incentive and funding policies are elaborated as they are essential to overcome challenging aspects that limit large scale application. The European Union, for example, has invested in projects such as LIFE MySOIL: a fungi-based soil decontamination from petroleum derivatives, achieving more than 90% reduction in TPHs (European Dissemination Agency 2023). It can serve as references for developing countries, such as Brazil, to contribute to an evolution into large-scale field application of mycoremediation.

It is also expected that regulatory barriers will receive attention, as in a large array of countries, legal gaps regarding the application of this kind of remediation still exist. Standardized risk assessment protocols need to be developed to consider not only degradation efficiency but also occupational safety and social acceptance, for example. Community involvement and clear communication of the benefits and risks associated with fungi-based remediation are essential to ensure the legitimacy of public remediation policies (OECD 2012; EPA 2012).

Finally, biotechnological tools can be used as catalysts for innovation. Genetic engineering techniques, such as CRISPR gene editing, the development of microbial consortia, and the application of omics tools (genomics, proteomics, and metabolomics), can enhance fungal robustness in extreme environments and improve their efficiency in treating complex contaminants (Miglani et al. 2022; Pham et al. 2024). However, such advances must be linked to regulatory frameworks that ensure biosafety and socio-environmental protection.

Thus, the integration of mycoremediation into public policies depends not only on scientific and technological advances but also on clear regulatory frameworks, governmental incentives, and social engagement. By aligning with principles of sustainability and environmental governance, this biotechnology can be consolidated as a strategic tool to tackle pollution and promote both environmental and human health.

Authors' Contribution Conceptualization: Fernanda Maria Policarpo Tonelli, and Flávia Cristina Policarpo Tonelli; writing—original draft preparation: Fernanda Maria Policarpo Tonelli, Flávia Cristina Policarpo Tonelli, Joice Cristina de Paula, Iasmin Fernandes Emediato, Luiz Felipe Elver Santana Soares, Vinicius Marx Silva Delgado, and Naony Sousa Costa Martins; writing—review and editing: Fernanda Maria Policarpo Tonelli, Flávia Cristina Policarpo Tonelli, and Naony Sousa Costa Martins; supervision: Fernanda Maria Policarpo Tonelli. All authors have read and agreed to the submitted version of the chapter.

References

AbuQamar SF, El-Fattah HIA, Nader MM, Zaghloul RA, El-Mageed TAA, Selim S, Omar BA, Mosa WF, Saad AM, El-Tarabily KA, El-Saadony MT (2023) Marine Environmental Research Exploiting fungi in bioremediation for cleaning-up emerging pollutants in aquatic ecosystems. Mar Environ Res 190:106068.

Akhtar N, Mannan MA (2020) Mycoremediation: expunging environmental pollutants. Biotechnol Rep (Amst) 26:e00452. https://doi.org/10.1016/j.btre.2020.e00452

Akpasi SO, Anekwe IMS, Tetteh EK, Amune UO, Shoyiga HO, Mahlangu TP, Kiambi SL (2023) Mycoremediation as a potentially promising technology: current status and prospects—a review. Appl Sci 13(8):4978. https://doi.org/10.3390/app13084978

Antón-Herrero R, Chicca I, García-Delgado C, Crognale S, Lelli D, Gargarello RM, Herrero J, Fischer A, Thannberger L, Eymar E, Petruccioli M, D'Annibale A (2023) Main factors deter-

mining the scale-up effectiveness of mycoremediation for the decontamination of aliphatic hydrocarbons in soil. J Fungi 9(12):1205. https://doi.org/10.3390/jof9121205
Bou-Mitri C, Kermasha S (2018) Lyoprotection and stabilization of laccase extract from Coriolus hirsutus, using selected additives. AMB Expr 8:152. https://doi.org/10.1186/s13568-018-0683-3
Brazil (1981) Law N° 6,938. Available at: https://www.planalto.gov.br/ccivil_03/leis/l6938.htm
Brazil (1988) Constitution of the Federative Republic of Brazil. Available at: https://www.planalto.gov.br/ccivil_03/constituicao/constituicao.htm
Brazil (1990) Law N° 8,080. Available at: https://www.planalto.gov.br/ccivil_03/leis/l8080.htm
Brunn H, Arnold G, Körner W, Rippen G, Steinhäuser KG, Valentin I (2023) PFAS: forever chemicals—persistent, bioaccumulative and mobile. Reviewing the status and the need for their phase out and remediation of contaminated sites. Environ Sci Eur 35:20. https://doi.org/10.1186/s12302-023-00721-8
Canada (2025) Microbial risk assessment framework under the New Substances Notification Regulations (Organisms) of the Canadian Environmental Protection Act, 1999. Available at: https://www.canada.ca/en/environment-climate-change/services/managing-pollution/evaluating-new-substances/biotechnology-living-organisms/microbial-risk-assessment-framework.html
Chaurasia PK, Yadav RSS, Yadava S (2013) A review on mechanism of laccase action. Res Rev BioSci 7(2):66–71
Chen Z, Yao J, Knudsen TS, Li H, Liu B, Ma B, Pang W (2024) Application of a laccase-catalyzed oxidative coupling for the removal and detoxification of typical flotation reagent salicylhydroxamic acid. Chem Eng J 500:157012. https://doi.org/10.1016/j.cej.2024.157012
Cheute VMS, Backes E, Pateis VDO, de Oliveira Junior VA, Uber TM, dos Santos Filho JR, dos Santos LFO, Castoldi R, de Souza CGM, Polonio JC, Contato AG, Bracht A, Peralta RM (2025) Optimization of immobilization, characterization, and environmental applications of laccases from *Pycnoporus sanguineus* UEM-20. Processes 13(6):1800. https://doi.org/10.3390/pr13061800
Cordner A, Goldenman G, Birnbaum LS, Brown P, Miller MF, Mueller R, Patton S, Salvatore DH, Trasande L (2021) The true costs of PFAS and the benefits of acting now. Environ Sci Technol 55(14):9630–9633. https://doi.org/10.1021/acs.est.1c03565
Crittenden J, Raudabaugh D, Gunsch CK (2025) Isolation, characterization, and mycostimulation of fungi for the degradation of polycyclic aromatic hydrocarbons at a superfund site. Biodegradation 36:1–15. https://doi.org/10.1007/s10532-024-10106-0
Dinakarkumar Y, Ramakrishnan G, Gujjula KR, Vasu V, Balamurugan P, Murali G (2024) Fungal bioremediation: an overview of the mechanisms, applications and future perspectives. Environ Chem Ecotoxicol 6:293–302. https://doi.org/10.1016/j.enceco.2024.07.002
Domanovac MV, Volf M, Runjavec MS, Terzić I (2025) Sustainable and advanced strategies for bioremediation of highly contaminated wastewater. Processes 13(7):2250. https://doi.org/10.3390/pr13072250
EPA - United States Environmental Protection Agency (2012) Microbial risk assessment guideline: Pathogenic microorganisms with focus on food and water. Available at: https://www.epa.gov/risk/microbial-risk-assessment-guideline-pathogenic-microorganisms-focus-food-and-water
European Dissemination Agency (2023) LIFE MySOIL: a-step further in bioremediation—mycoremediation for soil recovery. LIFE Programme. Available at: https://lifemysoil.eu/download/a-step-further-bioremediation-mycoremediation-for-soil-recovery/
Farzin OA, Godini H, Noorisepehr M, Mahmoudi E (2023) Assessment of fungal aerosols dispersion from municipal solid waste disposal site: a case study of Karaj, Iran. J Environ Health Sustain Dev 8(3):2039–2049
Gao B, Yu B, Huang X, Li H, Jia Y, Wang M, Lu Y, Zhang X, Li W (2024) Cadmium and calcium ions' effects on the growth of Pleurotus ostreatus mycelia are related to phosphatidylethanolamine content. Fungal Biol 128(7):2190–2196. https://doi.org/10.1016/j.funbio.2024.08.012
Ghosh S, Rusyn I, Dmytruk OV, Dmytruk KV, Onyeaka H, Gryzenhout M, Gafforov Y (2023) Filamentous fungi for sustainable remediation of pharmaceutical compounds, heavy

metal and oil hydrocarbons. Front Bioeng Biotechnol 11:1–14. https://doi.org/10.3389/fbioe.2023.1106973

Habig M, Patneedi SK, Stam R, Licht HHDF (2025) Horizontal transfer of accessory chromosomes in fungi—a regulated process for exchange of genetic material? Heredity. (in press). https://doi.org/10.1038/s41437-025-00746-0

Jin X, Yu X, Zhu G, Zheng Z, Feng F, Zhang Z (2016) Conditions optimizing and application of laccase-mediator system (LMS) for the laccase-catalyzed pesticide degradation. Sci Rep 6:35787. https://doi.org/10.1038/srep35787

Kanakaraju D, Glass BD, Goh PS (2025) Advanced oxidation process-mediated removal of pharmaceuticals from water: a review of recent advances. Environ Sci Pollut Res Int 32:14316–14350. https://doi.org/10.1007/s11356-025-36547-5

Kępińska-Pacelik J, Biel W (2021) Alimentary risk of mycotoxins for humans and animals. Toxins 13(11):822. https://doi.org/10.3390/toxins13110822

Khan MF, Hof C, Niemcová P, Murphy CD (2023) Recent advances in fungal xenobiotic metabolism: enzymes and applications. World J Microbiol Biotechnol 39(11):296. https://doi.org/10.1007/s11274-023-03737-7

Kontro MH, Kirsi M, Laitinen SK (2022) Exposure to bacterial and fungal bioaerosols in facilities processing biodegradable waste. Front Public Health 10:789861. https://doi.org/10.3389/fpubh.2022.789861

Li Y, Fan H, Li B, Liu X (2024) Environmental impact of xenobiotic aromatic compounds and their biodegradation potential in *Comamonas testosteroni*. Int J Mol Sci 25(24):13317. https://doi.org/10.3390/ijms252413317

Ling AL (2024) Estimated scale of costs to remove PFAS from the environment at current emission rates. Sci Total Environ 918:170647. https://doi.org/10.1016/j.scitotenv.2024.170647

Liu Y, Zhang Y, Shi Y, Shen F, Yang Y, Wang M, Zhang G, Deng T, Lai S (2021) Characterization of fungal aerosol in a landfill and an incineration plants in Guangzhou, Southern China: the link to potential impacts. Sci Total Environ 764:142908. https://doi.org/10.1016/j.scitotenv.2020.142908

Magan N, Fragoeiro S, Bastos C (2010) Environmental factors and bioremediation of xenobiotics using white rot fungi. Mycobiology 38(4):238–248. https://doi.org/10.4489/MYCO.2010.38.4.238

Mahmoud YA, El-Halmouch YH, Nasr EE, Al-Sodany YM, El-Nogoumy BA, Ali SS (2024) Exploring the potential of biosurfactants produced by fungi found in soil contaminated with petrochemical wastes. Sci Rep 14(1):25733. https://doi.org/10.1038/s41598-024-75865-5

Mate DM, Alcalde M (2015) Laccase engineering: from rational design to directed evolution. Biotechnol Adv 33(1):25–40. https://doi.org/10.1016/j.biotechadv.2014.12.007

Mayr SA, Rennhofer H, Gille L, Schwaiger N, Nyanhongo GS, Weiss R, Guebitz GM (2024) Effect of salts on laccase-catalyzed polymerization of Lignosulfonate. ChemSusChem 17(14):e202301134. https://doi.org/10.1002/cssc.202301134

Miglani R, Parveen N, Kumar A, Ansari MA, Khanna S, Rawat G, Panda AK, Bisht SS, Upadhyay J, Ansari MN (2022) Degradation of xenobiotic pollutants: an environmentally sustainable approach. Metabolites 12(9):818. https://doi.org/10.3390/metabo12090818

Morogovsky A, Handelman M, Kandil AA, Shadkchan Y, Osherov N (2022) Horizontal gene transfer of Triazole resistance in Aspergillus fumigatus. Microbiol Spectr 10(3):e0111222. https://doi.org/10.1128/spectrum.01112-22

Nascimento ACB, Nhampossa NA, Félix TP, Itabaiana I Jr, Nascimento RP (2025) Exploring fungal biodegradation pathways of 2,4-D: enzymatic mechanisms, synergistic actions, and environmental applications. ACS Omega 10(35):39398–39414. https://doi.org/10.1021/acsomega.5c05161

Navina BK, Velmurugan NK, Kumar PS, Rangasamy G, Palanivelu J, Thamarai P, Vickram AS, Saravanan A, Shakoor A (2024) Fungal bioremediation approaches for the removal of toxic pollutants: mechanistic understanding for biorefinery applications. Chemosphere 350:141123. https://doi.org/10.1016/j.chemosphere.2024.141123

Ningombam L, Mana T, Pradhan S, Apum G, Singh YD (2025) Fungal bioremediation in environmental pollution and recent strategies. Discov Environ 3(100):1–14. https://doi.org/10.1007/s44274-025-00267-x

OECD - Organisation for Economic Co-operation and Development (2012) Guidance to the environmental safety evaluation of microbial biocontrol agents. Available at: https://www.oecd.org/content/dam/oecd/en/publications/reports/2014/09/oecd-guidance-to-the-environmental-safety-evaluation-of-microbial-biocontrol-agents_g1g4856f/9789264221659-en.pdf

Oliveira CV d S, Pepe VLE (2024) Política de saúde e Autoridades Reguladoras: desafios da regulação de tecnologias nas Emergências em Saúde Pública. Ciênc Saúde Coletiva 29(7):e03442024. https://doi.org/10.1590/1413-81232024297.03442024

Omotayo OP, Omotayo AO, Mwanza M, Babalola OO (2019) Prevalence of mycotoxins and their consequences on human health. Toxicol Res 35(1):1–7. https://doi.org/10.5487/TR.2019.35.1.001

Oziegbe O, Oziegbe EJ, Ojo-Omoniyi O (2025) Bioremediation of heavy metals in aquatic environment: a review. Cleaner Chem Eng 11:100193. https://doi.org/10.1016/j.clce.2025.100193

Pham LTM, Deng K, Choudhary H, Northen TR, Singer SW, Adams PD, Simmons BA, Sale KL (2024) An engineered laccase from *Fomitiporia mediterranea* accelerates lignocellulose degradation. Biomolecules 14(3):324. https://doi.org/10.3390/biom14030324

Rathankumar AK, Saikia K, Cabana H, Kumar VV (2022) Surfactant-aided mycoremediation of soil contaminated with polycyclic aromatic hydrocarbons. Environ Res 209:112926. https://doi.org/10.1016/j.envres.2022.112926

van Rhijn N, Arikan-Akdagli S, Beardsley J, Bongomin F, Chakrabarti A, C-A Chen S, Chiller T, Colombo AL, Govender NP, Alastruey-Izquierdo A, Kidd SE, Lackner M, Li R, Hagen F (2024) Beyond bacteria: the growing threat of antifungal resistance. Lancet 404(10457):1017–1018. https://www.thelancet.com/journals/lancet/article/PIIS0140-6736(24)01695-7/fulltext

Rodrigues MA (2025) Coleção Esquematizado - Direito Ambiental - 12ª Edição, 12th edn. SRV, Rio de Janeiro, p 331. E-book ISBN 9788553628209. Available at: https://app.minhabiblioteca.com.br/reader/books/9788553628209/

Sahay S (2022) Extremophilic fungi: ecology, physiology and applications. Springer. https://doi.org/10.1007/978-981-16-4907-3

Saikia S, Yadav M, Hoque RA, Yadav HS (2023) Bioremediation mediated by manganese peroxidase—an overview. Biocatal Biotransform 41(3):161–173. https://doi.org/10.1080/10242422.2022.2113517

Salem MM, Mohamed TM, Shaban AM, Mahmoud YA, Eid MA, El-Zawawy NA (2024) Optimization, purification and characterization of laccase from a new endophytic *Trichoderma harzianum* AUMC14897 isolated from *Opuntia ficus-indica* and its applications in dye decolorization and wastewater treatment. Microb Cell Fact 23(1):266. https://doi.org/10.1186/s12934-024-02530-x

Shourie A, Vijayalakshmi U (2022) Fungal diversity and its role in mycoremediation. Geomicrobiol J 39(3–5):426–444. https://doi.org/10.1080/01490451.2022.2032883

da Silva JA (2019) Direito ambiental constitucional, 11th edn. Malheiros, São Paulo

Singh M, Mishra RC, Shah I, Wadhwa V, Mor V (2023) Xenobiotics: sources, pathways, degradation, and risk associated with major emphasis on pharmaceutical compounds. In: Singh R, Singh P, Tripathi S, Chandra KK, Bhadouria R (eds) Xenobiotics in urban ecosystems sources, distribution and health impacts. Springer, Cham, pp 87–106. https://doi.org/10.1007/978-3-031-35775-6_5

Singh AK, Abellanas-Perez P, de Andrades D, Cornet I, Fernandez-Lafuente R, Bilal M (2025) Laccase-based biocatalytic systems application in sustainable degradation of pharmaceutically active contaminants. J Hazard Mater 485:136803. https://doi.org/10.1016/j.jhazmat.2024.136803

Sośnicka A, Kózka B, Makarova K, Giebułtowicz J, Klimaszewska M, Turło J (2022) Optimization of white-rot fungi mycelial culture components for bioremediation of pharmaceutical-derived pollutants. Water 14(9):1374. https://doi.org/10.3390/w14091374

Soucek P (2011) Xenobiotics. In: Schwab M (ed) Encyclopedia of cancer. Springer, Berlin/Heidelberg, pp 4869–4873. https://doi.org/10.1007/978-3-662-46875-3_6276

Štefanac T, Grgas D, Dragičević TL (2021) Xenobiotics—division and methods of detection: a review. J Xenobiot 11(4):130–141. https://doi.org/10.3390/jox11040009

Strong PJ, Claus H (2011) Laccase: a review of its past and its future in bioremediation. Crit Rev Environ Sci Technol 41(4):373–434. https://doi.org/10.1080/10643380902945706

Thirumalaivasan N, Gnanasekaran L, Kumar S, Durvasulu R, Sundaram T, Rajendran S, Nangan S, Kanagaraj K (2024) Utilization of fungal and bacterial bioremediation techniques for the treatment of toxic waste and biowaste. Front Mater 11:1416445. https://doi.org/10.3389/fmats.2024.1416445

Torres-Farradá G, Thijs S, Rineau F, Guerra G, Vangronsveld J (2024) White rot fungi as tools for the bioremediation of Xenobiotics: a review. J Fungi 10(3):167. https://doi.org/10.3390/jof10030167

United Nations (1987) UN World Commission on Environment and Development. Our Common Future Report Brundtland GH (Coordinator) 1987. Available at: https://www.are.admin.ch/dam/are/en/dokumente/nachhaltige_entwicklung/dokumente/bericht/our_common_future-brundtlandreport1987.pdf.download.pdf/our_common_futurebrundtlandreport1987.pdf

United Nations (1992) Rio declaration on environment and development. Available at: https://www5.pucsp.br/ecopolitica/projetos_fluxos/doc_principais_ecopolitica/Declaracao_rio_1992.pdf

United Nations (2025) What are the sustainable development goals?. Available at: https://www.undp.org/sustainable-development-goals

Viana CEM, Lima VDS, Rodrigues K, Pereira L, Silva GMM (2025) Bioremediation of endocrine disruptors (EDs): a systematic review of fungal application in ED removal from wastewater. Water 17(5):640. https://doi.org/10.3390/w17050640

Wang F, Fang L, Shi Z (2024) Bioremediation of Contaminated Soil by Fungi: A Call for Research. J. Fungi 10(10):684. https://doi.org/10.3390/jof10100684

WHO – World Health Organization (2022) World health statistics 2022: monitoring health for the SDGs. Available at: https://www.who.int/publications/i/item/9789240051157

Xiangyu B, Chao L, Shilong H, Jiping Z, Hu J (2025) Combining advanced oxidation processes with biological processes in organic wastewater treatment: recent developments, trends, and advances. Desalin Water Treat 323:101263. https://doi.org/10.1016/j.dwt.2025.101263

Xie S (2024) Biosorption of heavy metal ions from contaminated wastewater: an eco-friendly approach. Green Chem Lett Rev 17(1):1–15. https://doi.org/10.1080/17518253.2024.2357213

Yang Y, Liu R, Zhou Y, Tang Y, Zhang J, Wang Y, Dai T, Zou P, Bi X, Li S (2024) Screening and performance optimization of fungi for heavy metal adsorption in electrolytes. Front Microbiol 15:1371877. https://doi.org/10.3389/fmicb.2024.1371877

Zhang Z, Wang Y, Chen B, Lei C, Yu Y, Xu N, Zhang Q, Wang T, Gao W, Lu T, Gillings M, Qian H (2022) Xenobiotic pollution affects transcription of antibiotic resistance and virulence factors in aquatic microcosms. Environ Pollut 306:119396. https://doi.org/10.1016/j.envpol.2022.119396

Zhang W, Li Q, Wang J, Wang Z, Zhan H, Yu X, Zheng Y, Xiao T, Zhou LW (2023) Biodegradation of benzo[a]pyrene by a white-rot fungus *Phlebia acerina*: surfactant-enhanced degradation and possible genes involved. J Fungi 9(10):978. https://doi.org/10.3390/jof9100978

Bioreactor-Based Mycoremediation: Design, Applications, and Advances

11

Debraj Maji and Subhasish Dutta

Abstract

Microbial remediation to sequester environmental pollutants has emerged as a promising biotechnological strategy for a sustainable way to deal with recalcitrant pollutants. While bacteria are commonly used in microbial bioremediation, fungi offer a competitive edge due to their production of powerful extracellular enzymes. Fungi-derived laccases, peroxidases, and cellulases can break down a wide range of persistent pollutants, including polycyclic aromatic hydrocarbons (PAHs), persistent organic pollutants, toxic dyes, pesticides, antibiotics, and plastics. Mycoremediation can be performed either directly at the contamination site (in situ) or away from the site (ex situ). However, in situ remediation is mostly slow, and it is not possible to control and optimize the factors affecting the remediation process. To address these bottlenecks, bioreactors are used for optimal growth and pollutant degradation, improving the efficiency of bioremediation. Bioreactors provide controlled conditions, enhancing fungal growth, enzyme production, and degradation rates. Different bioreactor designs, such as stirred tank, packed bed, and fluidized bed reactors, offer specific advantages depending on the type of contaminant and fungal species used. This approach is useful for treating various environmental pollutants. Improvements in bioreactor designs and fungal strains can make mycoremediation even more effective in the future. This chapter discusses the design and applications of different types of bioreactors involved in the mycoremediation process.

D. Maji
Department of Bioscience and Biotechnology, Indian Institute of Technology (IIT), Kharagpur, India

S. Dutta (✉)
Biosafety Support Unit (BSU), Regional Centre for Biotechnology (RCB), Faridabad, India

A. Roy et al. (eds.), *Mycoremediation of Xenobiotics*, Emerging Paradigms in Pharmaceutical Research, https://doi.org/10.1007/978-981-95-5743-1_11

Keywords

Mycoremediation · Bioreactors · Recalcitrant pollutants · Bioprocess · Biodegradation

11.1 Introduction

Environmental conditions are vital for maintaining ecosystem stability. However, industrial and technical progress has accelerated severe environmental degradation and biodiversity loss. Pollution caused by persistent organic pollutants (POPs), petroleum-derived compounds, and heavy metals has become a serious global challenge, affecting both terrestrial and aquatic systems. Like, polycyclic aromatic hydrocarbons (PAHs) contaminate over 2.3 million hectares of arable land worldwide, causing ~15–40% decline in crop yields, leading to significant risk to agricultural productivity and food security (Ningombam et al. 2025).

Contaminants released into the air, water, and soil by human activities have the potential to enter the human body via absorption, ingestion, or inhalation. Heavy metals, such as mercury (Hg), cadmium (Cd), arsenic (As), and lead (Pb), are extremely toxic, leading to cancer as well as reproductive, cardiovascular, neurological, and kidney diseases. Effective monitoring and management of these environmental toxins is crucial, as prolonged exposure through contaminated food, water, or dust can lead to serious long-term effects, including neurotoxicity, organ damage, and heightened vulnerability to respiratory and cardiovascular diseases. (Shetty et al. 2023).

The development of sustainable and efficient remediation strategies is essential to address these persistent pollutants. Conventional physical and chemical methods for eliminating these contaminants are often expensive, produce hazardous by-products, and prove ineffective when handling highly toxic substances present at low concentrations. Mycoremediation provides an efficient, cost-effective, and environmentally responsible approach to in-situ remediation. Fungi represent excellent candidates for contaminant degradation owing to their rapid growth, adaptability to changing environmental circumstances, extensive hyphal networks, high surface area to volume-ratio, tolerance to heavy metals, and diverse repertoire of extracellular enzymes. In controlled bioreactors, regulated fungal biomass and metabolites can accelerate the degradation of persistent pollutants, such as polycyclic aromatic hydrocarbons (PAHs), dyes, medicines, insecticides, and phthalates, and can be applied directly to contaminated sites (Akhtar and ul Mannan 2020).

Fungi have shown great success in the bioremediation of persistent organic pollutants like polycyclic aromatic hydrocarbons (PAHs) and heavy metals. They secrete metabolites and organic acids that can chelate metals and transform them into less harmful or immobilized forms. Additionally, fungal cell walls contain functional groups, like amino, carboxyl, and hydroxyl groups, which facilitate detoxification and metal adsorption. The extensive mycelia networks of fungi not only provide a larger surface area for pollutant interaction but also aid in the transportation of nutrients and enzymes, enabling penetration into deeper soil layers that are typically inaccessible to bacteria. Moreover, these networks facilitate co-metabolism, wherein the breakdown of one chemical enhances the degradation of others. This enables fungi to immobilize heavy metals and degrade organic contaminants like PAHs simultaneously. For instance, *Aspergillus niger* has been reported to reduce cadmium concentration by ~87%, whereas *Phanerochaete chrysosporium* can break down up to 95% of benzo[*a*]pyrene, a polycyclic aromatic hydrocarbon, under laboratory conditions.

Due to their multifunctionality, fungi represent a robust and efficient biological tool for remediating contaminated water and soils, effectively eliminating multiple contaminants with minimal chemical input and environmental disturbance. This process not only lowers the bioavailability and toxicity of the metal but also prevents further mobility in the environment. Overall, fungi constitute a versatile and sustainable resource for environmental management, with metabolic adaptability, ecological ubiquity, and diverse enzymatic systems endowing them with an inherent capacity to detoxify environments contaminated with organic and inorganic pollutants (Dinakarkumar et al. 2024).

This chapter provides a brief overview of the fungal species commonly used in mycoremediation, and simultaneously highlights the role of bioreactor systems in enhancing the process. Bioreactor-based remediation offers a fast and well-controlled method for treating contaminants ex situ. The chapter also describes various types of bioreactors used in mycoremediation and evaluates their effectiveness in pollutant removal.

11.2 Fungal Species Involved in Mycoremediation

Mycoremediation is a bioremediation process that uses fungi to remove toxic substances from various environmental matrices, whether the fungi are alive or dead. It is a cost-effective approach that doesn't generate any hazardous by-products. The success of mycoremediation largely depends on the selection of an appropriate fungal species for the pollutant of concern. Fungi are ubiquitous microorganisms and can be found in a vast range of ecological niches across terrestrial and aquatic environments. Fungal taxonomy has classified these organisms into nineteen major phyla, which include the following: *Chytridiomycota, Rozellomycota, Glomeromycota, Kickxellomycota, Entorrhizomycota, Basidiomycota, Neocallimastigomycota, Olpidiomycota, Monoblepharomycota, Calcarisporiellomycota, Mortierellomycota, Aphelidiomycota, Zoopagomycota, Caulochytriomycota, Entomophthoromycota, Blastocladiomycota, Mucoromycota, Ascomycota,* and *Basidiobolomycota* (Wijayawardene et al. 2024). Both filamentous fungi (molds) and macrofungi (mushrooms) can contribute to mycoremediation, allowing the biological degradation, conversion, or immobilization of environmental pollutants (Dinakarkumar et al. 2024).

Fungi are key agent in mycoremediation due to their enzymatic diversity and ability to degrade a broad spectrum of pollutants. Complex organic compounds, like polycyclic aromatic hydrocarbons and other persistent contaminants, are degraded into simpler, less toxic forms by reactive species produced by their extracellular enzymes, which include oxidoreductases like laccases and peroxidases. Intracellular enzymes, such as cytochromes, facilitate the oxidation and transformation of pollutants within fungal cells. The combined action of intracellular and extracellular enzymes provides the ability to efficiently digest both organic and inorganic contaminants. Current research indicates that fungi offer significant advantages when compared to conventional remediation techniques, as they help in the holistic ecosystem detoxification (Vaksmaa et al. 2023).

11.2.1 Rot Fungi and Roles in Mycoremediation

Among fungi, rot fungi hold particular importance in mycoremediation due to their ability to colonize and degrade plant biomass. Rot fungi are broadly classified into three types based on decay pattern: white rot, brown rot, and soft rot. White-rot fungi have garnered attention in biodegradation due to their nonspecific ligninolytic enzymes, laccase, manganese peroxidase (MnP), and lignin peroxidase (LiP), which allow them to break down a variety of xenobiotics, such as phenols, hydrocarbons, dyes, medications, paper and pulp effluents, and pesticides (Echezonachi 2022).

11.2.1.1 White-Rot Fungi (WRF)

White-rot fungi are considered the most efficient natural lignin degraders, imparting a characteristic bleached appearance to wood by breaking down lignin and lignin-like compounds. They can be broken down by both internal cytochrome P450 (CYP450) systems and external ligninolytic enzymes (including laccase, lignin peroxidase, manganese peroxidase, and versatile peroxidase). WRF can break down a diverse range of refractory organic pollutants, including endocrine disruptors, pesticides, PAHs, pharmaceuticals, and herbicides, due to the synergistic function of their enzymatic machinery. Specifically, cytochrome P450 enzymes (CYP450s) play a crucial role in the oxidation, detoxification, and transformation of hydrophobic contaminants through various processes, including hydroxylation, dealkylation, epoxidation, and dehalogenation. This coordinated enzymatic machinery makes WRF a highly versatile and efficient agent for mycoremediation, with significant potential for addressing environmental contamination. Examples are as follows: *Pleurotus ostreatus*, *Trametes versicolor*, and *Phanerochaete chrysosporium* (Lin et al. 2022).

11.2.1.2 Brown-Rot Fungi

Brown-rot fungi belong to the phylum Basidiomycota and are typically found on coniferous trees. They are different from white-rot fungi by modifying lignin through non-enzymatic breakdown and play a significant role in the mycoremediation of heavy metals and other environmental pollutants. Brown-rot fungi secrete organic acids, mainly oxalic acid, which increases metal solubility, enabling their removal from contaminated substances. The acids also chelate metal ions, shielding fungal cells from toxicity and facilitating safe detoxification. Additionally, in order to accelerate the breakdown of contaminants, they also use oxidative processes, like Fenton-type reactions, to decompose complex organic compounds found in soil and wood. Species like *Fomitopsis palustris*, *Coniophora puteana*, and *Laetiporus sulphureus* are commonly applied in mycoremediation (Kartal et al. 2004).

In mycoremediation practice, white-rot fungi have been most widely exploited due to their ability to mineralize complex organics. Brown-rot species are increasingly investigated for pollutant degradation via oxidative radical systems, while soft-rot fungi contribute under specialized conditions. In general, rot fungi significantly expand the toolkit of bioremediation: their nonspecific oxidative mechanisms allow them to tackle a broad spectrum of environmental contaminants, often outperforming bacteria in degrading aromatic pollutants (Rosa et al. 2024). Table 11.1 describes the major environmental pollutants and respective fungus species involved in their biodegradation processes.

11.3 Mechanisms of Mycoremediation

Fungi are highly adaptable microorganisms capable of thriving across diverse ecosystems, including those with extreme environmental conditions. They are drawing attention due to their unique biochemical and physiological capabilities, especially their robust oxidative metabolism and production of diverse secondary metabolites. They can break down complex hydrocarbons and toxic substances by transforming them into more nontoxic biodegradable forms, making them good candidates for bioremediation. Fungi exhibit exceptional capabilities in interacting with metal ions, including uptake, intracellular accumulation, transformation, and detoxification. A range of well-developed biological mechanisms (Fig. 11.1) drives their effectiveness in remediating heavy metals and organic pollutants (Shourie and Vijayalakshmi 2022).

11.3.1 Metal Bioprecipitation

One of the major mechanisms employed by filamentous fungi is the transformation of soluble heavy metal ions into insoluble precipitates. Fungi produce low-molecular-weight organic acids, such as gluconic, oxalic, and citric acids, which play a significant role in heavy metal removal. Among these, oxalic acid is considered the most effective chelating agent, forming insoluble metal complexes that reduce heavy metal bioavailability (Priyanka and Dwivedi 2023). For instance, oxalic acid can react with divalent cations such as Pb^{2+} or Cu^{2+} to form insoluble metal oxalates, shown below,

$$Pb^{2+} + C_2O_4^{2-} \rightarrow PbC_2O_4(s) \quad (11.1)$$

$$Cu^{2+} + C_2O_4^{2-} \rightarrow CuC_2O_4(s) \quad (11.2)$$

Table 11.1 Major environmental pollutants and fungi involved in their biodegradation

Fungi	Phyla	Pollutants	Biodegradation mechanism	References
Trametes versicolor	Basidiomycota	*Antibiotics*	Laccase and the cytochrome P450 system	Akhtar and ul Mannan (2020), Amobonye et al. (2023), Gao et al. (2018), Tan et al. (2025)
Lentinula edodes	Basidiomycota		Cytochrome 450 system	
Irpex lacteus	Basidiomycetes		Peroxidase	
Pleurotus ostreatus	Basidiomycetes		Ligninolytic enzymatic attack (laccase/MnP)	
Phanerochaete chrysosporium	Basidiomycota	*Polycyclic aromatic hydrocarbon*	Cytochrome P-450 monooxygenase	Akhtar and ul Mannan (2020), Kijpornyongpan et al. (2022), Rodrigues Barbosa et al. (2020), Villani et al. (2025)
Trametes versicolor	Basidiomycota		Dioxygenase	
Pleurotus ostreatus	Basidiomycota		FAD-dependent monooxygenases and laccase enzymes	
Pleurotus eryngii	Basidiomycota		Glutathione transferase and epoxide hydrolase activity	
Cochliobolus lunatus	Ascomycota		Dehydrogenases and oxidative enzymes	
Botryosphaeria laricina	Ascomycota	*Pesticide and herbicide*	Ligninolytic enzymes	Amobonye et al. (2023), Justo and Hibbett (2011)
Aspergillus glaucus	Ascomycota		Esterification	
Trametes pavonia	Basidiomycota		Ligninolytic enzymes, dioxygenation	
Penicillium spiculisporus	Ascomycota		Demethylation-mediated degradation	

(continued)

Table 11.1 (continued)

Fungi	Phyla	Pollutants	Biodegradation mechanism	References
Aspergillus niger	Ascomycota	*Heavy metals*	Cell wall biosorption (chitin, carboxyl groups)	Akhtar and ul Mannan (2020), Dusengemungu et al. (2020), Liu et al. (2017)
Penicillium chrysogenum (spp.)	Ascomycota		EPS chelation and metabolic	
Trichoderma spp.	Ascomycota		Biosorption and oxidative stress tolerance	
Rhizomucor species	Mucoromycota		Ligninolytic enzymes for heavy metals	
Pleurotus ostreatus	Basidiomycota		Ligninolytic enzymes for heavy metals	
Trametes versicolor	Basidiomycota	*Pharmaceuticals*	Laccase-mediated degradation of analgesics/ hormones	Akhtar and ul Mannan (2020) Amobonye et al. (2023)
Pleurotus ostreatus	Basidiomycota		MnP-assisted transformation of pharmaceuticals	
Mucor hiemali	Mucoromycota		Ligninolytic enzymes	
Phanerochaete chrysosporium	Basidiomycota		Ligninolytic enzymes	
Phlebia acerina	Basidiomycota	*Dyes*	Manganese peroxidase (MnP), laccase	Akhtar and ul Mannan (2020), Yuan et al. (2021), Zhang et al. (2021)
Bjerkandera adusta	Basidiomycota		Lignin peroxidase (LiP), manganese peroxidase (MnP)	
Antrodia sp. *P5*	Basidiomycota		Ligninolytic enzymes	
Trametes gibbosa	Basidiomycota		Extracellular ligninolytic enzymes: MnP, laccase, lignin peroxidase (LiP)	
Marasmius cladophyllus	Basidiomycota		Laccase	

(continued)

Table 11.1 (continued)

Fungi	Phyla	Pollutants	Biodegradation mechanism	References
Pleurotus ostreatus	Basidiomycota	*Phthalates*	Mycelium-associated enzymes: esterases, oxygenase, and oxidases/dehydrogenases	Ahuactzin-Pérez et al. (2018), Puranik et al. (2023)
Aspergillus flavus	Ascomycota		Laccase	
Penicllium brocae			Cutinase	
Fusarium subglutinans	Ascomycota		Cutinase	
Aspergillus fumigates	Ascomycota	*Detergents*	Utilizes extracellular enzymes to break down detergent components, facilitating their mineralization	Amobonye et al. (2023), Singh et al. (2021)
Penicillium verrrucosum	Ascomycota		–	
Geotrichum candidum	Ascomycota		–	

The precipitation of Pb^{2+} as PbC_2O_4 by *Aspergillus niger* has been well reported as an effective mechanism for lead detoxification (Huang et al. 2023). Similarly, white-rot fungi such as *Bjerkandera fumosa* and *Trametes versicolor* also form oxalate crystals in response to metal exposure. Some fungal strains, including *Aspergillus* sp. and *Fusarium oxysporum*, are reported to produce calcite and co-precipitate heavy toxic heavy metals, like lead, and radionuclides, like strontium, as carbonates (Singh and Singh 2024).

11.3.2 Biosorption and Bioaccumulation of Heavy Metals

Similar to bioprecipitation, one of the primary processes by which fungi remove heavy metal contamination is biosorption. Biosorption is a passive physicochemical process by which metal ions are adsorbed onto the surface of fungal biomass. Fungal cell wall contains several active functional groups like carboxyl (-COOH), amine ($-NH_2$), thiol (-SH), and phosphate ($-PO_4^{3-}$). The functional groups interact with metal ions through processes, like ion exchange, complexation, coordination, electrostatic interaction, and micro-precipitation (Robalds et al. 2016). The biosorption of heavy metals, including Cu, Zn, Cd, Mn, Pb, Fe, Ni, Ag, Th, Ra, Sr, and U,

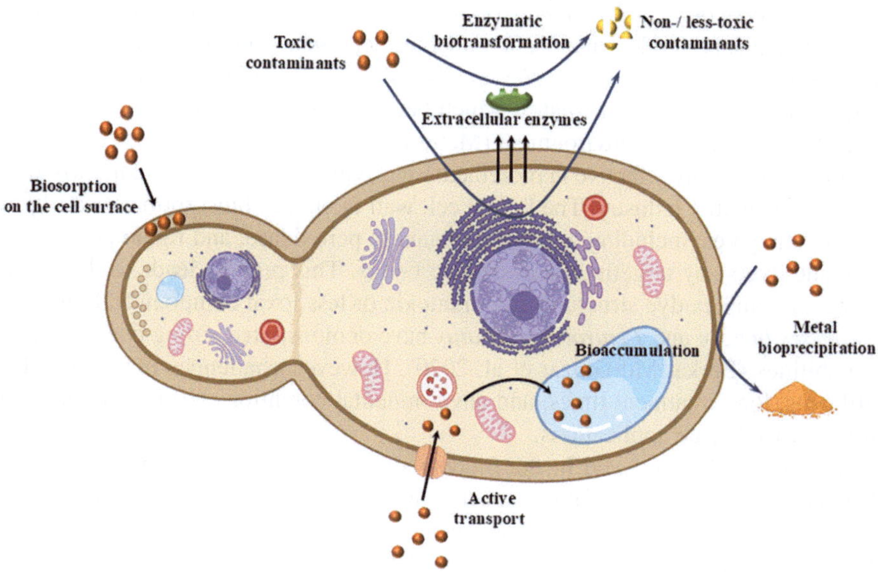

Fig. 11.1 Fungal mechanisms for pollutant removal through surface adsorption, extracellular and intracellular enzymatic degradation, active transport into the cell, bioprecipitation, and bioaccumulation

by fungal biomass has been extensively studied and documented by numerous researchers. Scientists have reported effective biosorption of heavy metals using calcium alginate-immobilized cells of *Aspergillus niger* B77, which demonstrated the ability to remove Mn, Fe, Ni, Cu, Zn, Cd, and Pb from contaminated environments. (Tsekova et al. 2010). On the other hand, bioaccumulation is an active, metabolism-driven process, wherein metal ions are transported across the cell membrane via metal transporters and subsequently sequestered in intracellular compartments (e.g., vacuoles) or bound to metallothioneins and phytochelatins. (Priyanka and Dwivedi 2023).

11.3.3 Enzymatic Degradation of Organic Pollutants

Fungi are capable of producing a wide range of hydrolytic and oxidative enzymes. These enzymes can break down recalcitrant organic pollutants, such as pesticides, synthetic dyes, pharmaceuticals, polycyclic aromatic hydrocarbons (PAHs), and petroleum hydrocarbons.

Fungi are capable of degrading pesticides through minor modification of their chemical structure by hydrolysis, oxidation, and dechlorination. Degradation starts with the adsorption of pesticides on the fungal hyphae, followed by enzymatic alteration. Different extracellular and intracellular enzymes, such as laccase, oxidase, and peroxidase, play important roles in degradation. Fungi like *Pleurotus ostreatus*,

Coriolus versicolor, and *Trametes versicolor* have been reported to degrade pesticides, like phenylamide, triazine, organochlorine, organophosphate, and carbamate. *Aspergillus and Penicillium species* are also widely reported for transforming pesticides into nontoxic intermediates, which become innocuous and can undergo further degradation in the environment (Matúš et al. 2023).

Fungi are highly effective in bioremediating synthetic dyes, particularly azo dyes used in the textile industry. The fungal cell wall facilitates biosorption, while ligninolytic enzymes, including laccase, manganese peroxidase, and lignin peroxidase, degrade dye(s) by targeting their specific bonds. This process leads to the breakdown of complex dye structures into nontoxic or less toxic compounds. Fungi like *Trametes hirsute* and *Pleurotus ostreatus* have demonstrated strong decolorization capabilities (Barkat Md Gulzar et al. 2020). However, challenges like high pH of effluents, long treatment times, and environmental conditions need to be managed for industrial-scale applications.

Fungi are also involved in the biodegradation of plastics like polyhydroxyalkanoates (PHAs). Some fungal species are known to synthesize depolymerase enzymes that degrade PHAs into lower monomers like 3-hydroxybutyrate. These enzymes catalyze oxidative cleavage of C–C and C–H bonds in aromatic rings through single-electron transfers or radical-mediated mechanisms. For instance, manganese peroxidases (MnP) catalyze the oxidation of Mn^{2+} to Mn^{3+}, which is a diffusible oxidant for phenolic substrates in PAHs. The PHA-degrading potential has been observed in *Aspergillus, Penicillium*, and *Fusarium* species, particularly under nutrient-scarce conditions (Akhtar and ul Mannan 2020).

Fungi degrade pharmaceutical waste primarily through the action of extracellular ligninolytic enzymes, such as laccase, manganese peroxidase, and other versatile peroxidases, along with intracellular enzymes like cytochrome P450 (Lin et al. 2022). These enzymes oxidize and transform complex drug molecules into simpler, nontoxic compounds. *Trametes versicolor* efficiently breaks down drugs naproxen, ketoprofen, diclofenac, and carbamazepine. *Mucor hiemalis* and *Phanerochaete chrysosporium* also eliminate drugs, like acetaminophen and some anti-inflammatory drugs. In bioreactor systems, *T. versicolor* was able to degrade naproxen to undetectable levels within 6 hours (Esterhuizen-Londt et al. 2017). Similarly, *Phanerochaete chrysosporium* and *Lentinula edodes* are capable of breaking down anti-inflammatory drugs like piroxicam and even endocrine disruptors such as synthetic testosterone. Similarly, the antifungal drugs bifonazole and clotrimazole were effectively absorbed and degraded by the mycelia of the edible fungi *Lentinula edodes*.

The degradation process depends on both the chemical structure of the organic pollutant and the specific fungal species used. While some compounds are degraded mainly by oxidative enzymes, others are removed through biosorption and transformation. Bioreactors using fungal strains are emerging as promising systems for large-scale treatment of toxic recalcitrant waste like synthetic dyes, pesticides, and pharmaceuticals. However, optimization of environmental conditions such as pH, temperature, and retention time is essential to improve degradation efficiency.

11.4 Types of Bioreactors Used in Mycoremediation

Fungi can remove recalcitrant and resistant pollution, such as PAHs, pesticides, synthetic dyes, explosives, and pharmaceutical residue. When fungal bioremediation is implemented in a bioreactor system, environmental parameters can be precisely controlled and optimized, thereby enhancing process reproducibility and overall efficiency.

11.4.1 Stirred Tank Bioreactors (STRs)

Stirred tank bioreactors (STRs) are commonly used for submerged fungal cultures in liquid media and provide efficient mixing and aeration through mechanical agitation. These reactors are ideal for the treatment of hydrocarbon-rich industrial effluents, phenolic compounds, and dyes, using fungi, like *Aspergillus terreus*, *Trichoderma harzianum*, and *Phanerochaete chrysosporium*. The continuous stirring ensures uniform distribution of cells, nutrients, and oxygen, allowing precise control of pH and temperature. STRs support high reaction rates and scalability but may impose shear stress on delicate fungal hyphae and require high energy for mixing. Foam generation and oxygen transfer limitations can also affect performance if not properly controlled. A fungal isolate, *Aspergillus terreus* SA3, previously obtained from the wastewater of a local textile industry, was effectively employed for the removal of sulfur black dye from textile effluent. In a stirred tank reactor (STR) operated at a dye concentration of 50 ppm with a hydraulic retention time (HRT) of 24 hours, the reported removal efficiencies were 84.53% for color, 66.50% for biochemical oxygen demand (BOD), and 75.24% for chemical oxygen demand (COD) (Andleeb et al. 2010).

11.4.2 Packed Bed Bioreactors (PBRs)

PBRs are fixed-film systems where fungal cultures grow on solid supports, like polyurethane foam, sawdust, ceramic beads, or inert polymers. These reactors are especially effective for treating wastewater containing dyes, pharmaceuticals, and pesticides. Fungi such as *Trametes versicolor*, *Aspergillus niger*, and *Pleurotus ostreatus* form biofilms on the packed materials, degrading pollutants as the contaminated liquid flows through the bed. The design offers high biomass retention, eliminating the need to separate fungi from the treated effluent. However, limitations include diffusion gradients within the biofilm, risk of clogging due to overgrowth, and hydraulic channelling, which can affect treatment efficiency. A packed bed reactor (PBR) system fed with a laccase/mediator solution demonstrated effective mycoremediation of PAH-contaminated soil, achieving degradation rates of up to 100% for pyrene and 96% for the highly recalcitrant benzo[a]pyrene. The continuous flow of fungal laccase significantly enhanced PAH degradation compared to batch systems, highlighting the potential of PBRs for efficient and rapid enzymatic soil bioremediation (Vipotnik et al. 2021).

11.4.3 Slurry Phase Bioreactors (SBs)

Slurry phase bioreactors (SBs) are widely used in fungal bioremediation for the treatment of contaminated soils, sediments, and solid wastes. In this system, the solid matrix is mixed with water to form a slurry, improving the bioavailability of hydrophobic pollutants such as PAHs, pesticides, and petroleum hydrocarbons. The reactor operates in batch or semicontinuous mode, and mixing ensures better contact between fungal biomass, pollutants, and nutrients. This type of reactor allows controlled aeration and pH adjustment, but it may require high energy input for mixing and generate large volumes of slurry that need further treatment. Recent trends also explore the use of simultaneous electron acceptors within SB systems for improved bioremediation of hydrocarbons and chlorinated organics. While fungal applications are gaining attention, further research is needed to deepen our understanding of fungal community dynamics within SBs using molecular tools and to assess treatment success through more representative ecotoxicity assays (Robles-González et al. 2008).

11.4.4 Airlift Bioreactors

Airlift bioreactors are pneumatically operated systems where air or gas is introduced at the base to create circulation and aeration without mechanical stirring. These bioreactors are well-suited for fungi sensitive to shear stress and are used for the treatment of volatile organic compounds (VOCs), phenols, and other oxygen-demanding pollutants. The movement of liquid between the riser and downcomer zones ensures mild mixing and enhances gas-liquid mass transfer. Fungi such as *Aspergillus flavus* and white-rot fungi can efficiently degrade pollutants in these systems. Airlift reactors are energy-efficient and simple in design, but their performance may be affected at high cell densities due to limited mixing and oxygen distribution. The study done by (Aragão et al. 2020) evaluated the biodegradation of sugarcane vinasse using *Pleurotus sajor-caju* CCB020 in air-lift bioreactors, achieving up to 58% reduction in COD and 71% in BOD. The process also enabled simultaneous biomass and enzyme (laccase and manganese peroxidase) production, demonstrating that bioreactor configuration significantly influences treatment efficiency.

11.4.5 Membrane Bioreactors (MBRs)

Membrane bioreactors (MBRs) combine biological fungal treatment with membrane filtration to retain fungal biomass and produce a clear, high-quality effluent. These systems are particularly useful for the removal of textile dyes, pharmaceuticals, and endocrine-disrupting compounds from wastewater. Fungi like *Trametes hirsuta* and *Pleurotus ostreatus* secrete powerful oxidative enzymes that break down complex pollutants. The membrane acts as a barrier, preventing biomass washout and allowing extended operation with high degradation efficiency. However, membrane fouling is a significant drawback, requiring regular cleaning and replacement, which increases operational cost. The compact design and ability to treat high-strength wastewater make MBRs a promising tool for advanced fungal bioremediation. Another study described the potential of a submerged microfiltration membrane bioreactor using the white-rot fungus *Coriolus versicolor* for treating textile dye wastewater. Operating at 29 ± 1 °C and pH 4.5 ± 2 with a hydraulic retention time of 15 hours, the system achieved stable decolorization of approximately 98% and TOC removal over 95%. Despite membrane fouling due to fungal growth, the integrated fungi–membrane system proved highly effective, with the fungal culture alone contributing 35–50% to color removal and 70% to TOC reduction.

11.4.6 Partitioning Bioreactors (Two-Phase Systems)

Two-phase partitioning bioreactors (TPPBs) are designed for the treatment of pollutants that are poorly soluble or highly toxic to fungi by introducing a second, nonaqueous, biocompatible phase (typically an organic solvent). The pollutant preferentially partitions into the solvent, reducing its concentration in the aqueous phase and protecting the fungal cells from toxicity (Daugulis 2001). The dual-phase approach increases pollutant solubility and maintains microbial activity, but careful solvent selection and reactor design are crucial to prevent solvent toxicity or volatilization. These reactors are promising for industrial effluents containing recalcitrant or hydrophobic organic compounds. In a study, a continuous two-phase partitioning bioreactor (C-TPPB) using coiled Hytrel 8206 tubing was employed for the bioremediation of 4-chlorophenol, achieving near-complete removal (87–95%) across influent concentrations of 750–1500 mg/L. The system effectively buffered toxic substrate levels through diffusion across the polymer wall while maintaining microbial activity and enhancing resistance to substrate inhibition, demonstrating the advantages of controlled, subinhibitory exposure in TPPB systems.

Different bioreactor systems offer unique advantages for fungal bioremediation, depending on the type of pollutant, fungal species, and environmental conditions. Selecting the appropriate bioreactor design enhances fungal activity, biomass growth, and degradation efficiency (Fig. 11.2). With ongoing innovations, mycoremediation using bioreactors holds great promise for sustainable and eco-friendly pollution control in soil, water, and air environments.

11.5 Factors Affecting the Mycoremediation of Recalcitrant Pollutants in Bioreactors

The efficiency of enzymatic degradation of pollutants in fungi-based bioreactor systems depends on a range of environmental, biological, and engineering factors. These factors influence both fungal growth and the activity of their extracellular

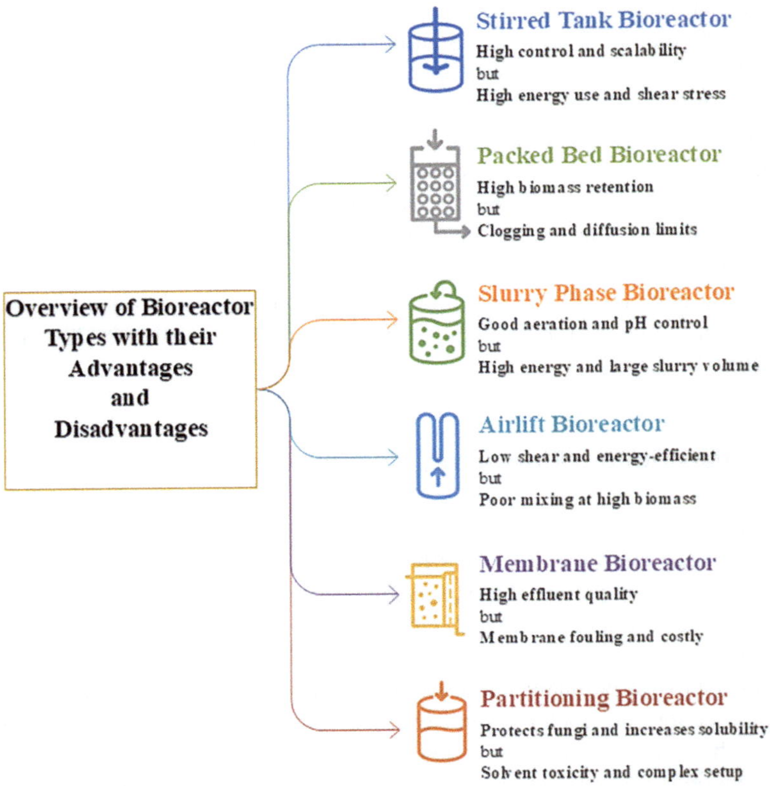

Fig. 11.2 Overview of Bioreactor types with their advantages and disadvantages

enzymes, which are crucial for breaking down complex pollutants. Understanding and optimizing these factors are critical for improving the degradation performance and ensuring process stability in bioreactor operations.

11.5.1 Choice of Fungal Strain

Different fungi produce different enzymes with varying efficiency. For example, white-rot fungi like *Phanerochaete chrysosporium* produce ligninolytic enzymes that can degrade a wide range of pollutants. Screening and selection of strains that function effectively under certain conditions must be carried out for effective treatment (Dinakarkumar et al. 2024).

In some bioreactor systems, fungal cultures may be used in combination with bacteria or other fungi. These microbial consortia can exhibit synergistic interactions, where one species degrades a pollutant into an intermediate that another species can further metabolize. This can broaden the spectrum of pollutants targeted and improve overall degradation rates.

11.5.2 Environmental Conditions

Temperature and pH play a key role in fungal metabolism and enzyme activity. Most fungi prefer slightly acidic environments, and each strain has an optimal temperature and pH range for growth and enzyme production. Extreme conditions can denature enzymes or inhibit fungal activity. Aeration is also important, especially for white-rot fungi that degrade pollutants using oxygen-dependent enzymes (Dhakar and Pandey 2013).

11.5.3 Nutrient Availability

Fungi need carbon, nitrogen, and other nutrients to grow and produce enzymes. A balanced nutrient supply enhances fungal metabolism and enzyme synthesis. The carbon-to-nitrogen (C:N) ratio is particularly influential in regulating enzyme expression. A high C:N ratio often favors ligninolytic enzyme production over biomass synthesis (Yu et al. 2005). In some cases, fungi may require an additional carbon source to co-metabolize pollutants more effectively. Nutrient limitation can downregulate enzyme synthesis and lead to metabolic stress, while excessive nutrient supplementation may suppress ligninolytic enzyme production due to catabolite repression. Therefore, maintaining a balanced nutrient profile in the bioreactor medium is critical for achieving sustained enzymatic degradation.

11.5.4 Moisture and Water Activity

Fungi require adequate moisture for enzyme production and nutrient transport. Low water activity can reduce fungal activity, while excess water may limit oxygen availability, especially in solid-state bioreactors.

11.5.5 Reactor Design and Operation

The type and configuration of the bioreactor affect mass transfer, mixing, and oxygen distribution. Parameters such as mixing, aeration, substrate feeding, and temperature control are all affected by the reactor type. Efficient mixing promotes uniform distribution of nutrients, temperature, and oxygen and prevents the formation of any concentration gradients. It also enhances contact between fungal mycelia and the target pollutants. Poor mixing can result in dead zones, stratification, and uneven degradation performance. Temperature and pH control systems also help maintain optimal conditions for fungal activity (Tekere and Tekere 2019).

11.5.6 Hydraulic Retention Time (HRT)

The residence time of the substrate in the reactor must be sufficient to allow complete enzymatic degradation. Short HRTs may lead to incomplete treatment, while too long HRTs can reduce throughput and affect fungal biomass maintenance. HRT must be optimized based on degradation kinetics, pollutant complexity, and fungal growth rate to achieve effective and stable pollutant removal.

11.5.7 Pollutant Characteristics

The chemical structure, solubility, and concentration of pollutants influence their degradability. Hydrophobic or high-molecular-weight compounds are often more resistant. Fungal enzymes may degrade simple pollutants faster than complex or highly toxic ones. High concentrations of xenobiotics can inhibit fungal growth or interfere with enzyme production. Moreover, certain pollutants may generate toxic intermediates during partial degradation, necessitating the selection of fungi capable of complete mineralization or detoxification.

11.6 Recent Advances in the Degradation of Pollutants by Fungi

Recent advances in genetic and enzyme engineering have significantly improved the efficiency of fungal strains in pollutant breakdown. Researchers are now able to identify and enhance specific enzymes, such as laccases, peroxidases, and cytochrome P450s, that play key roles in degrading harmful substances, like polycyclic aromatic hydrocarbons (PAHs), pesticides, synthetic dyes, and heavy metals.

One of the major developments is the use of gene editing tools like CRISPR-Cas, TALEN, and site-directed mutagenesis to modify fungal genomes for better pollutant degradation (Jaiswal et al. 2019). These tools allow scientists to enhance enzyme activity, stability under harsh environmental conditions, and specificity towards certain contaminants (Garg et al. 2025). Metagenomic studies have also contributed to fungal bioremediation by identifying new genes and metabolic pathways involved in pollutant degradation. These studies reveal how fungi interact with other microbes in the environment, often forming synergistic microbial consortia that are more effective than single strains. Such consortia can degrade complex pollutants like plastics, hydrocarbons, and pharmaceutical residues. Transcriptomics provides further insight by showing how fungi respond at the gene level when exposed to pollutants. This helps researchers understand which genes are activated during stress and how detoxification pathways are regulated, allowing for the development of targeted bioremediation strategies.

11.7 Conclusion

Mycoremediation represents a promising and environmentally sustainable approach for the removal of diverse environmental pollutants. Fungi possess unique enzymatic and metabolic capabilities that enable the degradation of complex and recalcitrant contaminants. The integration of mycoremediation with bioreactor systems significantly enhances its efficiency and scalability, as bioreactors provide controlled operational conditions that promote fungal growth, enzyme production, and degradation kinetics. Various reactor configurations, including stirred tank, packed bed, slurry phase, and fluidized bed systems, can be employed, depending on the pollutant type and fungal species utilized. These systems not only accelerate pollutant removal but also ensure greater consistency and reproducibility of treatment outcomes. The convergence of fungal biotechnology with bioreactor engineering thus opens new avenues for the development of efficient and sustainable bioremediation strategies. Furthermore, future advancements in reactor design and fungal strain engineering are expected to further expand the potential of mycoremediation in addressing global environmental pollution challenges.

References

Ahuactzin-Pérez M, Tlécuitl-Beristain S, García-Dávila J, Santacruz-Juárez E, González-Pérez M, Gutiérrez-Ruíz MC, Sánchez C (2018) Kinetics and pathway of biodegradation of dibutyl phthalate by Pleurotus ostreatus. Fungal Biol 122:991–997. https://doi.org/10.1016/J.FUNBIO.2018.07.001

Akhtar N, ul Mannan MA (2020) Mycoremediation: expunging environmental pollutants. Biotechnol Rep (Amst) 26:e00452. https://doi.org/10.1016/J.BTRE.2020.E00452

Amobonye A, Aruwa CE, Aransiola S, Omame J, Alabi TD, Lalung J (2023) The potential of fungi in the bioremediation of pharmaceutically active compounds: a comprehensive review. Front Microbiol 14:1207792. https://doi.org/10.3389/FMICB.2023.1207792/XML

Andleeb S, Atiq N, Ali MI, Razi-Ul-Hussnain R, Shafique M, Ahmad B, Ghumro PB, Hussain M, Hameed A, Ahmad S (2010) Biological treatment of textile effluent in stirred tank bioreactor. Int J Agric Biol 12:256–260

Aragão MS, Menezes DB, Ramos LC, Oliveira HS, Bharagava RN, Romanholo Ferreira LF, Teixeira JA, Ruzene DS, Silva DP (2020) Mycoremediation of vinasse by surface response methodology and preliminary studies in air-lift bioreactors. Chemosphere 244:125432. https://doi.org/10.1016/J.CHEMOSPHERE.2019.125432

Barkat Md Gulzar A, Kumar Vandana U, Paul P, B. Mazumder P (2020) The role of mushrooms in biodegradation and decolorization of dyes. In: An introduction to mushroom. https://doi.org/10.5772/INTECHOPEN.90737

Daugulis AJ (2001) Two-phase partitioning bioreactors: a new technology platform for destroying xenobiotics. Trends Biotechnol 19:457–462. https://doi.org/10.1016/S0167-7799(01)01789-9

Dhakar K, Pandey A (2013) Laccase production from a temperature and pH tolerant fungal strain of Trametes hirsuta (MTCC 11397). Enzyme Res 2013:869062. https://doi.org/10.1155/2013/869062

Dinakarkumar Y, Ramakrishnan G, Gujjula KR, Vasu V, Balamurugan P, Murali G (2024) Fungal bioremediation: an overview of the mechanisms, applications and future perspectives. Environ Chem Ecotoxicol 6:293–302. https://doi.org/10.1016/J.ENCECO.2024.07.002

Dusengemungu L, Kasali G, Gwanama C, Ouma KO (2020) Recent advances in biosorption of copper and cobalt by filamentous fungi. Front Microbiol 11:582016. https://doi.org/10.3389/FMICB.2020.582016/XML

Echezonachi SO (2022) The role of white rot fungi in bioremediation. In: Microbes and microbial biotechnology for green remediation, pp 305–320. https://doi.org/10.1016/B978-0-323-90452-0.00034-7

Esterhuizen-Londt M, Hendel AL, Pflugmacher S (2017) Mycoremediation of diclofenac using Mucor hiemalis. Toxicol Environ Chem 99:795–808. https://doi.org/10.1080/02772248.2017.1296444

Gao N, Liu CX, Xu QM, Cheng JS, Yuan YJ (2018) Simultaneous removal of ciprofloxacin, norfloxacin, sulfamethoxazole by co-producing oxidative enzymes system of Phanerochaete chrysosporium and Pycnoporus sanguineus. Chemosphere 195:146–155. https://doi.org/10.1016/J.CHEMOSPHERE.2017.12.062

Garg S, Kim M, Romero-Suarez D (2025) Current advancements in fungal engineering technologies for Sustainable Development Goals. Trends Microbiol 33:285–301. https://doi.org/10.1016/J.TIM.2024.11.001

Huang Y, Zhang L, Yuan S, Liu W, Zhang C, Tian D, Ye X (2023) The production of oxalate by Aspergillus niger under different Lead concentrations. Agronomy 13. https://doi.org/10.3390/AGRONOMY13041182

Jaiswal S, Singh DK, Shukla P (2019) Gene editing and systems biology tools for pesticide bioremediation: a review. Front Microbiol 10:87. https://doi.org/10.3389/FMICB.2019.00087

Justo A, Hibbett DS (2011) Phylogenetic classification of Trametes (Basidiomycota, Polyporales) based on a five–marker dataset. Taxon 60:1567–1583. https://doi.org/10.1002/TAX.606003

Kartal SN, Munir E, Kakitani T, Imamura Y (2004) Bioremediation of CCA-treated wood by brown-rot fungi Fomitopsis palustris, Coniophora puteana, and Laetiporus sulphureus. J Wood Sci 50:182–188. https://doi.org/10.1007/S10086-003-0544-8/METRICS

Kijpornyongpan T, Schwartz A, Yaguchi A, Salvachúa D (2022) Systems biology-guided understanding of white-rot fungi for biotechnological applications: A review. iScience 25. https://doi.org/10.1016/J.ISCI.2022.104640/ASSET/6107FF59-1BFA-479E-8C82-1E3636644B09/MAIN.ASSETS/GR6_LRG.JPG

Lin S, Wei J, Yang B, Zhang M, Zhuo R (2022) Bioremediation of organic pollutants by white rot fungal cytochrome P450: the role and mechanism of CYP450 in biodegradation. Chemosphere 301:134776. https://doi.org/10.1016/J.CHEMOSPHERE.2022.134776

Liu SH, Zeng GM, Niu QY, Liu Y, Zhou L, Jiang LH, Tan X f, Xu P, Zhang C, Cheng M (2017) Bioremediation mechanisms of combined pollution of PAHs and heavy metals by bacteria and fungi: a mini review. Bioresour Technol 224:25–33. https://doi.org/10.1016/J.BIORTECH.2016.11.095

Matúš P, Littera P, Farkas B, Urík M (2023) Review on performance of Aspergillus and Penicillium species in biodegradation of organochlorine and organophosphorus pesticides. Microorganisms 11:1485. https://doi.org/10.3390/MICROORGANISMS11061485

Ningombam L, Mana T, Pradhan S, Apum G, Singh YD (2025) Fungal bioremediation in environmental pollution and recent strategies. Discov Environ 3:1–13. https://doi.org/10.1007/S44274-025-00267-X/TABLES/2

Priyanka, Dwivedi SK (2023) Fungi mediated detoxification of heavy metals: insights on mechanisms, influencing factors and recent developments. J Water Process Eng 53:103800. https://doi.org/10.1016/J.JWPE.2023.103800

Puranik S, Shukla L, Kundu A, Kamil D, Paul S, Venkadasamy G, Salim R, Singh SK, Kumar D, Kumar A (2023) Exploring potent fungal isolates from sanitary landfill soil for in vitro degradation of dibutyl phthalate. J Fungi 9:125. https://doi.org/10.3390/JOF9010125

Robalds A, Naja GM, Klavins M (2016) Highlighting inconsistencies regarding metal biosorption. J Hazard Mater 304:553–556. https://doi.org/10.1016/J.JHAZMAT.2015.10.042

Robles-González IV, Fava F, Poggi-Varaldo HM (2008) A review on slurry bioreactors for bioremediation of soils and sediments. Microb Cell Factories 7:5. https://doi.org/10.1186/1475-2859-7-5

Rodrigues Barbosa J, dos Santos Freitas MM, da Silva Martins LH, de Carvalho RN (2020) Polysaccharides of mushroom Pleurotus spp.: new extraction techniques, biological activities and development of new technologies. Carbohydr Polym 229:115550. https://doi.org/10.1016/J.CARBPOL.2019.115550

Rosa FM, Mota TFM, Busso C, de Arruda PV, Brito PEM, Miranda JPM, Trentin AB, Dekker RFH, da Cunha MAA (2024) Filamentous fungi as bioremediation agents of industrial effluents: a systematic review. Fermentation 10:143. https://doi.org/10.3390/FERMENTATION10030143/S1

Shetty SS, Deepthi D, Harshitha S, Sonkusare S, Naik PB, Madhyastha H (2023) Environmental pollutants and their effects on human health. Heliyon 9:e19496. https://doi.org/10.1016/J.HELIYON.2023.E19496/ASSET/008DDC1C-5AB7-41B0-84F1-A6F6C7CF98C5/MAIN.ASSETS/GR5.JPG

Shourie A, Vijayalakshmi U (2022) Fungal diversity and its role in mycoremediation. Geomicrobiol J 39:426–444. https://doi.org/10.1080/01490451.2022.2032883;SUBPAGE:STRING:ABSTRACT;JOURNAL:JOURNAL:UGMB20;WGROUP:STRING:PUBLICATION

Singh VK, Singh R (2024) Role of white rot fungi in sustainable remediation of heavy metals from the contaminated environment. Mycology 15:585. https://doi.org/10.1080/21501203.2024.2389290

Singh S, Mangla J, Singh S (2021) Evaluation of Aspergillus fumigatus NTCC1222 as a source of enzymes for detergent industry. Resour Environ Sustain 5:100030. https://doi.org/10.1016/J.RESENV.2021.100030

Tan Z, Beltrán-Flores E, Ramos-Meza GD, Alonso LL, Sarrà M (2025) Eliminating antibiotics by white-rot-fungi Trametes versicolor from manure solids and synthetic wastewater. Environ Pollut 378:126504. https://doi.org/10.1016/J.ENVPOL.2025.126504

Tekere M, Tekere M (2019) Microbial bioremediation and different bioreactors designs applied. In: Biotechnol Bioeng. https://doi.org/10.5772/INTECHOPEN.83661

Tsekova K, Todorova D, Ganeva S (2010) Removal of heavy metals from industrial wastewater by free and immobilized cells of Aspergillus niger. Int Biodeterior Biodegradation 64:447–451. https://doi.org/10.1016/J.IBIOD.2010.05.003

Vaksmaa A, Guerrero-Cruz S, Ghosh P, Zeghal E, Hernando-Morales V, Niemann H (2023) Role of fungi in bioremediation of emerging pollutants. Front Mar Sci 10:1070905. https://doi.org/10.3389/FMARS.2023.1070905/XML

Villani A, Fanelli F, Mulè G, Moretti A, Loi M (2025) Shedding light on Pleurotus: an update on taxonomy, properties, and photobiology. Microbiol Res 295:128110. https://doi.org/10.1016/J.MICRES.2025.128110

Vipotnik Z, Michelin M, Tavares T (2021) Development of a packed bed reactor for the removal of aromatic hydrocarbons from soil using laccase/mediator feeding system. Microbiol Res 245:126687. https://doi.org/10.1016/J.MICRES.2020.126687

Wijayawardene NN, Hyde KD, Mikhailov KV, Péter G, Aptroot A, Pires-Zottarelli CLA, Goto BT, Tokarev YS, Haelewaters D, Karunarathna SC, Kirk PM, André ALCM, Saxena RK, Schoutteten N, Wimalasena MK, Aleoshin VV, Al-Hatmi AMS, Ariyawansa KGSU, Assunção AR, Bamunuarachchige TC, Baral HO, Bhat DJ, Błaszkowski J, Boekhout T, Boonyuen N, Brysch-Herzberg M, Cao B, Cazabonne J, Chen XM, Coleine C, Dai DQ, Daniel HM, da Silva SBG, de Souza FA, Dolatabadi S, Dubey MK, Dutta AK, Ediriweera A, Egidi E, Elshahed MS, Fan X, Felix JRB, Galappaththi MCA, Groenewald M, Han LS, Huang B, Hurdeal VG, Ignatieva AN, Jerônimo GH, de Jesus AL, Kondratyuk S, Kumla J, Kukwa M, Li Q, Lima JLR, Liu XY, Lu W, Lumbsch HT, Madrid H, Magurno F, Marson G, McKenzie EHC, Menkis A, Mešić A, Nascimento ECR, Nassonova ES, Nie Y, Oliveira NVL, Ossowska EA, Pawłowska J, Peintner U, Pozdnyakov IR, Premarathne BM, Priyashantha AKH, Quandt CA, Queiroz MB, Rajeshkumar KC, Raza M, Roy N, Samarakoon MC, Santos AA, Santos LA, Schumm F, Selbmann L, Selçuk F, Simmons DR, Simakova AV, Smith MT, Sruthi OP, Suwannarach N, Tanaka K, Tibpromma S, Tomás EO, Ulukapı M, Van Vooren N, Wanasinghe DN, Weber E, Wu Q, Yang EF, Yoshioka R, Youssef NH, Zandijk A, Zhang GQ, Zhang JY, Zhao H, Zhao RL, Zverkov OA, Thines M, Karpov SA (2024) Classes and phyla of the kingdom Fungi. Fungal Divers 128:1–165. https://doi.org/10.1007/S13225-024-00540-Z

Yu G, Wen X, Qian Y (2005) Production of the ligninolytic enzymes by immobilized Phanerochaete chrysosporium in an air atmosphere. World J Microbiol Biotechnol 21:323–327. https://doi.org/10.1007/S11274-004-3571-8
Yuan T, Zhang S, Chen Y, Zhang R, Chen L, Ruan X, Zhang S, Zhang F (2021) Enhanced reactive blue 4 biodegradation performance of newly isolated white rot fungus Antrodia P5 by the synergistic effect of herbal extraction residue. Front Microbiol 12:644679. https://doi.org/10.3389/FMICB.2021.644679/BIBTEX
Zhang J, Chi Y, Feng L (2021) The mechanism of degradation of alizarin red by a white-rot fungus Trametes gibbosa. BMC Biotechnol 21:1–17. https://doi.org/10.1186/S12896-021-00720-8/FIGURES/8

Mycoremediation of Xenobiotics: Harnessing Fungi for Environmental Detoxification

12

Juwita Ratna Dewi, Reny Tiarantika, Devy Ulandari, and Hendrix Yulis Setyawan

Abstract

Mycoremediation, an application of fungi for the degradation of xenobiotics and other environmental pollutants, has been proven to be a multi-perspective, environmentally friendly, and sustainable biotechnology for water and soil treatment. In this chapter, the technological, ecological, and economical features of mycoremediation are emphasized, and the latest advances and future directions are targeted. Major progress is novel fungal enzyme development and design, transcriptomics and metabolomics pathway mapping of biodegradation, and CRISPR/Cas9-based strain improvement. Hybrid technologies that integrate mycoremediation with phytoremediation, electrokinetic remediation, and nanobioremediation are presented as mechanisms to maximize contaminant removal and scalability. Challenges persist in translating laboratory success to heterogeneous field environments, optimizing the production of the inoculum, and minimizing ecological risk. The chapter is also keen to highlight the significance of interdisciplinarity, regulatory measures, and public engagement in guaranteeing safe adoption. In addition to pollution cleanup, mycoremediation helps sequester carbon, enhance soil health, and promote climatic resilience, all of which are aligned with the United Nations' sustainability agenda. The argument not only places mycoremediation as a green technology but also as a driver for the development of a circular bioeconomy and mainstreaming climate policy.

J. R. Dewi · R. Tiarantika
Department of Environmental Sciences, Graduate School, Universitas Brawijaya, Malang, East Java, Indonesia

D. Ulandari · H. Y. Setyawan (✉)
Department of Agroindustrial Technology, Faculty of Agroindustrial Technology and Biosystem, Universitas Brawijaya, Malang, East Java, Indonesia
e-mail: hendrix@ub.ac.id

A. Roy et al. (eds.), *Mycoremediation of Xenobiotics*, Emerging Paradigms in Pharmaceutical Research, https://doi.org/10.1007/978-981-95-5743-1_12

Keywords

Xenobiotics · Mycoremediation · Circular bioeconomy · CRISPR/Cas9

12.1 Introduction

The massive rise of industrialization, vehicle emissions, overuse of chemicals in the agricultural system, and human activities are the major contributors to global pollution. These activities released numerous hazardous substances into the environment, including xenobiotics, human-made compounds, such as pharmaceuticals, pesticides, dioxin, and polychlorinated biphenyls (PCBs). Additionally, poor waste management by industries or the government intensifies this condition. Among the strategies employed to address these pollutants is the application of fungi for their degradation, a process known as myco-remediation.

Mycoremediation has been widely used to mitigate contaminated soil and wastewater. Employing fungi in mycoremediation facilitates ecological restoration by diversifying microbial communities and enforcing ecological balance (Akpasi et al. 2023; Hacıoğlu et al. 2024). Lignin-degrading isoenzymes of *P. ostreatus* and *A. bisporus* showed a potential utilization for LDPE depolymerization (Maisto et al. 2024). While suspended and immobilized, *Geotrichum candidum* profoundly decolorized textile effluent after 6 h of application (Rajhans et al. 2021b). In addition, white fungus, *Trametes* versicolor, in a closed system could effectively remove antibiotics up to 90% (Tan et al. 2025), and other fungal species have shown similar capabilities against a wide spectrum of xenobiotics (Anand et al. 2023; Roy et al. 2025). These findings suggest that mycoremediation is a versatile, eco-friendly, and a promising biotechnological approach for the sustainable management of environmental pollutants.

Although previous studies have shown that mycoremediation is a promising approach in managing pollutants, however, the current understanding remains largely restricted to controlled condition or laboratory scale. In reality, these pollutants exist in a complex environmental system where the efficacy of the fungal degradation is subject to temperature, pH, moisture content, and the presence of other contaminants (Dasgupta et al. 2024; Roy et al. 2025). Furthermore, the incomplete knowledge of fungal metabolism pathway, particularly the enzymatic mechanism essential for xenobiotics degradation, remains a challenge. The potential secretion of toxic fungal by-product secretion and their subsequent ecological impact, as well as the overall effectiveness of the degradation process under real environmental condition, also remain understudied (Pujiati et al. 2024; Sośnicka et al. 2022).

12.2 Technological Advancements

The advances of late in biotechnology that have improved the capacity of mycoremediation tenfold to degrade xenobiotics are enhanced enzyme performance, identification and development of novel enzymes (Khan 2025b), and the application of genomic and gene editing technology to improve fungi performance (Ganesh Kumar et al. 2021). Transcriptomic and metabolic analyses have also been employed to identify a new regulatory network to control the degradation of pollutants (Peidro-Guzmán et al. 2021). Through the method of traditional random mutagenesis to enhance genetic manipulation, fungal strains have been enhanced to develop more effective ligninolytic and degradative enzymes, thus increasing their capacity to break down various pollutants (Kryczyk-Poprawa et al. 2020; Roy et al. 2025; Sharma et al. 2018; Venice et al. 2020). *Aspergillus sydowii* genome studies displayed distinctive pathways accountable for high-molecular-weight hydrocarbon degradation, paving the way for the coming gene (Ganesh Kumar et al. 2021). Additionally, genome editing is now possible with ultrahigh accuracy using next-generation tools like CRISPR/Cas9, allowing target improvement of fungal metabolic processes and advancement of xenobiotic degradation even further (Stein et al. 2018).

Genetically modified recombinant enzyme is now manufactured on a larger scale with enhanced stability and potency for efficient mycoremediation (Sharma et al. 2018). The utilization of advanced continuous bioreactor systems is also favorable for fungal growth environment throughout xenobiotic degradation (Subbulakshmi et al. 2021). These technologies enhance the reusability and stability of fungal biomass, producing increased rates of degradation and reducing ecological disturbance. As mycoremediation is based on fungal natural processes, it has a smaller carbon footprint compared to conventional procedures, which cater to the increasing need of clean technologies in curbing xenobiotic pollution in different ecosystems (Bhandari et al. 2021; Dhagat and Jujjavarapu 2022). Overall, they enhance the effectiveness, dependability, and sustainability of mycoremediation.

While biotechnology has far improved xenobiotics bioremediation, the discipline is also improving day by day and is still far from needing drastic challenges and research direction to be undertaken. Among them, one of high promises is genomic and other omics technologies application. NGS and bioinformatic development tool provide us deeper knowledge about the fungal metabolic pathway and how they interact with xenobiotics (Achudhan et al. 2023). Omics dataset, such as genomics, metabolomics, proteomics, and transcriptomics, might also be used to deconvolute toxicity mechanism and biodegradation pathway (Achudhan et al. 2023). Alternatively, synthetic biology, wherein collaboration with design of synthetic biological model and engineered microbial consortia offers a new platform to improve recalcitrant contaminants. In this case, gene-editing tools and synthetic genetic networks are used for more selective and efficient degradation of recalcitrant pollutants (Jaiswal and Shukla 2020).

Enhancing enzyme performance remained paramount in controlling xenobiotics. Thus, studies need to aim at augmenting stability, catalytic activity, and yield, particularly laccase and peroxidases, via genetic and protein engineering methodologies (Rahman et al. 2024). Concurrently, nanobioremediation, being a boundary of biotechnology and nanotechnology, is gradually emerging as a viable strategy. This method focuses on nanomaterial biosynthesis for pollution abatement and life cycle assessment (LCA) for environmental impact assessment (Pandey et al. 2022).

Nevertheless, intensive research on environmental impact as well as sustainability remains extremely scarce, especially to estimate the ecological impact of applying genetically altered fungi and nanomaterials in mycoremediation, demonstrating the need for prolonged studies (Pandey et al. 2022; Perera and Hemamali 2022).

12.3 Synergy with Other Environmental Technologies

Various bioremediation methods, such as mycoremediation, phytoremediation, and bioaugmentation, have been employed to tackle contaminated soil and water. Each method has its distinctive mechanism, benefits, and limitations. For instance, mycoremediation is capable of degrading a wide range of pollutants, including complex organic pollutants (Ali et al. 2017), extract heavy metals (Dou et al. 2024), and the fungi can survive under non-sterilized medium (Sadañoski et al. 2020). However, the effectiveness of this method strictly depends on the environmental conditions, such as pH, temperature, substrate, and mycelium age (Ali et al. 2017; Antón-Herrero et al. 2023; Khan 2025b). On the other hand, phytoremediation, which utilizes plants and its associated microbes, is capable of degrading, removing, or transforming various contaminants and is considered a cost-effective and environmentally friendly approach but is restricted by the plant's ability to absorb the contaminant and requires long-term commitment and monitoring (Badamasi et al. 2025; Zhu et al. 2023). These limitations of the methods and the complexity and diversity of the environment potentially reduced the effectiveness of the degradation process. Therefore, integrating these bioremediation procedures may provide better results since combining the strengths of each method may effectively remove a wider range of contaminants and enhance the general bioremediation outcome.

Myco-electrochemical remediation is a novel approach combining mycoremediation with electrokinetic remediation (EKR). This approach has shown a significant improvement in reducing petroleum hydrocarbon both in sandy and clay soil compared to mycoremediation or EKR alone (Lelli et al. 2025). Integrating these approaches brought a positive synergistic effect where the application of EKR contributed to improving the distribution and activity of the fungi in the soil, especially in the heterogenous soil when the pollutant is not evenly distributed (Crognale et al. 2020; Lelli et al. 2025). In addition, mycoremediation, which relies on the activity of fungal enzyme, the degrading effect of this process can be further amplified by applying EKR, which pronouncedly enhance the effectiveness of the degradation process (Choudhury et al. 2025; Lelli et al. 2025). Although research in integrating EKR with mycoremediation is limited to date, EKR has been widely integrated with other bioremediation approach which shown remarkable results in addressing heavy metals and organic pollutants (Bokde et al. 2025; Hamdi et al. 2025; He et al. 2018), indicating a similar potential to integrate EKR with mycoremediation. Hence, several recommendations are proposed to optimize the integration of these methods: (1) optimizing the operation condition under the combined EKR with mycoremediation, which include the fungal species, electrode material, and operational condition that support both the EKR and fungal growth (Fdez-Sanromán et al. 2021; Lelli et al. 2025); (2) enhancing the sustainability of this integrated approach, by utilizing renewable energy resources (Hassan 2021); and (3) scaling up the integrated approach to fit field application (Bokde et al. 2025; Fdez-Sanromán et al. 2021). The future research and development of this integrated approach may provide a solution in addressing the challenges to improve their effectiveness and broaden the application in diverse environmental conditions.

Mycoremediation and phytoremediation hybridity is extensively researched. The hybrid process has shown promise as a possible substitute for enhancing the efficiency of environmental remediation, for instance, xenobiotics removal. Macro fungal extracellular enzymes have been effectively used for the degradation of complex organic pollutants and heavy metal removal (Ali et al. 2017; Hidalgo et al. 2023). Fungi are also identified to transform toxic complex organic compounds into less toxic via mycelium. When applied to soils, mycoremediation will monitor soil erosion and water holding capacity that will in turn enhance the ecosystem quality (Maruška et al. 2024). Conversely, phytoremediation employing plants and the accompanying microbial as significant agents has been effective in degrading, removing, transforming, or immobilizing contaminants from water and soil (Farraji et al. 2020; Parveen et al. 2021; Paulo et al. 2023; Singh et al. 2024). The synergistic effect of mycoremediation and phytoremediation can guarantee increased degradation or removal of intricate pollutants by optimizing the strengths of plants and fungi. For example, research on the remediation of polycyclic aromatic hydrocarbon (PAH)-polluted soil by using willow (*Salix viminalis*) and white rot fungus (*Crucibulum laeve*) was found to exhibit extreme efficacy in the decontamination of phenanthrene, pyrene, and benzo(a)pyrene over single treatment (Ma et al. 2021). However, there are a number of challenges still present in applying this hybrid approach, including the following: (1) Contaminated soil nature—this includes pollutant concentration, soil type, vegetation, weather factors, and the occurrence of other microorganisms (Phang et al. 2024; Wang and Aghajani Delavar 2024). (2) Cost-effectiveness—phytoremediation is fashionable due to its economic viability; the application of this integrated approach should still remain economically viable when applied at a greater scale. The integration of remediation into economically viable operations, like bioenergy and agriculture, is possible (Haller and Jonsson 2020; Yadav et al. 2017). (3) Technology readiness level—most of the present available data are lab-scale experiments; thus, more effort is required to create a standard procedure and protocol for field application (Liang et al. 2024; Maisto et al. 2024).

Therefore, to overcome the limitation of this approach, several future directions are proposed, such as optimizing the interaction between the plant and fungi to enhance remediation efficacy. This can be done by studying or combining different plant and fungi species (Adeyemi et al. 2022; Li et al. 2022); developing an integrated system by adapting current technology, like utilizing an advanced phytoreactor, which is suitable for phyto- and mycoremediation processes (Samudro and Mangkoedihardjo 2023); conducting various comprehensive field trial and scaling up model for a broader application (Liang et al. 2024; Wang and Aghajani Delavar 2024); and using the emerging biotechnological advancement, such as genetic engineering and genomic approach to select a specific plant and fungi that yield the highest effectivity and manage to address various contaminants in one site can be further considered (Sahoo 2024; Takata et al. 2013).

Another emerging research interest area is nanobioremediation, an integrated approach that combines nanotechnology with bioremediation. Principally, this technology utilizes nanomaterials, including nanoparticles, such as titanium oxide, zero-valent iron, and other metal oxides, nanotubes, and nanocomposites, which combine with biological systems, including fungi, for remediation purposes. The implementation of nanobioremediation means harnessing the strength of nanotechnology and bioremediation to optimize the remediation process. Previous studies have shown that nanoparticles have the potential to stimulate the activity of microorganism, thus promoting pollutant absorption or serving as a catalytic agent to accelerate the degradation process, and facilitate site-specific delivery of bioremediation agents (Andrada Suarez et al. 2025; Indraratne et al. 2021; Sahu et al. 2025; Stein et al. 2018). Integrating nanomaterials with fungi has shown a significant degradation of xenobiotics. A short-term exposure of silver nanoparticles (AgNPs) significantly enhanced the enzymatic activity and viability of white-rot fungi in decolorizing and degrading pollutants such as azo dyes (Wang et al. 2022). Likewise, a combination of magnetite nanoparticles (Fe_3O_4) with *Aspergillus flavus* demonstrated a remarkable efficiency in dye removal and subsequently reduced the toxicity of textile wastewater (Polli et al. 2023). However, the compatibility of fungi and nanomaterials, environmental conditions, impact of the nanomaterials on the environment, sustainability of the process, the regulatory framework, and standard protocol for this integrated approach remain to be elucidated.

The interplay between fungi and nanomaterials is highly complex, as it may yield synergistic or inhibitory effects on the remediation process. Since mycoremediation involves fungi, maintaining the optimum environmental conditions, such as pH, temperature, and substrate or pollutant concentration, is essential to facilitate optimal remediation. The addition of nanomaterials may affect the physiology of the fungi and extracellular enzymes activity, which limits the remediation processes (Biswal 2023; Choudhury et al. 2025; Kryczyk-Poprawa et al. 2020). Thus, the stability and biosynthesis of fungal-assisted nanoparticles required further investigation to ensure consistent results of remediation (Biswal 2023; Yadav et al. 2017). The introduction of nanomaterials into the environment poses ecological risk, including toxicity to nontarget organisms and the potential for bioaccumulation. High concentration of nanoparticles reportedly leads to DNA damage and induce oxidative stress in plants (Sharma et al. 2025). Thus, a life cycle assessment (LCA) study is necessary to evaluate the impact of nanobioremediation (Pandey et al. 2022). Also, the existing regulations and protocols to address the health and safety aspects of nanoparticles are insufficient. Bridging this regulatory gap is essential to ensure their safety application (Badamasi et al. 2025; Singh et al. 2024).

Therefore, future efforts regarding the application of myco-nanoparticles remediation should focus on green synthesis, such as fungal-based green synthesis of silver nanoparticles using *Aspergillus brasiliensis,* to comply with sustainable and eco-friendly remediation (de Oliveira et al. 2025). Additionally, implementing an AI-assisted tool to predict toxicity and optimize myco-nanoparticles remediation provides mitigation strategies to reduce the risk and optimize the efficiency of remediation (Sharma et al. 2025). Ultimately, large-scale field trial and the development of comprehensive policy frameworks are crucial to ensuring the safety and effectiveness of this approach (Badamasi et al. 2025). Hence, the collaboration among policymakers, researchers, and industry is crucial in establishing a standard framework that facilitates innovations without compromising public safety (Khan 2025a).

12.4 Scaling and Commercialization

12.4.1 Challenges in Scaling Up Fungal Remediation Processes for Large-Scale Applications

There are serious problems faced in scaling up fungi-based remediation processes, mostly related to effectiveness and generalizability. Fungi, namely, *Aspergillus niger* and *Trichoderma pseudokoningii*, have also demonstrated a great potential to degrade hydrocarbons and other toxic organic pollutants (Amobonye et al. 2023; El-Bouhy et al. 2024; Khilji et al. 2021; Mahmud et al. 2022). The development of such waves varies, but their fast readiness is not the same under all field conditions. Humidity, temperature, pH, and others are equally of great effect on the ability of the fungi to reach their targets (El-Bouhy et al. 2024; Khilji et al. 2021). More dramatic weather can slow down or shut down the fungi's work of cleaning up contaminants. Thus, fungal bioremediation in nature hardly fulfills the expectation, which means it is a little bit hard but promising to further develop it.

Large-scale production of the fungal inoculum requires massive resources, and thus, the demand for fermenters is a difficult challenge, in addition to the long production history that often reaches up to 120 days (Maruška et al. 2024; Sama et al. 2023). Furthermore, the application of fungi for field inoculation often encounters difficulties in establishing due to an unfavorable environment. A major drawback comes from the competition with faster-growing microorganism. Moreover, low content of nutrients in the growth medium can restrict inoculum multiplication. Characteristics of some microbes can be seen to promote each other's impact. Even though meilleur site de rencontre gratuit en France identifying the best blend of species and quantities is still a challenge, a wrong recipe to be a drawback (Zulfiqar et al. 2023).

Hence, it is necessary to find out which combinations are the most effective. It is considered that microbial consortia contribute to increased efficiency by their unique characteristics. Another approach with great potential is genetic engineering, which can endow fungi with increased resistance in challenging environments. Both approaches will need further research before they can be readily adopted. The development of inoculum is still an important problem to address to apply fungal bioremediation successfully, and the high cost related to application might be a bottleneck for scaling up the technology. This is relatively greener compared to chemical or physical methods, but the initial cost is also usually high (Rajhans et al. 2021a; Zulfiqar et al. 2023). These costs include infrastructure, machinery, and high-quality inoculum, and these can be particularly challenging for developing countries that want to start the technologies (Peekate and Ogolo 2024; Sattar et al. 2022). Still, the long-term gains from cutting pollution could be significant. An intriguing new field is utilizing nanotechnology to increase the effectiveness of fungi. The synergistic action of fungi and pollutants can be enhanced by nanomaterials, allowing for faster and stronger remediation, as previously presented (Ghosh et al. 2023; Pete et al. 2021). Although the investment at the beginning of the project is high, the technology can be economical in the long term, making fungal bioremediation more attractive. These advancements are critical for the development of more accessible and economical strategies for fungal removal in the future.

The main problem we are addressing stems from the variability and fluctuations of impurities in industrial rubbish. Although fungi degrade diverse organic substances, their efficiency may be impeded when materials are combined (Alao and Adebayo 2022; Melati et al. 2022). For example, some fungi are only limited in their ability to degrade polycyclic aromatic hydrocarbons due to interference from competing substances. Contaminants of long-range transport impair fungal metabolism by inhibiting their activity (Bala et al. 2022; Gugel et al. 2024). Understanding the interplay of chemicals and fungi in their native habitat is essential. The complexity of field conditions makes it difficult to address those challenges; therefore, it is important to support the bench experiment by field studies to form better and more realistic remediation plans.

Table 12.1 shows that hybrid strategies that combine quantitative data from different areas, like bio-based innovation, environmental technology, and economics, could get rid of most of the big problems that stop businesses from growing. For example, using agricultural waste as a substrate is one important step that could significantly lower the overall cost of developing inoculum while still following the current rules for a circular economy.

The development of industrial mycoremediation on a large scale represents a challenge for the effective use of these technologies in industry. The efficacy of fungi in remediation processes may vary, depending on environmental conditions, nutrient type, or toxicity level. The commercial-scale production of the inoculum of fungal strains is more expensive in comparison to that of the conventional processes (Bala et al. 2022; Gugel et al. 2024). Improvements in fungal bioremediation processes are necessary, even though there are a lot of practical problems and challenges that come with fungal remediation. Recent advancements in computer simulation technology may enable precise predictions of fungal growth dynamics across various environmental contexts (El-Bouhy et al. 2024; Sama et al. 2023). These new technologies give scientists new and exciting options, especially for sites that are very polluted and pose the greatest risk to humans. However, it is important to remember that there are still many ethical and technological challenges to overcome. Before any new technology can be used in routine clinical practice, it must be thoroughly tested and validated to make sure it is safe and effective. In conclusion, the future of fungal bioremediation appears promising, yet it faces significant threats.

Table 12.1 Challenges and effective strategies for expanding mycoremediation initiatives

Key challenges	Impact	Solution strategies	Implementation examples
Variation in environmental conditions	Reduced fungal effectiveness	Selection of tolerant fungal strains	Use of *Pleurotus ostreatus* on agricultural waste
High inoculum production costs	Limited industrial scale	Optimization of low-cost culture media	Agricultural waste as a culture substrate
Bioreactor limitations	Unstable fungal growth	Development of specialized bioreactor designs	Liquid bioreactor for Trametes versicolor
Low viability in the field	Failure of long-term remediation	Long-lasting spore formulations	Fungal spore pellets for soil application
Complexity of pollutants	Fungi are less effective on mixed pollutants	Multi-fungal consortia	Combination of *Phanerochaete chrysosporium* and *Aspergillus niger*

12.4.2 Development of Bioreactors, Fungal Cultivation Techniques, and Delivery Systems

Improving bioreactors, fungal cultures, and microorganism delivery systems is considered a key part of the development of cost-effective technology-based bioremediation. Bioreactor design is very important for the industrial use of mycoremediation. The bioreactors enabled the degradation of the pollutants by the fungus to occur in a relatively controlled environment in comparison with direct use in field studies. The degradation environment can be improved by changes in environmental factors, including pH, temperature, humidity, levels of oxygen, and so forth, which are important to be considered, because they mainly affect the efficiency of pollutant breakdown. An efficient approach is to trap defined microbial consortia within microspheres.

Oleaginous fungi: recent advancements in the synthesis of fungal triacylglycerides. The potential of oleaginous fungi for biodiesel production utilizing industrial waste (Athoillah and Ahmad 2022). Their research indicated that *Yarrowia lipolytica* and *Lipomyces starkeyi* could eliminate 66–80% of COD from palm oil mill effluents. These results show how fungi can help with bioremediation and energy in two different ways. This combined method is a clean way to get rid of pollution and also helps with the world's energy shortage. We can use the unique traits of these filamentous fungi to make bioreactors that grow them efficiently and use less fossil fuel. These insights provide a foundation for the advancement of next-generation bioreactor systems that convert waste into value within a circular bioeconomy. This suggests the potential of oleaginous fungi as a sustainable technology in the future (Athoillah and Ahmad 2022).

A bioreactor needs a good environment to work well. The pH, temperature, and C:N (Carbon/Nitrogen) ratio are all very important for the bioremediation process. Changes in these factors affect how quickly mercury-resistant bacterial consortia grow (Amandita et al. 2023). This shows how important it is to customize each bioreactor for the specific microorganism being used. In an ambient environment, the temperature changes in the system slow down enough that the microbes can work at their best to keep the system running.

This stability is crucial if the microbes are to survive over the long term. If not properly controlled, the efficacy of bioremediation is dramatically decreased. Hence, adaptive designs that consider the needs of different organisms in bioreactor engineering are crucial, as comprehensive information of their desired environment is required (Amandita et al. 2023). Fungi culturing techniques play crucial roles in biotechnology, especially in bioconversion and production of high-value products, such as enzymes, secondary metabolites, and biomass. One common approach involves fermentation in bioreactors.

The microorganisms' support is the key to bioremediation success. Bioaugmentation is a common technology in which pollutant-metabolizing microorganisms are added to the soil. Zahir et al. (2024) studied microbial signal transduction in natural and engineered communities and reported that the mode in which the individual microorganisms are interacting can hamper the performance of the systems. Accordingly, such systems may possess higher sensitivity to environmental variations. Additionally, this approach could be employed in bioenergy generation, since bioaugmentation combined with the reconstruction of complex microbial populations offers new opportunities.

In addition, dedicated bioreactors have been evaluated for their potential to combat oil pollution (Sayed et al. 2021). It has been reported that the use of oil spill booms is suitable for containing bioremediation reactors and is also effective. This approach provides a greater capacity for bioaugmentation and biostimulation and results in increased TPH degradation. A further benefit is its ability to reduce emissions of pollutants. This method has also been used to streamline post-bioremediation recovery of the pollutants, thus improving the efficiency in recovering post-biodegradation of substances. In addition, natural biofilm consortia are also of great significance, as Bôto et al. (2021) highlighted the activity of native microbial communities in hydrocarbon degradation.

The effect on bioreactors and the new technologies is also very important for the future of bioremediation. Kaushal and Fulekar (2022) have put forth the theory of the contemporary era in the wake of nanotechnology, on a new way to use nano-rhizoremediation. The theory is based on a vision and aims to improve remediation policies by using the unique properties of nanomaterials and unusual plant root systems. The main goal of this method is to make remediation much more effective and successful in all kinds of settings.

12.4.3 Economic Feasibility and Commercialization of Mycoremediation Technologies

Mycoremediation is becoming more popular in many fields because of its potential to make money as a biotechnological practice that aims to reduce pollution in the environment. This new method is being recognized as one of the most cost-effective and eco-friendly options. Second, it is important that this tool helps solve the complicated problem of total release of pollutants, which Antón-Herrero et al. (2022) describe as human pollution (organic matter) and toxicants (heavy metals, hydrocarbons). This shows that this technology has many uses and is important for protecting the environment.

So, what is mycoremediation? Mycoremediation is the use of the natural properties of fungi to break down unwanted materials using biology. Fungi release enzymes, both extractable and nonextractable, and degrade and adsorb pollutants by means of biosorption and bioaccumulation. This development allows for environmental cleaning without causing harm to the environment, making mycoremediation especially appropriate for industries facing rigid environmental mandates. Furthermore, this method is believed to be more general than most chemical types, as many of them are costly and environmentally affecting. Consequently, mycoremediation has a number of both advantages and disadvantages, which makes it a good situation for future industrial applications (Efremenko et al. 2024).

One of the main obstacles to commercializing mycoremediation has been the relatively high cost of developing it first. Rapid production of large amounts of fungal biomass is unfeasible without specialized facilities and abundant substrates. Nevertheless, the use of local fungal species may save costs for the process, as reported by researchers, where the use of native fungi can decrease costs (Chang et al. 2021). These beneficial properties could make local fungal production more efficient. In addition, fungi possess biological versatility, enabling them to adapt to a wide range of pollutants. This malleable nature allows a wide application of the mycoremediation process. For example, fungal waste materials might be used to remove pollutants, such as phthalates. We are putting the circular economy into action by giving these waste streams a new purpose.

For now, this technological solution is one of the most critical for global pollution and can be an effective fuel to boost economies. To use the technology effectively, all innovations must be well-planned and well-organized (Mahmud et al. 2022; Silva et al. 2022). In addition, it does not only mean environmentally beneficial technologies but also social and economic benefits. As industries must develop more and more environmentally friendly production methods, eco-friendly ways of product development are becoming beneficial. Moreover, the products or materials manufactured with pollution cleaning processes have a more significant value and comply with the global developments of the green economy. The support of the government and public policies and private investment can promote these technologies. Therefore, the future of the development of mycoremediation is promising on two levels: environmental care and economic growth.

12.4.4 Industrial Applications in Agriculture, Mining, Wastewater Treatment, and Urban Pollution

Biotechnology is becoming more and more important to businesses, and it plays a big role in their success. Some areas that are moving away from traditional solutions and toward bio-based ones are agriculture, mining, waste management, and urban services. Research, including that of Deng and Wu (2024), investigates the integration of fungi into solar-assisted processes for the secure management of waste and the enhancement of environmental quality. Fungal treatment can also help with liquid waste, which is one of the main reasons why water gets dirty. This means that technology can work in both cities and rural areas. Fungi can break down organic pollutants that stay in waste streams for a long time. This can also help cities. Some fungi change harmful chemicals into biomolecules that can be used in cosmetics, medicine, and food. This is good for both the world and business. This two-in-one benefit shows how useful mycoremediation can be in cities today.

12.5 Climate Change Resilience and Mitigation

12.5.1 The Role of Mycoremediation in Carbon Sequestration and Soil Health

Mycoremediation uses fungi to store carbon and make soil better. Fungi break down pollutants and change the structure and fertility of the soil by making stable organic compounds that help the soil while breaking down pollutants. Chemicals from the breakdown help plants grow and store carbon without messing up the natural balance of nutrients.

Mycoremediation also plays a significant role in managing industrial pollutants. The ability of fungi to degrade recalcitrant organic pollutants (ROPs), a group of polluting substances that are resistant to biodegradation due to their structural complexity and resistance of these compounds to degradation, is well established. Fungi also clean the soil and increase the amount of organic carbon in ecosystems (Qin et al. 2023). Because fungi can do both of these things, they help keep the ecosystems healthy and keep important ecological functions going. The process also affects the natural carbon cycle, which shows how important fungi for environment balance.

The payoff is a stronger, more resilient soil, making fungi an important part of sustainability-based remediation. Apart from hydrocarbons, fungi are also involved in heavy metal detoxification. It has been studied that metal-resistant fungi indigenous to a region could uptake and detoxify the harmful heavy metals (Akram et al. 2023). These fungi not only decrease the level of poisonous metals but also diversify the soil microbial community, leading to improved soil fertility. In general, it was found that soils that were less fertile had a lower potential for carbon storage, indicating the relationship between metal clean-up and carbon sequestering.

Mycoremediation is also being applied to the degradation of synthetic compounds, including plastics. Some fungi could colonize low-density polyethylene. The process of colonization creates new organic compounds, which benefit the soil by improving soil structure and increasing porosity. Thus, the soil is better aerated and fertile (Khruengsai et al. 2021), growing the plant for maximum growth and the ability to store more carbon. This means that we are never more than 25 fungi from at least one object that gets digested axiomatically, and the uncertainty once again affirms the role that fungi play in carbon sequestration, as mycoremediation becomes an ecological answer to the plastic question, and our worlds of environmental restoration begin to expand.

Mycoremediation is pertinent not only in the control of pollutants but also as part of global climate resilience (Table 12.2), such as the following fungi: mycorrhiza and white rot—counter above-ground and below-ground traffic through soil carbon sequestration. Fungi contribute to C soil sequestration by increasing C storage and stabilization as organic matter. It's a valuable contribution to increased soil fertility and expansion of the carbon cycle (and less CO_2 in the air). In addition, fungi are also important for soil aggregation, reducing erosion and water infiltration, all are essential for adaptation to the changing climate. Recent research shows that fungi may be able to capture as much carbon as a third of all the fossil fuel emissions in the world.

Table 12.2 Contribution of mycoremediation to carbon sequestration and soil health

Indicators	Quantitative data	Climate impact	Source
Plant carbon transfer to mycorrhizal fungi	1–13% of plant carbon	Increases soil carbon storage	Hawkins et al. (2023)
Global carbon transfer to *mycorrhizal* fungi	≈ 13 gigatons CO_2/year	Equivalent to 36% of fossil fuel emissions	Huxtable (2023)
Increased soil organic carbon (SOC)	+17.2%	Improves fertility and carbon sequestration	Wang et al. (2016)
Increased plant photosynthesis (Pn) due to AMF	+15.3–33.1%	Supports plant productivity in extreme climates	Wang et al. (2016)
Degradation of hazardous compounds (PCDD/Fs) by *Pleurotus pulmonarius* fungi	Up to 96% in 72 days	Reduces carbon loads from persistent pollution	Stiebeling and Labes (2022)
Degradation of hazardous compounds (PCDD/Fs) by *Pleurotus pulmonarius* fungi	US$ 3.12 per gram per day (complex medium) and US$ 0.69 g^{-1} d^{-1} (N-limited medium)	For marine fungi, bioremediation methods for complex waste/media	Stiebeling and Labes (2022)

12.5.2 Contribution of Mycoremediation to Climate Change Adaptation and Resilience

Fungi can live in many different places because they can change to fit their surroundings. They may live in a wide range of places, from dirty cities to farmland. Because they are so strong, fungi are very helpful in solving today's environmental problems (Maček et al. 2022). Fungi are quite flexible, so they can be used in various kinds of cleanup plans in the United States and around the world. Changes in the climate, on the other hand, are also to blame for difficulties, such the rise in the number of harmful fungus populations. Jenks et al. (2023) have posited that such modifications could adversely affect plant life and destabilize soil microbial ecosystems. In view of this concern, it is clear that mycoremediation methods that deal with both contaminants and pathogens are very important. A coordinated strategy would help ecosystems stay healthy even if the weather changed. Because of this, it's evident that we need to make mycoremediation technologies to deal with new problems.

Mycoremediation can make farming better for the environment. Fungal inoculants help crops grow better by making nutrients more available in the soil (Gupta 2025). This means that less synthetic pesticides and fertilizers are needed. This, in turn, means that less pollution is emitted into the air, which contributes to climate change. Adding mycoremediation to farming systems helps keep the soil healthy and makes sure there is enough food. Mycoremediation could help people deal with the effects of climate change if professionals from different professions work together. The regeneration of land can be expedited by integrating fungal studies with traditional land management practices (Qin et al. 2023). To get more people to use this technology, the community needs to be involved as well (Hamm et al. 2022). When the general population gets involved, it not only makes bioremediation take less time, but it also creates new chances for businesses to make money in the field. If more research is done, the right rules are put in place, and if more people learn about mycoremediation, it might become an important part of making ecosystems more adaptable to climate change.

12.5.3 Integration of Mycoremediation into Climate Policy and Environmental Restoration

Adding mycoremediation to climate policies and initiatives for restoring the environment creates a lot of chances for cleaning up pollution and making ecosystems stronger. Fungi-assisted technology can help the soil hold more carbon, break down organic contaminants, and make it more fertile. In the past, rehabilitation attempts have mostly used mechanical or chemical procedures, which typically come with substantial costs for both people and the environment. Mycoremediation is a better option for the environment that costs a lot less and has both bioremediation and carbon storage benefits. Fungi are becoming more and more known as a tool in "nature-based solutions" because they can do two things at once. International organizations like the IPCC (Intergovernmental Panel on Climate Change) and the UNFCCC (United Nations Framework Convention on Climate Change) are starting to talk about the possible uses of fungi. However, these ideas haven't yet been widely adopted in public policy because there isn't enough information or real-world proof. Pilot projects spearheaded by governments, universities, and business entities can yield empirical data that transitions the notion from theoretical research to actual, policy-relevant implementations.

From the standpoint of environmental governance, the integration of mycoremediation into climate policy fosters sustainable land utilization. Kalogiannidis et al. (2024) stress how important it is to include climate change scenarios in planning at the regional and local levels. Using fungi in environmental planning models helps local governments get the most out of their soil recovery efforts and boost land productivity while stopping additional damage from intensive farming. This decentralized model lets global policies be changed to fit local needs and connects climate initiatives directly to ecosystem restoration.

In this way, fungi not only improve soil quality but also strengthen the ecosystem's carrying capacity for flora and fauna. Policies that integrate mycoremediation with tree planting programs will be more effective than focusing solely on reforestation. This is because nutrient-poor soil is often a major obstacle to restoration success. The role of fungi in improving soil quality makes rehabilitation programs more sustainable. Fungi can be considered natural partners in any ecosystem restoration effort. This fact reinforces the urgency of integrating mycoremediation into the environmental restoration agenda. Thus, fungi become a strategic component in maintaining the sustainability of global ecosystems.

Mycoremediation also contributes to urban water management. Nature-based solutions, including fungi, can strengthen stormwater management (Leone et al. 2022). This technology functions as a bioremediation for urban runoff. Its application can reduce flood risk and improve water quality in city canals. Urban aquatic ecosystems become healthier and support biodiversity. This approach demonstrates the relevance of fungi not only in agriculture but also in urban planning. Integrating mycoremediation into spatial planning policies offers both ecological and social benefits. Therefore, this technology can be part of urban climate adaptation strategies. The dual benefits of flood control and environmental improvement underscore its urgency, strengthening the position of fungi as a cross-sectoral solution.

Climate policy must also consider public health aspects. Urban pollution has significant impacts on human health. Integrating mycoremediation into spatial planning can reduce pollution and improve air quality (Allam and Soomauroo 2024). Reducing pollution can reduce the risk of environmentally related diseases. This strategy aligns with the urban public health agenda. Integrating environmental and health policies offers dual benefits to residents. Fungal-based remediation can improve long-term quality of life. Furthermore, such inclusive policies can reduce public health costs. Thus, fungi act as a bridge between environmental and public health policies. This fact reinforces the importance of an interdisciplinary approach in climate policy. Mycoremediation has a direct impact on both human well-being and ecosystem sustainability.

From an economic perspective, the integration of mycoremediation can also be linked to a carbon credit mechanism. The carbon storage process by fungi can be measured, verified, and calculated as a tangible contribution to emissions reduction. This opens up opportunities for companies to gain financial benefits through the carbon market. Thus, mycoremediation provides not only ecological benefits but also economic ones. This technology must be developed efficiently to avoid burdening financing (Gupta 2025). If standards for quantifying carbon storage by fungi are available, the legitimacy of mycoremediation within the carbon credit scheme will be strengthened. This will accelerate its adoption in the industry and international levels. Companies can leverage this technology to meet their sustainability targets. Policies that combine environmental aspects with financial incentives will be more readily accepted. This fact demonstrates that balancing ecology and economics is key to successful integration.

Despite its immense potential, the integration of mycoremediation still faces challenges. One major obstacle is the limited understanding of policymakers. Many still view fungal technology as a laboratory experiment. The lack of published success stories on an industrial scale means this technology has not yet become a priority on the environmental agenda. Thus, improved science communication is critical. We as researchers need to do a better job communicating our research to policymakers.

To ensure the secure and efficient integration of fungal technology into environmental policy, it is essential to establish legal frameworks. This is an unequivocal requirement. Governments must implement methods to guarantee that the deployed fungal strains are safe, do not harm local ecosystems, and lack invasive traits. To reach this goal, the business needs to put in place systems for certification, monitoring, and verification. If these strategies are used, it will be clear and responsible what the remedial tasks are. One of the most important things that may help people and businesses trust technology is having clear, well-followed rules. The absence of clear legal frameworks could impede public acceptance and reduce excitement for the deployment of mycoremediation. Several developed nations have implemented legislation governing the use of living animals for environmental remediation. These frameworks can be modified to address the unique circumstances prevalent in other countries. Robust regulatory frameworks establish the necessary basis for the responsible and widely accepted expansion of fungal remediation techniques. This is essential to attain the intended outcomes.

The social dimension significantly influences the success of mycoremediation integration. Community involvement in remediation projects promotes more sustainable success. A sense of ownership over restoration outcomes grows when communities participate from the early stages of implementation. Project sustainability is more assured if local communities feel they are a vital part of the process. Good policies should provide space for community participation at every stage of the program. A community-based approach can also strengthen residents' adaptive capacity to face the pressures of climate change. The relevance of environmental technology cannot be separated from the social aspects surrounding it. The effectiveness of ecosystem restoration projects increases significantly when communities are involved as partners. Public participation also creates shared solidarity in preserving the environment. This fact confirms that inclusive policies are a prerequisite for successful mycoremediation integration.

The private sector plays a strategic role in supporting the implementation of mushroom technology. Companies play a role through funding, research support, and direct implementation in the field. Social responsibility obligations encourage many companies to seek ways to reduce their ecological footprint. Mycoremediation can be a practical strategy to fulfill this commitment. Policies that provide incentives will attract broader involvement from industry. Reliance on public funding alone will limit the scope of implementation, making private sector contributions crucial. Policy design should be geared toward facilitating cross-sector collaboration. Multi-stakeholder involvement accelerates large-scale technology adoption. Such collaboration is key to accelerating the transition to sustainable development. This fact demonstrates that the successful integration of mycoremediation depends on partnerships between various actors.

Mycoremediation offers a complementary combination of ecological, social, and economic benefits. This technology can reduce pollution while strengthening the ecosystem's carrying capacity. The carbon sequestration process facilitated by fungi significantly supports climate mitigation efforts. Soil and water quality can be restored, making ecosystems better prepared to cope with the pressures of climate change. The resulting ecological benefits also impact public health. Economic value arises through potential integration into emerging carbon credit mechanisms. Widespread implementation can only be achieved through regulatory support, incentives, and community engagement. Collaboration between government, industry, academia, and communities is a critical factor in the success of this strategy. Fungi can ultimately function as natural partners in maintaining climate balance and the sustainability of global ecosystems. This fact confirms mycoremediation's position as an integral part of future solutions for sustainable development.

12.6 Global Implementation and Public Awareness

The application of fungi for ecological detoxification, referred to as mycoremediation, has increasingly become popular as a low-cost and environmentally friendly option for remediating contaminated sites. Despite the many promising prospects, levels of uptake of mycoremediation vary with regions, differences in infrastructure, inputs, and technological capabilities. Keeping these regional variances in mind; incorporating measures to enhance public awareness, education, and acceptance; and large-scale fungal-related remediation technologies are crucial, especially in areas with scarce resources.

Geographic differences as the primary drivers for mycoremediation technology adoption have been extensively researched. Adoption of mycoremediation technologies is extremely diverse by regions, and that is stimulated most by regional economy differences, environmental policy differences, and availability of scientific resources. Antón-Herrero et al. (2023) theorize that the selection of species of fungi and environmental factors, like economic resources, technological equipment, and environmental policy, differ significantly region-by-region since the main drivers for scalability of mycoremediation. In others, such as the United States and Germany, the technologies are an inherent part of overall environmental management practices, attracted large research grants, and possess highly evolved regulatory mechanisms (Geris et al. 2024). Experiments have proven the efficiency of species, such as *Aspergillus niger* and *Pleurotus ostreatus*, in breaking down persistent organic pollutants and heavy metals in polluted soil (Yadav et al. 2021). On the other hand, poor resource areas experience bottlenecks in using mycoremediation due to limited access to scientific information, lower budgets, and inadequate infrastructure (Geris et al. 2024).

Low research areas, especially in Africa, Asia, and Latin America, are experiencing increased levels of environmental pollution due to industrial activities, drainage from agriculture, and poor waste management. Rani et al. (2025) state that mycoremediation is a new solution with a green and affordable substitute for the traditional remediation treatment. For instance, large oil spills in the Amazon of Ecuador have led to enormous hydrocarbon pollution, but research has determined that indigenous strains possess enzymatic mechanisms to degrade petroleum hydrocarbons with promise for recultivation of the soil (Cipriani-Avila et al. 2025). Likewise, where there are war-torn regions, e.g., Eastern European and Middle Eastern countries, soils contaminated with residues of explosives and heavy metals are of great ecological and public health concern. The research exhibits that mycoremediation can effectively degrade such pollutants, offering inexpensive and viable solutions to soil decontamination (Geris et al. 2024; Khatua et al. 2024).

The public's lack of information on the advantages of remediation employing fungi in many locations can discourage large-scale implementation. Raising awareness of the local communities uses educational programs, seminars, and even research participation. Collaborations between universities, civil society organizations, and local governments allow the sharing of information and promote community involvement in environmental management practices (Rani et al. 2025). Further, integrating the principles of mycoremediation into the education body at different levels can stimulate students' awareness of environmental issues and of sustainable development. Public acceptance is also influenced by perceiving risk and benefit. Effective communication of the safety, efficacy, and ecological effects of mycoremediation is needed. The generation of pilot trial and actual case data can reinforce public confidence and emphasize the achievability of fungus-based remediation as a viable and cost-effective methodology to control the environment (Cipriani-Avila et al. 2025).

12.7 Ethical, Societal, and Legal Considerations

The implementation of genetically modified (GMOs) fungi in environmental remediations presents significant ethical, societal, and legal challenges. While such approaches offer substantial environmental gains, their ethical adoption depends heavily on public perceptions and societal acceptance. Regulatory GMOs differ widely, highlighting the need for considerations of both legal obligations and ethical responsibilities throughout research and implementation.

The use of GM fungi for environmental remediation may disrupt native ecosystems or the transfer of modified traits to nontarget species. These risks underscore the necessity of conducting thorough ecological risk assessments prior to introducing genetically modified fungi into natural environments (Pollution Sustainability Directory 2025c). Moreover, ethical discussions have arisen around the deliberation of fungal genome alteration, particularly regarding the long-term effects on biodiversity and the natural genetic diversity of fungal populations. Such modifications may trigger unforeseen ecological interactions, emphasizing the importance of careful ethical evaluation and engagement with the public during the development and implementation of GM fungi (Choudhury et al. 2025).

Regulations governing the use of GMOs, including fungal species, differ across countries but share a common goal of safeguarding both the environment and public health. In the United States, oversight is provided through the Coordinated Framework for Biotechnology, with agencies such as the USDA and EPA, responsible for evaluating potential risks to agriculture and ecosystems (US Department of Agriculture 2025). Within the European Union, GMO governance is shaped by the precautionary principle, requiring rigorous safety evaluations prior to environmental release. These assessments consider possible impacts on human and animal health, as well as ecological consequences, while policies on traceability and labeling ensure transparency and consumer awareness (European Commission 2025).

Public perception as a key determinant in the successful adoption of fungal-based bioremediation technologies indicates that understanding and trust among the public significantly affect the uptake of bioremediation methods. Misunderstanding or limited awareness of it can create resistance, whereas open communication and active community involvement can help promoting acceptance and support (Pollution Sustainability Directory 2025d). Additionally, societal acceptance is closely tied to perceived benefits, including environmental sustainability and cost efficiency. When communities recognize positive outcomes of fungal remediation, such as reducing reliance on chemicals and enhanced soil quality, they are more likely to endorse and engage these initiatives (Pollution Sustainability Directory 2025c).

Although considerable progress has been made in terms of mycoremediation, gaps in knowledge remain important. These include insufficient insight into the long-term ecological effects of fungal-based remediation, limitations of the biochemical mechanism understanding driving fungal degradation of complex contaminants, and inconsistent performance of fungi under various environmental conditions (Geris et al. 2024). Closing the gaps will require dedicated research aimed at clarifying the interactions between fungi and pollutants, identifying the ecological factors that influence fungal activity across diverse habitats, and evaluating potential unintended impacts in natural ecosystems (Pollution Sustainability Directory 2025a). Interdisciplinary collaborations play a critical role in advancing the efficiency of mycoremediation technologies. Partnerships involving mycologists, environmental researchers or engineers, chemists, and engineers can foster the creation of more effective fungal strains, the refinement of remediation techniques, and the development of scalable applications (Thirumalaivasan et al. 2024). A combination of perspectives from multiple fields can be a comprehensive approach of mycoremediation, considering not only the biological dimensions but also the chemical, physical, and socioeconomic factors that determines overall success of remediation initiatives (Pollution Sustainability Directory 2025a).

Recent progress in research techniques has greatly advanced the understanding of fungal mechanisms of bioremediation. The use of omics-based approaches, including genomics, proteomics, and metabolomics, provides detailed insights into fungal responses to pollutants at the molecular level (Thirumalaivasan et al. 2024). In addition, advances in computational modeling and bioinformatics make it possible to predict fungal interactions with environmental contaminants, supporting the development of more targeted and effective bioremediation strategies (Pollution Sustainability Directory 2025b).

12.8 Future Research Directions

Substantial progress has been made in mycoremediation research; however, key knowledge gaps remain. Targeted research initiatives are needed to improve the efficacy, scalability, and sustainability of fungal remediation strategies. Future efforts should adopt an interdisciplinary approach, integrating insights from microbiology, environmental engineering, policy studies, and social sciences to maximizing fungi's role in addressing environmental contamination.

A major limitation in advancing mycoremediation lies in the incomplete characterization of fungal enzymatic pathways that drive the breakdown of persistent xenobiotics, including pesticides, plastics, and polycyclic aromatic hydrocarbons (Akpasi et al. 2023). Many of the existing research has concentrated on individual contaminants; however, environmental pollution typically consists of complex mixtures. The synergistic or antagonistic effects of these chemical combinations during fungal degradation remain insufficiently explored (Kumar et al. 2021). Another unresolved issue is the transition from lab-scale studies to practical applications. Although fungi such as *Pleurotus ostreatus* and *Trametes versicolor* have demonstrated strong degradation potential under controlled conditions, their efficacy in heterogeneous and large-scale environments is far less predictable (Singh and Singh 2024). Concerns also arise regarding the long-term ecological consequences of deploying fungal species, especially genetically engineered strains, into natural ecosystems, as this may disrupt ecological balance and raise biosafety challenges (Goncalves et al. 2022). Additionally, research on fungal-based remediation within aquatic systems is relatively underdeveloped compared to soil-focused investigations, especially about fungal survival, adaptability, and pollutant transformation in dynamic water environments (Behera & Das 2023). These knowledge gaps emphasize the need for more comprehensive ecological, molecular, and environmental studies to improve the reliability and scalability of fungal detoxification technologies.

Addressing these gaps requires the application of omics technologies (such as genomics, proteomics, and metabolomics) to elucidate the metabolic pathways that support fungal degradation of xenobiotics (AbuQamar et al. 2023). These approaches hold the potential to uncover new enzymes and regulatory networks that can be engineered or optimized for bioremediation purposes. Large-scale pilot projects also needed to evaluate fungal bioremediation performance in real environmental conditions. Establishing collaborative frameworks among academic institutions, government bodies, and industrial stakeholders could give support, such as demonstrations, while simultaneously providing valuable data on ecological safety, cost efficiency, and long-term viability (Akhtar and Mannan 2020). Furthermore, engaging with local communities ensures that mycoremediation practices are implemented in ways that are socially acceptable and culturally sensitive (Choudhury et al. 2025). International collaborations may accelerate knowledge exchange and technology transfer while enabling the evaluation of indigenous fungal strains across diverse ecosystems. Such collaborations would enhance the global relevance and applicability of fungal-based remediation strategies (Geris et al. 2024).

The advancement of mycoremediation will increasingly rely on interdisciplinary collaboration. Integrating mycological expertise with environmental engineering offers some opportunities to refine fungal delivery systems, such as immobilized matrices, fungal mats, or engineered bioreactors that can enhance pollutant accessibility and degradation efficiency (Anverkhan and Pandya 2022). Contributions from materials science are equally valuable, particularly in developing composite carriers designed to improve fungal stability and persistence in polluted environments (Thirumalaivasan et al. 2024). Moreover, integrating ecological modeling with environmental chemistry can strengthen predictions regarding contaminant behavior and fungal activity under variable ecological conditions (Ningombam et al. 2025). The role of social sciences should not be overlooked, as they provide critical insights into public perception, regulatory frameworks, and community engagement, which ultimately shape the acceptance and practical adoption of mycoremediation strategies (Maglione et al. 2024). Ultimately, the future success of fungal-based remediation rests on a multidisciplinary convergence that not only advances mechanistic understanding but also fosters inclusive, scalable, and socially responsive environmental solutions.

12.9 Conclusion

Mycoremediation is an exciting area of green environmental management, a biologically driven, low-carbon remedy to traditional remediation technologies. Its potential is not limited to pollutant degradation but extends to the restoration of ecosystem health, soil aggregation, and carbon sequestration, making it of immense worth in the domain of climate change mitigation and adaptation. However, scale-up to large applications from laboratory-controlled tests will be limited by restrictions in fungal inoculum production, environmental heterogeneity, and possible biosafety issues. The innovation will rely on an interlinked multidisciplinary effort incorporating biotechnology, materials science, ecological modelling, and social science with the support of enabling regulations and public goodwill. Academic-industrial-policymaker interface will be crucial to evolve standard procedures, scale-up demonstration, and generate public trust. Through sustained innovation and integration of systems, mycoremediation can emerge as a central driver in worldwide strategies for pollution control, climatic adaptability, and circular economy transformation.

References

AbuQamar SF, Abd El-Fattah HI, Nader MM, Zaghloul RA, Abd El-Mageed TA, Selim S, Omar BA, Mosa WF, Saad AM, El-Tarabily KA, El-Saadony MT (2023) Exploiting fungi in bioremediation for cleaning-up emerging pollutants in aquatic ecosystems. Mar Environ Res 190(June):106068. https://doi.org/10.1016/j.marenvres.2023.106068

Achudhan AB, Masilamani M, Kannan P, Saleena LM (2023) In-silico integration in environmental remediation. In: Recent trends in constructed wetlands for industrial wastewater treatment. Springer Nature Singapore, pp 249–263. https://doi.org/10.1007/978-981-99-2564-3_13

Adeyemi NO, Sakariyawo OS, Soremi PAS, Atayese MO (2022) Phytoremediation using arbuscular mycorrhizal fungi. In: Current developments in biotechnology and bioengineering. Elsevier, pp 73–92. https://doi.org/10.1016/B978-0-323-99907-6.00016-5

Akhtar N, Mannan MA u (2020) Mycoremediation: expunging environmental pollutants. Biotechnol Rep (Amst) 26:e00452. https://doi.org/10.1016/j.btre.2020.e00452

Akpasi SO, Anekwe IMS, Tetteh EK, Amune UO, Shoyiga HO, Mahlangu TP, Kiambi SL (2023) Mycoremediation as a potentially promising technology: current status and prospects—a review. Appl Sci 13(8):4978. https://doi.org/10.3390/app13084978

Akram MB, Khan I, Rehman MU, Sarwar A, Ullah N, Rahman SU, Aziz T, Alharbi M, Alshammari A, Alasmari AF (2023) Mycoremediation of heavy metals contaminated soil by using indigenous metallotolerant fungi. Pol J Chem Technol 25(3):1–13. https://doi.org/10.2478/pjct-2023-0019

Alao MB, Adebayo EA (2022) Fungi as a veritable tool in bioremediation of polycyclic aromatic hydrocarbons-polluted wastewater. Z Allg Mikrobiol 62(3–4):223–244. https://doi.org/10.1002/jobm.202100376

Ali A, Guo D, Mahar A, Wang P, Shen F, Li R, Zhang Z (2017) Mycoremediation of potentially toxic trace elements—a biological tool for soil cleanup: a review. Pedosphere 27(2):205–222. https://doi.org/10.1016/S1002-0160(17)60311-4

Allam Z, Soomauroo Z (2024) Inaction in urban climate policy undermines health. Environ Health Insights 18:11786302241246909. https://doi.org/10.1177/11786302241246909

Amandita FY, Efadeswarni, Ikhwani, A. Z. N., Idris I, Sugiharto A, Fatma YS, Alfiansah YR, Sudiana IM (2023) The growth of mercury-resistant bacterial consortium at different levels of pH, temperature, and C:N Ratio. IOP Conf Ser Earth Environ Sci 1201(1):12068. https://doi.org/10.1088/1755-1315/1201/1/012068

Amobonye A, Aruwa CE, Aransiola SA, Omame J, Alabi TD, Lalung J (2023) The potential of fungi in the bioremediation of pharmaceutically active compounds: a comprehensive review. Front Microbiol 14:1207792. https://doi.org/10.3389/fmicb.2023.1207792

Anand U, Dey S, Bontempi E, Ducoli S, Vethaak AD, Dey A, Federici S (2023) Biotechnological methods to remove microplastics: a review. Environ Chem Lett 21(3):1787–1810. https://doi.org/10.1007/s10311-022-01552-4

Andrada Suarez EE, Roca Jalil ME, Fernandez Baldo MA, Cuozzo SA (2025) Nanobiotechnology approaches for the remediation of persistent and emerging organic pollutants: strategies, interactions, and effectiveness. Environ Sci Nano 12(2):979–1011. https://doi.org/10.1039/D4EN00424H

Antón-Herrero R, García-Delgado C, Baena N, Mayans B, Delgado-Moreno L, Eymar E (2022) Assessment of different spent mushroom substrates to bioremediate soils contaminated with petroleum hydrocarbons. Appl Sci 12(15):7720. https://doi.org/10.3390/app12157720

Antón-Herrero R, Chicca I, García-Delgado C, Crognale S, Lelli D, Gargarello RM, Herrero J, Fischer A, Thannberger L, Eymar E, Petruccioli M, D'Annibale A (2023) Main factors determining the scale-up effectiveness of mycoremediation for the decontamination of aliphatic hydrocarbons in soil. J Fungi 9(12):1205. https://doi.org/10.3390/jof9121205

Anverkhan PS, Pandya A (2022) Mycoremediation- a powerful planetary healer. Int J All Res Educ Sci Methods 10(6):1929–1941

Athoillah AZ, Ahmad FB (2022) Biodiesel production from bioremediation of palm oil mill effluent via oleaginous fungi. Clean-Soil Air Water 50(4):2200025. https://doi.org/10.1002/clen.202200025

Badamasi H, Aliyu Abdullahi U, Praveen Kumar A, Durumin Iya NI, Varra V, Ademola Olaleye A, Falalu Hamza M (2025) Nanotechnology-assisted phytoremediation of heavy metal contaminated soils: a state-of-the-art review on recent progress, challenges, and future directions. Soil and sediment contamination: an international journal, pp 1–44. https://doi.org/10.1080/15320383.2025.2536035

Bala S, Garg D, Thirumalesh BV, Sharma M, Sridhar K, Inbaraj BS, Tripathi M (2022) Recent strategies for bioremediation of emerging pollutants: a review for a green and sustainable environment. Toxics 10(8):484. https://doi.org/10.3390/toxics10080484

Behera AD, Das S (2023) Ecological insights and potential application of marine filamentous fungi in environmental restoration. Reviews in Environmental Science and Biotechnology 22(2):281–318. https://doi.org/10.1007/s11157-023-09655-2

Bhandari S, Poudel DK, Marahatha R, Dawadi S, Khadayat K, Phuyal S, Shrestha S, Gaire S, Basnet K, Khadka U, Parajuli N (2021) Microbial enzymes used in bioremediation. J Chem 2021:1–17. https://doi.org/10.1155/2021/8849512

Biswal T (2023) Nanobioremediation and its application for sustainable environment. In: Phytoremediation. Springer, pp 463–485. https://doi.org/10.1007/978-3-031-17988-4_22

Bokde AA, Subash A, Gholap V, Kandasubramanian B (2025) Electrokinetic soil-water remediation: a comprehensive exploration. Separation & Purification Reviews, pp 1–22. https://doi.org/10.1080/15422119.2025.2537073

Bôto ML, Magalhães C, Perdigão R, Alexandrino DAM, Fernandes JP, Bernabéu AM, Ramos S, Carvalho MF, Semedo M, LaRoche J, Almeida CMR, Mucha AP (2021) Harnessing the potential of native microbial communities for bioremediation of oil spills in the Iberian Peninsula NW coast. Front Microbiol 12:633659. https://doi.org/10.3389/fmicb.2021.633659

Chang B, Yang C-P, Yang C (2021) Application of fungus enzymes in spent mushroom composts from edible mushroom cultivation for phthalate removal. Microorganisms 9(9):1989. https://doi.org/10.3390/microorganisms9091989

Choudhury M, Adhikari MD, Agarwal S, Samanta P, Sharma A, Kundu D, Kumar S (2025) Pharmaceuticals and personal care products in soil: sources, impacts and myco-remediation strategies. Emerging Contam 11(2):100488. https://doi.org/10.1016/j.emcon.2025.100488

Cipriani-Avila I, Decock C, Zambrano-romero A, Zaldumbide K, Garcés-ruiz M, Caiza-olmedo J, Gordillo A, Luna V, Gerin PA (2025) Mycoremediation of petroleum-contaminated soil using native Ganoderma and Trametes strains from the Ecuadorian Amazon. J Fungi 11(9):651

Crognale S, Cocarta DM, Streche C, D'Annibale A (2020) Development of laboratory-scale sequential electrokinetic and biological treatment of chronically hydrocarbon-impacted soils. New Biotechnol 58:38–44. https://doi.org/10.1016/j.nbt.2020.04.002

Dasgupta D, Barman S, Sarkar J, Mridha D, Labrousse P, Roychowdhury T, Acharya K, Sarkar J, Chakraborty N (2024) Mycoremediation of different wastewater toxicants and its prospects in developing value-added products: a review. J Water Process Eng 58:104747. https://doi.org/10.1016/j.jwpe.2023.104747

de Oliveira RJV, Sousa FLN, Freitas DV, Silva FAC, de Almeida TS, Aguilera P, Machado G, Araújo BGP (2025) Synthesis of silver nanoparticles co-stabilized by carboxymethylcellulose using a sugarcane endophytic Aspergillus brasiliensis. The Microbe 6:100223. https://doi.org/10.1016/j.microb.2024.100223

Deng Z, Wu Z (2024) Design and evaluation of solar photocatalytic reclamation for purification of agricultural wastewater. Ijepes 2(2):27–44. https://doi.org/10.62051/ijepes.v2n2.04

Dhagat S, Jujjavarapu SE (2022) Utility of lignin-modifying enzymes: a green technology for organic compound mycodegradation. J Chem Technol Biotechnol 97(2):343–358. https://doi.org/10.1002/jctb.6807

Dou R, Xie Y, Liu FX, Wang B, Xu F, Xiao K (2024) In situ mycoremediation of acid rain and heavy metals co-contaminated soil through microbial inoculation with Pleurotus ostreatus. Sci Total Environ 912:169020. https://doi.org/10.1016/j.scitotenv.2023.169020

Efremenko E, Stepanov N, Senko O, Aslanli A, Maslova O, Lyagin I (2024) Using fungi in artificial microbial consortia to solve bioremediation problems. Microorganisms 12(3):470. https://doi.org/10.3390/microorganisms12030470

El-Bouhy Z, Reda RM, Mohamed F, Elashhab M, Rahman ANA (2024) Polycyclic aromatic aydrocarbons (PAHs) pollution approaches in aquatic ecosystems: perils and remedies using green technologies. Zagazig Vet J 52(1):49–87. https://doi.org/10.21608/zvjz.2024.239217.1221

European Commission (2025) Commission welcomes provisional political agreement on new genomic techniques for plants [Press release]. https://ec.europa.eu/commission/presscorner/detail/en/ip_25_2912

Farraji H, Robinson B, Mohajeri P, Abedi T (2020) Phytoremediation: green technology for improving aquatic and terrestrial environments. Nippon J Environ Sci 1(1):1–30. https://doi.org/10.46266/njes.1002

Fdez-Sanromán A, Pazos M, Rosales E, Sanromán MÁ (2021) Prospects on integrated electrokinetic systems for decontamination of soil polluted with organic contaminants. Curr Opin Electrochem 27:100692. https://doi.org/10.1016/j.coelec.2021.100692

Ganesh Kumar A, Manisha D, Sujitha K, Magesh Peter D, Kirubagaran R, Dharani G (2021) Genome sequence analysis of deep sea Aspergillus sydowii BOBA1 and effect of high pressure on biodegradation of spent engine oil. Sci Rep 11(1):9347. https://doi.org/10.1038/s41598-021-88525-9

Geris R, Malta M, Soares LA, de Souza Neta LC, Pereira NS, Soares M, Reis V d S, Pereira M d G (2024) A review about the mycoremediation of soil impacted by war-like activities: challenges and gaps. J Fungi 10(2):94. https://doi.org/10.3390/jof10020094

Ghosh S, Rusyn I, Dmytruk OV, Dmytruk KV, Onyeaka H, Gryzenhout M, Gafforov Y (2023) Filamentous fungi for sustainable remediation of pharmaceutical compounds, heavy metal and oil hydrocarbons. Front Bioeng Biotechnol 11:1106973. https://doi.org/10.3389/fbioe.2023.1106973

Gugel I, Summa D, Costa S, Manfredini S, Vertuani S, Marchetti F, Tamburini E (2024) Mycoremediation of synthetic azo dyes by white-rot fungi grown on diary waste: a step toward sustainable and circular bioeconomy. Fermentation 10(2):80. https://doi.org/10.3390/fermentation10020080

Gonçalves MFM, Esteves AC, Alves A (2022) Marine Fungi: Opportunities and Challenges. Encyclopedia 2(1):559–577. https://doi.org/10.3390/encyclopedia2010037

Gupta AK (2025) Climate change, natural disasters, and cutaneous fungal infections. Int J Dermatol 64(8):1349–1355. https://doi.org/10.1111/ijd.17908

Hacıoğlu B, Dupaul G, Paladino G, Edman M, Hedenström E (2024) Unlocking the biodegradative potential of native white-rot fungi: a comparative study of fiberbank organic pollutant mycoremediation. Bioengineered 15(1):2396642. https://doi.org/10.1080/21655979.2024.2396642

Haller H, Jonsson A (2020) Growing food in polluted soils: a review of risks and opportunities associated with combined phytoremediation and food production (CPFP). Chemosphere 254:126826. https://doi.org/10.1016/j.chemosphere.2020.126826

Hamdi FM, Altaee A, Aedan Y, Zhou J, Zaidi SJ, Alsaka L, Almalki R, Al-Askar A, Samal AK (2025) Black tea waste/iron slag reactive filter media-electrokinetic for mixed heavy metals treatment from contaminated site. J Contam Hydrol 270:104517. https://doi.org/10.1016/j.jconhyd.2025.104517

Hamm PS, Montoya KN, Montoya L, Cook K, Liphardt S, Taylor JW, Cook JA, Natvig DO (2022) Breathing can be dangerous: opportunistic fungal pathogens and the diverse community of the small mammal lung mycobiome. Front Fungal Biol 3:996574. https://doi.org/10.3389/ffunb.2022.996574

Hassan I (2021) Applications of electrokinetic remediation in environmental restoration. https://doi.org/10.11159/iceptp21.lx.103

Hawkins H-J, Cargill RIM, Van Nuland ME, Hagen SC, Field KJ, Sheldrake M, Soudzilovskaia NA, Kiers ET (2023) Mycorrhizal mycelium as a global carbon pool. Curr Biol 33(11):R560–R573. https://doi.org/10.1016/j.cub.2023.02.027

He J, He C, Chen X, Liang X, Huang T, Yang X, Shang H (2018) Comparative study of remediation of Cr(VI)-contaminated soil using electrokinetics combined with bioremediation. Environ Sci Pollut Res 25(18):17682–17689. https://doi.org/10.1007/s11356-018-1741-8

Hidalgo J, Epelde L, Anza M, Becerril JM, Garbisu C (2023) Mycoremediation with Agaricus bisporus and Pleurotus ostreatus growth substrates versus phytoremediation with Festuca rubra and Brassica sp. for the recovery of a Pb and γ-HCH contaminated soil. Chemosphere 327:138538. https://doi.org/10.1016/j.chemosphere.2023.138538

Huxtable A (2023) Fungi stores a third of carbon from fossil fuel emissions and could be essential to reaching net zero. The University of Sheffield. https://sheffield.ac.uk/news/fungi-stores-third-carbon-fossil-fuel-emissions-and-could-be-essential-reaching-net-zero

Indraratne SP, Pierzynski GM, Baker LR, Vara Prasad PV (2021) Nano-oxides immobilize cadmium, lead, and zinc in mine spoils and contaminated soils facilitating plant growth. Can J Soil Sci 101(3):543–554. https://doi.org/10.1139/cjss-2020-0127

Jaiswal S, Shukla P (2020) Alternative strategies for microbial remediation of pollutants via synthetic biology. Front Microbiol 11:808. https://doi.org/10.3389/fmicb.2020.00808

Jenks JD, Prattes J, Wurster S, Sprute R, Seidel D, Oliverio M, Egger M, Río C d, Sati H, Cornely OA, Thompson GR, Kontoyiannis DP, Hoenigl M (2023) Social determinants of health as drivers of fungal disease. EClinicalMedicine 66:102325. https://doi.org/10.1016/j.eclinm.2023.102325

Kalogiannidis S, Kalfas D, Papaevangelou O, Chatzitheodoridis F, Katsetsiadou K-N, Lekkas E (2024) Integration of climate change strategies into policy and planning for regional development: a case study of Greece. Land 13(3):268. https://doi.org/10.3390/land13030268

Kaushal RS, Fulekar MH (2022) Rhizo-nanoremediation: overview on practices: advances and perspective for the treatment of contaminants. Int J Front Biol Pharm Res 2(2):16–28. https://doi.org/10.53294/ijfbpr.2022.2.2.0037

Khan J (2025a) Bioremediation of contaminated soil and ground water: a review. Annu Methodol Arch Res Rev 3(7):233–243. https://doi.org/10.63075/tzx02z07

Khan MF (2025b) Fungi for sustainable pharmaceutical remediation: enzymatic innovations, challenges, and applications—a review. Processes 13(4):1034. https://doi.org/10.3390/pr13041034

Khatua S, Simal-Gandara J, Acharya K (2024) Myco-remediation of plastic pollution: current knowledge and future prospects. Biodegradation 35(3). Springer Netherlands. https://doi.org/10.1007/s10532-023-10053-2

Khilji SA, Aqeel M, Maqsood MF, Khalid N, Tufail A, Sajid ZA, AL-surhanee AA, Hashem M, Alamri S, Al-Mutairi KA, Noman A (2021) Hemarthria compressa—Aspergillus niger—Trichoderma pseudokoningii mediated trilateral perspective for bioremediation and detoxification of industrial paper sludge. Sustainability 13(21):12266. https://doi.org/10.3390/su132112266

Khruengsai S, Sripahco T, Pripdeevech P (2021) Low-density polyethylene film biodegradation potential by fungal species from Thailand. J Fungi 7(8):594. https://doi.org/10.3390/jof7080594

Kryczyk-Poprawa A, Piotrowska J, Żmudzki P, Opoka W, Muszyńska B (2020) Feasibility of the use of Lentinula edodes mycelium in terbinafine remediation. 3 Biotech 10(4):184. https://doi.org/10.1007/s13205-020-02177-6

Kumar A, Yadav AN, Mondal R, Kour D, Subrahmanyam G, Shabnam AA, Khan SA, Yadav KK, Sharma GK, Cabral-Pinto M, Fagodiya RK, Gupta DK, Hota S, Malyan SK (2021) Myco-remediation: a mechanistic understanding of contaminants alleviation from natural environment and future prospect. Chemosphere 284(April):131325. https://doi.org/10.1016/j.chemosphere.2021.131325

Lelli D, Pasvantu IC, Crognale S, Streche C, Cocârtă DM (2025) Sequential use of electrochemical and myco-remediation for sustainable depollution of hydrocarbon contaminated soils. Int J Environ Sci Technol 22:16111–16122. https://doi.org/10.1007/s13762-025-06616-w

Leone A, Grassini L, Balena P (2022) Urban planning and sustainable storm water management: gaps and potential for integration for climate adaptation strategies. Sustainability 14(24):16870. https://doi.org/10.3390/su142416870

Li X, Zhang Z, Luo J, Cui X, Xu J, Hou Y, Hao B, Guo W (2022) Arbuscular mycorrhizal fungi and nitrilotriacetic acid regulated Suaeda salsa growth in Cd-contaminated saline soil by driving rhizosphere bacterial assemblages. Environ Exp Bot 193:104669. https://doi.org/10.1016/j.envexpbot.2021.104669

Liang Z, Neményi A, Kovács GP, Gyuricza C (2024) Incorporating functional traits into heavy metals phytoremediation: the future of field-based phytoremediation. Ecol Indic 166:112262. https://doi.org/10.1016/j.ecolind.2024.112262

Ma X, Li X, Liu J, Cheng Y, Zou J, Zhai F, Sun Z, Han L (2021) Soil microbial community succession and interactions during combined plant/white-rot fungus remediation of polycyclic aromatic hydrocarbons. Sci Total Environ 752:142224. https://doi.org/10.1016/j.scitotenv.2020.142224

Maček I, Pintarič S, Šibanc N, Rajniš T, Kastelec D, Leštan D, Suhadolc M (2022) Plants play a crucial role in the development of soil fungal communities in the remediated substrate after EDTA washing of metal-contaminated soils. Front Environ Sci 10:978850. https://doi.org/10.3389/fenvs.2022.978850

Maglione G, Zinno P, Tropea A, Mussagy CU, Dufossé L, Giuffrida D, Mondello A (2024) Microbes' role in environmental pollution and remediation: a bioeconomy focus approach. AIMS Microbiology 10(3):723–755. https://doi.org/10.3934/microbiol.2024033

Mahmud T, Sabó I, Lambu ZN, Danlami D, Shehu AA (2022) Hydrocarbon degradation potentials of fungi: a review. J Environ Bioremediat Toxicol 5(1):50–56. https://doi.org/10.54987/jebat.v5i1.681

Maisto M, Zuzolo D, Tartaglia M, Prigioniero A, Ranauda MA, Germinario C, Falzarano A, Castelvetro V, Sciarrillo R, Guarino C (2024) Advances in plastic mycoremediation: focus on the isoenzymes of the lignin degradation complex. Sci Total Environ 948:174554. https://doi.org/10.1016/j.scitotenv.2024.174554

Maruška A, Mickienė R, Kaškonienė V, Grigiškis S, Stankevičius M, Drevinskas T, Kornyšova O, Donati E, Tiso N, Mikašauskaitė J, Zacchini M, Levišauskas D, Ragažinskienė O, Bimbiraitė-Survilienė K, Kanopka A, Dūda G (2024) Searching for chemical agents suppressing substrate microbiota in white-rot fungi large-scale cultivation. Microorganisms 12(6):1242. https://doi.org/10.3390/microorganisms12061242

Melati I, Rahayu G, Henny C (2022) The recent status of synthetic dyes mycoremediation: a review. IOP Conf Ser Earth Environ Sci 1062(1):12029. https://doi.org/10.1088/1755-1315/1062/1/012029

Ningombam L, Mana T, Pradhan S, Apum G, Singh YD (2025) Fungal bioremediation in environmental pollution and recent strategies. Discov Environ 3(1):100. https://doi.org/10.1007/s44274-025-00267-x

Pandey G, Chauhan R, Yadav AS, Bajpai S (2022) LCA of nanomaterials for bioremediation. In: Nanotechnology for environmental remediation. Wiley, pp 413–431. https://doi.org/10.1002/9783527834143.ch22

Parveen S, Bhat IUH, Khanam Z, Rak AE, Yusoff HM, Akhter MS (2021) Phytoremediation: in situ alternative for pollutant removal from contaminated natural media: a brief review. Biointerface Res Appl Chem 12(4):4945–4960. https://doi.org/10.33263/BRIAC124.49454960

Paulo AM, Caetano NS, Marques APGC (2023) The potential of bioaugmentation-assisted phytoremediation derived maize biomass for the production of biomethane via anaerobic digestion. Plants 12(20):3623. https://doi.org/10.3390/plants12203623

Peekate LP, Ogolo JM (2024) The potential of Serratia marcescens in bioremediation of crude-oil polluted soil. Umyu J Microbiol Res 9(1):75–83. https://doi.org/10.47430/ujmr.2491.008

Peidro-Guzmán H, Pérez-Llano Y, González-Abradelo D, Fernández-López MG, Dávila-Ramos S, Aranda E, Hernández DRO, García AO, Lira-Ruan V, Pliego OR, Santana MA, Schnabel D, Jiménez-Gómez I, Mouriño-Pérez RR, Aréchiga-Carvajal ET, del Rayo Sánchez-Carbente M, Folch-Mallol JL, Sánchez-Reyes A, Vaidyanathan VK et al (2021) Transcriptomic analysis of polyaromatic hydrocarbon degradation by the halophilic fungus Aspergillus sydowii at hypersaline conditions. Environ Microbiol 23(7):3435–3459. https://doi.org/10.1111/1462-2920.15166

Perera IC, Hemamali EH (2022) Genetically modified organisms for bioremediation: current research and advancements. In: Bioremediation of environmental pollutants. Springer, pp 163–186. https://doi.org/10.1007/978-3-030-86169-8_7

Pete AJ, Bharti B, Benton MG (2021) Nano-enhanced bioremediation for oil spills: a review. ACS ES&T Eng 1(6):928–946. https://doi.org/10.1021/acsestengg.0c00217

Phang L-Y, Mingyuan L, Mohammadi M, Tee C-S, Yuswan MH, Cheng W-H, Lai K-S (2024) Phytoremediation as a viable ecological and socioeconomic management strategy. Environ Sci Pollut Res 31(38):50126–50141. https://doi.org/10.1007/s11356-024-34585-z

Polli AD, de Oliveira Junior VA, Ribeiro MA d S, Polonio JC, Rosini B, Oliveira JA d S, Bini RD, Golias HC, Fávaro-Polonio CZ, Orlandelli RC, Vicentini VEP, Cotica LF, Peralta RM, Pamphile JA, Azevedo JL (2023) Synthesis, characterization, and reusability of novel nanobiocomposite of endophytic fungus aspergillus flavus and magnetic nanoparticles (Fe3O4) with dye bioremediation potential. Chemosphere 340:139956. https://doi.org/10.1016/j.chemosphere.2023.139956

Pollution Sustainability Directory (2025a) Mycoremediation applications — Pollution → Sustainability Directory. https://pollution.sustainability-directory.com/term/mycoremediation-applications/

Pollution Sustainability Directory (2025b) Mycoremediation in soil — Pollution → Sustainability Directory. https://pollution.sustainability-directory.com/term/mycoremediation-in-soil/

Pollution Sustainability Directory (2025c) What role does public perception play in bioremediation adoption? — Pollution → Sustainability Directory. https://pollution.sustainability-directory.com/question/what-roledoes-public-perception-play-in-bioremediation-adoption/

Pollution Sustainability Directory (2025d) Why is public acceptance crucial for bioremediation? — Pollution → Sustainability Directory. https://pollution.sustainability-directory.com/question/why-is-public-acceptancecrucial-for-bioremediation/

Pujiati P, Fatimah F, Ramadhan R, Ni'matuzahroh NM (2024) Mycoremediation of pesticide-contaminated soil: a review. BIO Web Conf 148:02020. https://doi.org/10.1051/bioconf/202414802020

Qin C, Pellitier PT, Nuland MEV, Peay K, Zhu K (2023) Niche modelling predicts that soil fungi occupy a precarious climate in boreal forests. Glob Ecol Biogeogr 32(7):1127–1139. https://doi.org/10.1111/geb.13684

Rahman M u, Ullah MW, Shah JA, Sethupathy S, Bilal H, Abdikakharovich SA, Khan AU, Khan KA, Elboughdiri N, Zhu D (2024) Harnessing the power of bacterial laccases for xenobiotic degradation in water: a 10-year overview. Sci Total Environ 918:170498. https://doi.org/10.1016/j.scitotenv.2024.170498

Rajhans G, Barik A, Sen SK, Raut S (2021a) Degradation of dyes by fungi: an insight into mycoremediation. Biotechnologia 102(4):445–455. https://doi.org/10.5114/bta.2021.111109

Rajhans G, Sen SK, Barik A, Raut S (2021b) De-colourization of textile effluent using immobilized Geotrichum candidum: an insight into mycoremediation. Lett Appl Microbiol 72(4):445–457. https://doi.org/10.1111/lam.13430

Rani S, Akash AU, Patel K (2025) Mycoremediation: leveraging fungi for ameliorating soil and water pollution. In: Bioremediation technologies for environmental pollution. IGI Global Scientific Publishing, pp 213–230

Roy A, Gogoi N, Haider FU, Farooq M (2025) Mycoremediation for sustainable remediation of environmental pollutants. Biocatal Agric Biotechnol 64:103526. https://doi.org/10.1016/j.bcab.2025.103526

Sadañoski MA, Tatarin AS, Barchuk ML, Gonzalez M, Pegoraro CN, Fonseca MI, Levin LN, Villalba LL (2020) Evaluation of bioremediation strategies for treating recalcitrant halo-organic pollutants in soil environments. Ecotoxicol Environ Saf 202:110929. https://doi.org/10.1016/j.ecoenv.2020.110929

Sahoo M (2024) Exploring the role of plants in bio-remediation: harnessing nature's clean-up agents. Int J Multidiscip Approach Res Sci 2(02):726–732. https://doi.org/10.59653/ijmars.v2i02.676

Sahu A, Manyapu V, Paliya S, Mandpe A (2025) Nanotechnology-assisted bioremediation: a multidisciplinary approach to environmental restoration. Soil Bioremediation, pp 403–430. https://doi.org/10.1007/978-3-031-76886-6_17

Sama DK, Agbor DT, Pangilinan PJM, Sunjo TE, Sama-Lang P (2023) Cutting edge use of microbial consortia for bioremediation of contaminated soil. Agric Sci 15(39):4–14. https://doi.org/10.22620/agrisci.2023.39.001

Samudro G, Mangkoedihardjo S (2023) Phytoreactor for arsenic biodetoxification: an integrated sequential process of phytoremediation involving multi-kingdom organisms. J Adv Res Appl Sci Eng Technol 36(2):21–35. https://doi.org/10.37934/araset.36.2.2135

Sattar S, Siddiqui S, Shahzad A, Bano A, Naeem M, Hussain R, Khan N, Jan BL, Yasmin H (2022) Comparative analysis of microbial consortiums and nanoparticles for rehabilitating petroleum waste contaminated soils. Molecules 27(6):1945. https://doi.org/10.3390/molecules27061945

Sayed K, Baloo L, Sharma NK (2021) Bioremediation of total petroleum hydrocarbons (TPH) by bioaugmentation and biostimulation in water with floating oil spill containment booms as bioreactor basin. Int J Environ Res Public Health 18(5):2226. https://doi.org/10.3390/ijerph18052226

Sharma B, Dangi AK, Shukla P (2018) Contemporary enzyme based technologies for bioremediation: a review. J Environ Manag 210:10–22. https://doi.org/10.1016/j.jenvman.2017.12.075

Sharma L, Dadhich A, Dhar I, Choudhary R, Dhiman M, Sharma MM (2025) Myco-nanotechnology at the crossroads: eco-friendly synthesis, agricultural applications, and AI-driven risk mitigation of fungal-derived nanoparticles. OpenNano 25:100254. https://doi.org/10.1016/j.onano.2025.100254

Silva AF d, Banat İM, Robl D, Giachini AJ (2022) Fungal bioproducts for petroleum hydrocarbons and toxic metals remediation: recent advances and emerging technologies. Bioprocess Biosyst Eng 46(3):393–428. https://doi.org/10.1007/s00449-022-02763-3

Singh VK, Singh R (2024) Role of white rot fungi in sustainable remediation of heavy metals from the contaminated environment. Mycology 15(4):585–601. https://doi.org/10.1080/21501203.2024.2389290

Singh A, Rawat S, Rajput VD, Minkina T, Mandzhieva S, Eloyan A, Singh RK, Singh O, El-Ramady H, Ghazaryan K (2024) Nanotechnology products in agriculture and environmental protection: advances and challenges. Egypt J Soil Sci 64(4):0–0. https://doi.org/10.21608/ejss.2024.300047.1802

Sośnicka A, Kózka B, Makarova K, Giebułtowicz J, Klimaszewska M, Turło J (2022) Optimization of white-rot fungi mycelial culture components for bioremediation of pharmaceutical-derived pollutants. Water 14(9):1374. https://doi.org/10.3390/w14091374

Stein HP, Navajas-Pérez R, Aranda E (2018) Potential for CRISPR genetic engineering to increase xenobiotic degradation capacities in model fungi. In: Nanotechnology in the life sciences. Springer, Cham, pp 61–78. https://doi.org/10.1007/978-3-030-02369-0_4

Stiebeling DA, Labes A (2022) Mycoremediation of sewage sludge and manure with marine fungi for the removal of organic pollutants. Front Mar Sci 9(September):1–11. https://doi.org/10.3389/fmars.2022.946220

Subbulakshmi G, Debbarma A, Sinha A, Panda S (2021) Bio remediation of xenobiotic compound: reclamation approach for environmental sustainability—a review. Mater Today Proc 47:1108–1113. https://doi.org/10.1016/j.matpr.2021.07.144

Takata N, Ghassemi-Armaki H, Terada Y, Takeyama M, Kumar S (2013) Effect of dislocation sources on slip in Fe 2 Nb laves phase with Ni in solution. MRS Proc 1516:269–274. https://doi.org/10.1557/opl.2012.1566

Tan Z, Beltrán-Flores E, Ramos-Meza GD, Alonso LL, Sarrà M (2025) Eliminating antibiotics by white-rot-fungi Trametes versicolor from manure solids and synthetic wastewater. Environ Pollut 378:126504. https://doi.org/10.1016/j.envpol.2025.126504

Thirumalaivasan N, Gnanasekaran L, Kumar S, Durvasulu R, Sundaram T, Rajendran S, Nangan S, Kanagaraj K (2024) Utilization of fungal and bacterial bioremediation techniques for the treatment of toxic waste and biowaste. Front Mater 11(July):1–20. https://doi.org/10.3389/fmats.2024.1416445

U.S. Department of Agriculture (2025) Regulation of biotech plants

Venice F, Davolos D, Spina F, Poli A, Prigione VP, Varese GC, Ghignone S (2020) Genome sequence of Trichoderma lixii MUT3171, a promising strain for Mycoremediation of PAH-contaminated sites. Microorganisms 8(9):1258. https://doi.org/10.3390/microorganisms8091258

Wang J, Aghajani Delavar M (2024) Modelling phytoremediation: concepts, methods, challenges and perspectives. Soil Environ Health 2(1):100062. https://doi.org/10.1016/j.seh.2024.100062

Wang Z-G, Bi Y-L, Jiang B, Zhakypbek Y, Peng S-P, Liu W-W, Liu H (2016) Arbuscular mycorrhizal fungi enhance soil carbon sequestration in the coalfields, Northwest China. Sci Rep 6(1):34336. https://doi.org/10.1038/srep34336

Wang Y, Chen S, Zhou J, Fan X, He L, Fan G (2022) Enhanced degradation capability of white-rot fungi after short-term pre-exposure to silver ion: performance and selectively antimicrobial mechanisms. Sci Total Environ 818:151672. https://doi.org/10.1016/j.scitotenv.2021.151672

Yadav KK, Singh JK, Gupta N, Kumar V (2017) A review of nanobioremediation technologies for environmental cleanup: a novel biological approach. J Mater Environ Sci 8(2):740–757

Yadav P, Rai SN, Mishra V et al (2021) Mycoremediation of environmental pollutants: a review with special emphasis on mushrooms. Environ Sustain 4:605–618

Zahir A, Okorie PA, Nwobasi VN, David E, Nwankwegu RO, Azi F (2024) Harnessing microbial signal transduction systems in natural and synthetic consortia for biotechnological applications. Biotechnol Appl Biochem 72(4):1079–1109. https://doi.org/10.1002/bab.2707

Zhu Y, Wang Y, He X, Li B, Du S (2023) Plant growth-promoting rhizobacteria: A good companion for heavy metal phytoremediation [Review]. Chemosphere, 338, Article 139475. https://doi.org/10.1016/j.chemosphere.2023.139475

Zulfiqar U, Haider FU, Maqsood MF, Mohy-Ud-Din W, Shabaan M, Ahmad M, Kaleem M, Ishfaq M, Aslam Z, Shahzad B (2023) Recent advances in microbial-assisted remediation of cadmium-contaminated soil. Plants 12(17):3147. https://doi.org/10.3390/plants12173147

Case Studies of Mycoremediation of Xenobiotics

13

Nisha Saxena and Noopur Srivastava

Abstract

Hazardous environmental contaminants, both naturally occurring and anthropogenically introduced, present significant ecological challenges. These pollutants include a wide range of chemical classes, such as polycyclic aromatic hydrocarbons (PAHs), dyes, detergents, plastics, dioxins, pharmaceuticals, and complex mixtures, like creosote and oil. A major concern arises when pollutants enter ecosystems through leaks or accidental discharges, overwhelming the natural microbial communities and hindering effective biodegradation. Unlike organic pollutants, metals are persistent, remaining in the environment for thousands of years due to their inability to be degraded. Although they undergo oxidation or complexation, they still accumulate in soil and water, posing long-term ecological and health risks. Bio-sorption, a process where microorganisms remove metals from aqueous environments, plays a vital role in remediation efforts. Myco-sorption, or fungal-mediated bio-sorption, has attracted considerable global interest owing to fungi's structural and biochemical versatility. Their expansive mycelial networks, nonspecific catabolic enzymes, and cell walls enriched with functional groups, such as amino, hydroxyl, carboxyl, and phosphate, allow them to bind and immobilize metals, metalloids, and radionuclides efficiently. Mycoremediation, the application of fungi for ecosystem restoration, is an emerging, sustainable strategy. Ongoing research explores fungal species, degradation pathways, and innovations like myco-nanotechnology to enhance remediation efficiency, offering promising solutions for polluted environments.

N. Saxena (✉)
Department of Chemistry, M.R.M. College, A Constituent Unit of L.N. Mithila University, Darbhanga, Bihar, India

N. Srivastava
Department of Chemistry and Biochemistry, Sharda School of Engineering & Sciences, Sharda University, Greater Noida, Uttar Pradesh, India

A. Roy et al. (eds.), *Mycoremediation of Xenobiotics*, Emerging Paradigms in Pharmaceutical Research, https://doi.org/10.1007/978-981-95-5743-1_13

Keywords

Mycoremediation · Environmental contaminants · Case studies

13.1 Introduction

13.1.1 Fungal Enzymes in Xenobiotic Bioremediation

Xenobiotics are natural or synthetic organic compounds foreign to biological systems and often pose serious ecological and health risks. They include industrial and agricultural products, such as fertilizers, pesticides, insecticides, synthetic dyes, plastics, and hydrocarbon derivatives (Olicón-Hernández et al. 2017; Sharma et al. 2018). Many xenobiotics are chemically stable aromatic compounds with substituted phenyl rings, making them resistant to natural degradation. Their persistence leads to bioaccumulation in soil and water, threatening humans, wildlife, and aquatic ecosystems. Several are known to be carcinogenic, teratogenic, or endocrine disruptors, exacerbating their impact on public health. The growing presence of these pollutants highlights the urgent need for cost-effective, environmentally sustainable remediation strategies to prevent long-term damage and restore ecological balance.

13.1.2 Bioremediation: Harnessing Biological Potential

Bioremediation employs biological organisms to transform hazardous substances into less toxic or harmless forms (Watanabe 2001). Microorganisms, such as bacteria and fungi, metabolize diverse xenobiotics, but fungi hold a distinct advantage due to their robust enzymatic systems and resilience in extreme environments. Unlike bacteria, whose ability to fully mineralize persistent pollutants, like polyaromatic hydrocarbons (PAHs), is often limited, fungi can degrade a wider range of recalcitrant compounds through both extracellular and intracellular mechanisms (Christian et al. 2005b; Mougin et al. 2009). This capacity makes fungi particularly valuable for achieving complete detoxification in challenging contamination scenarios.

13.1.3 Mycoremediation: Fungal-Based Bioremediation

Mycoremediation, the use of fungi for pollutant degradation, is gaining global recognition as an eco-friendly and cost-effective strategy. Filamentous fungi, especially white rot fungi (Basidiomycetes), are particularly valued for their exceptional ability to degrade lignin, a complex aromatic polymer in plant cell walls. This is achieved through potent extracellular oxidative enzymes, which also enable the breakdown of structurally similar xenobiotic pollutants (Morel et al. 2013). Comparative genomic studies have identified a "xenome"—a gene set dedicated to xenobiotic detoxification—in many white rot fungi, equipping them to naturally remediate a wide range of environmental contaminants with remarkable efficiency.

13.1.4 Key Enzymes in Xenobiotic Detoxification

The key fungal enzymes involved in xenobiotic transformation are oxidoreductases, which drive crucial oxidation–reduction reactions. The primary categories include laccases, peroxidases (lignin peroxidase, manganese peroxidase, versatile peroxidase), and oxygenases, such as cytochrome P450 monooxygenases, all of which play essential roles in degrading diverse environmental pollutants efficiently.

(i) *Peroxidases*: Peroxidases are heme-containing enzymes that utilize hydrogen peroxide as an oxidant. Key types include manganese peroxidase (MnP), lignin peroxidase (LiP), and versatile peroxidase (VP). These enzymes efficiently cleave complex aromatic structures, playing a crucial role in lignin breakdown and xenobiotic degradation, making them essential for effective fungal bioremediation processes (Doddapaneni et al. 2005).

(ii) *Laccases (LACs)*: Copper-containing oxidases, such as laccases, use molecular oxygen as the electron acceptor, producing water as a by-product. They can oxidize diverse phenolic and non-phenolic pollutants, making them vital for the detoxification of various environmental contaminants.

(iii) *Cytochrome P450 Monooxygenases (CYP450s)*: Intracellular oxygenases incorporate oxygen into hydrophobic xenobiotics, increasing their solubility and facilitating further metabolism or excretion, thus playing a crucial role in detoxification and biotransformation of persistent environmental pollutants.

These enzymes are distinguished for their nonspecific catalytic action, permitting them to act on varied pollutants. Furthermore, they continue to be active under a wide range of environmental conditions, including variable pH and temperature, making them robust tools for bioremediation (Verma and Madamwar 2002; Rastegari et al. 2019).

Enzymatic bioremediation provides distinct advantages, including avoiding the ecological risks associated with introducing live microbes, especially genetically modified organisms, into sensitive habitats (Ang et al. 2005). Isolated enzymes or mixtures enable precise, controlled, and targeted pollutant degradation, yielding safer and more predictable outcomes. Nonetheless, challenges persist, such as nutrient-poor conditions at contaminated sites, complex pollutant mixtures toxic to fungi, and the high cost of enzyme production and stabilization. Ongoing research addresses these issues by optimizing enzyme yields, developing efficient immobilization techniques, and integrating nutrient supplementation or co-metabolic strategies, thereby improving the overall efficiency and scalability of enzyme-based bioremediation systems.

Fungal enzymes, particularly those from white rot fungi, offer a powerful, sustainable solution for the detoxification of persistent xenobiotics. Their unique oxidative mechanisms, broad substrate specificity, and ability to function under diverse conditions make them ideal candidates for large-scale bioremediation projects. As industrialization continues to generate complex pollutants, harnessing and optimizing these biological tools will remain critical to mitigating environmental toxicity and protecting public health.

13.1.5 Critical Highlights of the Mycoremediation of Xenobiotics

1. *Xenobiotics*: Toxic, recalcitrant pollutants from industrial and agricultural sources; many are carcinogenic and environmentally persistent.
2. *Bioremediation*: Use of microbes to detoxify pollutants; fungi play a crucial role due to their powerful enzymes (Fig. 13.1); in short, the diagrammatic representation gives a brief layout for understanding the enzymatic and nonenzymatic processes.
3. *Mycoremediation*: Fungal-based pollutant degradation, especially effective with white rot fungi (*Basidiomycetes*).
4. *Key enzymes*: Peroxidases (MnP, LiP, VP), laccases, and CYP450 monooxygenases.

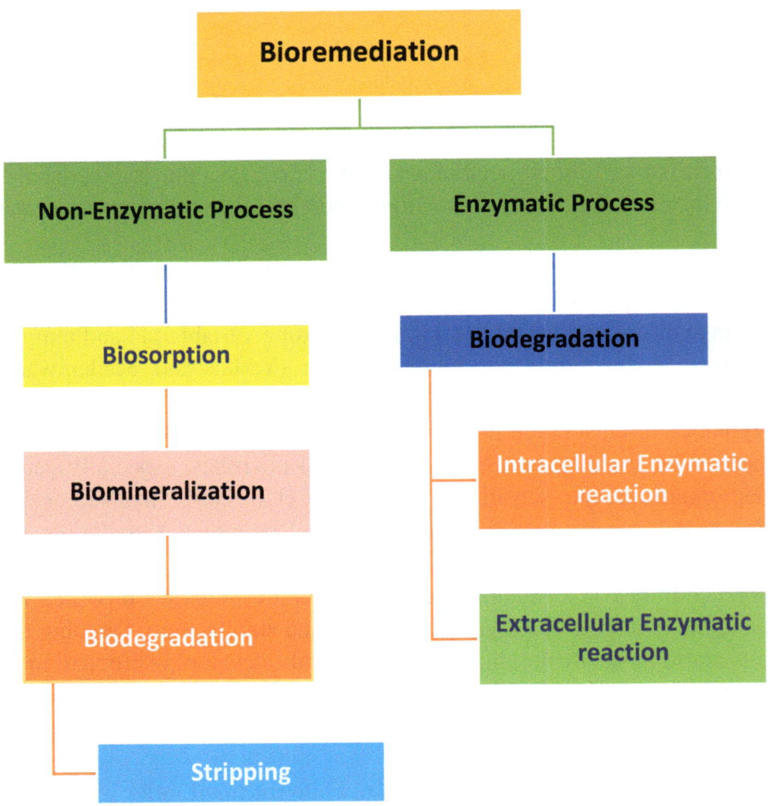

Fig. 13.1 Bioremediation mechanism

13.2 Xenobiotic Contamination Sources

Major sources of xenobiotics include synthetic dyes released from paper, textile, plastic, cosmetic, food, and pharmaceutical industries, many of which are toxic and carcinogenic (Asgher et al. 2008; Levin et al. 2005). Polycyclic aromatic hydrocarbons (PAHs), such as anthracene, pyrene, benzopyrene, and naphthalene, arise from fuel combustion, gas plants, and industrial operations and are potent carcinogens. Endocrine-disrupting chemicals (EDCs), such as nonylphenol, octylphenol, and bisphenol A (BPA), interfere with hormonal balance and threaten human and animal health. Additional sources include bleach plant effluents from the pulp and paper industry containing polychlorinated phenols (PCPs), toxic dyes (azo dyes, crystal violet), and chlorinated aromatic pesticides, such as DDT, chlordane, and lindane, all of which pose significant environmental and health risks (Fig. 13.2).

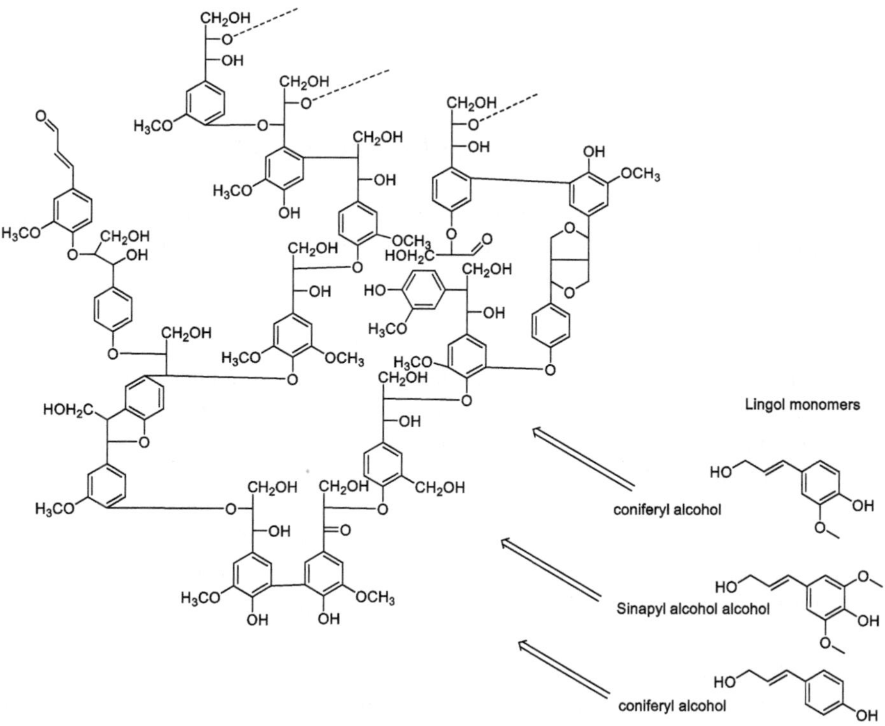

Fig. 13.2 Lignin shares structural features with certain xenobiotics, such as alkylated phenolics and the geminal phenyl motif of BPA, which resembles its monomeric alcohol-derived cross-linked network. However, nitrogen-containing compounds, fused polycyclic aromatics, and polychlorinated phenols are absent in lignin, making the enzymatic degradation of these xenobiotics less straightforward (Laurichesse and Avérous 2014)

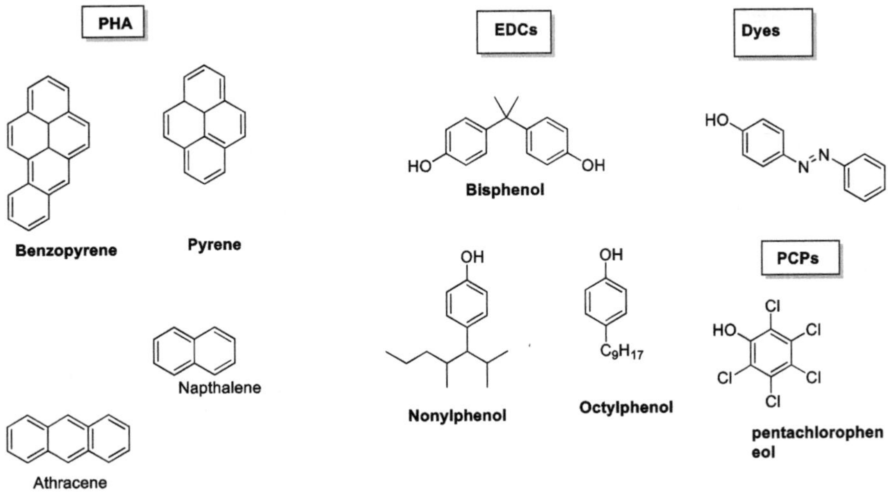

Fig. 13.2 (continued)

13.3 Metabolism of Xenobiotic Pollutants by Fungi

Fungi play a vital role in xenobiotic bioremediation by metabolizing pollutants through several pathways. The most desirable pathway is direct metabolism, where fungi completely degrade xenobiotic compounds into innocuous end products, such as carbon dioxide, water, and simple inorganic molecules (Mougin et al. 2009; Kumari et al. 2021). This process is especially likely under nutrient-limited conditions, where the xenobiotic serves as a source of carbon and/or energy for the fungal organism. Direct metabolism is preferred because it results in the near-complete elimination of the toxic compound from the environment.

More commonly, fungal detoxification occurs via co-metabolism. In this process, a secondary cosubstrate is used, allowing the fungus to transform the xenobiotic without deriving energy or growth benefits from it. Co-metabolism usually leads to partial structural modifications, such as the generation of substrate-free radicals that cross-link with themselves or with environmental components, like soil particles, producing adducts with reduced bioavailability and toxicity (Bollag 1992). These partially transformed compounds can subsequently undergo further degradation by other organisms, resulting in progressive detoxification (Prajapati et al. 2022).

A third mechanism is conjugation or oligomerization, wherein enzymatic activity modifies or aggregates the pollutant into less bioactive forms. Conjugation involves methylation, acetylation, alkylation, or binding with sugars or amino acids, followed by sequestration or excretion. Oligomerization, on the other hand, couples xenobiotics with identical or different molecules to form higher molecular weight complexes, reducing toxicity through immobilization or reduced bioavailability (Šimášková 2020).

Together, these metabolic strategies highlight fungi as potent agents for sustainable detoxification of hazardous environmental pollutants.

13.3.1 Essential Insights for Metabolism

- *Direct Metabolism*—Complete degradation to CO_2, water, and simple compounds.
- *Co-Metabolism*—Partial transformation using a cosubstrate; reduces bioavailability and toxicity.
- *Conjugation*—Methylation, acetylation, alkylation, or sugar/amino acid binding to detoxify.
- *Oligomerization*—Coupling of xenobiotics into larger, less bioactive complexes.
- *Outcome*—Reduced toxicity, bioavailability, and environmental impact.

13.4 Enzyme-Mediated Xenobiotic Degradation by Fungi

Fungi have emerged as powerful bioremediation agents due to their unique ability to secrete extracellular and intracellular enzymes capable of transforming a wide range of xenobiotics. The majority of fungal enzymes implicated in xenobiotic degradation are oxidoreductases, which catalyze electron transfer reactions that transform complex, toxic molecules into less harmful derivatives (Sharma et al. 2018). These enzymes can act both extracellularly, where they initiate the breakdown of recalcitrant pollutants that cannot enter the fungal cell, and intracellularly, where they further metabolize partially degraded compounds, contributing to detoxification and mineralization (Sharma et al. 2018; Khan et al. 2023) (Fig. 13.3).

13.4.1 Oxidoreductases: Key Catalysts in Bioremediation

Among the most studied fungal oxidoreductases are laccases (LACs), cytochrome P450 monooxygenases (CYPs), and heme peroxidases, including manganese peroxidase (MnP), lignin peroxidase (LiP), and versatile peroxidase (VP) (Doddapaneni et al. 2005; Baker et al. 2019). Laccases and CYPs typically utilize molecular oxygen as the terminal electron acceptor, whereas peroxidases employ hydrogen peroxide to drive oxidative reactions. The relatively low substrate specificity of these extracellular enzymes is advantageous for bioremediation, as it enables them to act on a wide variety of structurally diverse xenobiotic compounds, including persistent organic pollutants, dyes, pesticides, and endocrine-disrupting chemicals (Harms et al. 2011).

Most of these secreted oxidoreductases are heavily glycosylated proteins, which enhance their stability and activity under variable environmental conditions. For example, laccases may contain up to 16% carbohydrate content (Salony et al. 2006). In addition to enzymes, fungi produce redox mediators, such as veratryl alcohol,

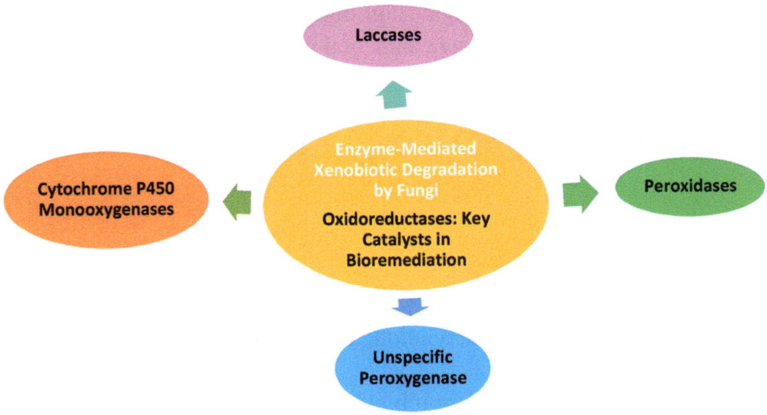

Fig. 13.3 Oxidoreductases involved in bioremediation

N-hydroxyacetanilide, and acetosyringone, which facilitate electron shuttling during oxidation reactions. These mediators are particularly critical when the xenobiotic molecule is too large to directly access the active site of the enzyme (Mougin et al. 2009; Nousiainen et al. 2012).

Once extracellular enzymes partially degrade xenobiotics, intracellular CYP monooxygenases further metabolize the resulting intermediates. CYPs are ubiquitous heme-containing enzymes that catalyze a variety of reactions, including hydroxylation, dealkylation, and sulfoxidation, frequently by inserting oxygen atoms into hydrophobic substrates (Morel et al. 2013). This stepwise combination of extracellular oxidation followed by intracellular transformation often results in near-complete detoxification of hazardous pollutants (Fig. 13.3 and Table 13.1).

13.4.1.1 Laccases: The Frontline Enzymes in Mycoremediation

Laccases (EC 1.10.3.2) are blue multicopper oxidases originally discovered for their role in lignin degradation. They are widely distributed among ascomycetes, basidiomycetes, and deuteromycetes and are typically encoded by gene families with multiple isoforms (Harms et al. 2011). Laccases catalyze the single-electron oxidation of phenolic and aromatic compounds, generating free radicals while simultaneously reducing molecular oxygen to water (Sharma 2025).

The radical products undergo self-coupling or cross-coupling reactions, resulting in polymerization, cross-linking, or further decomposition of the substrate through decarboxylation, dechlorination, or demethoxylation. These transformations reduce the bioavailability and toxicity of the pollutants. Laccase-catalyzed reactions often produce higher molecular weight end products, immobilizing pollutants and reducing their biological impact (Junghanns et al. 2005; Tsutsumi et al. 2001). Unlike peroxidases, laccases function across a broad pH range (2–10) and exhibit wide substrate specificity, enabling the degradation of diverse xenobiotics, including PAHs, organophosphorus insecticides, phenolic dyes, chlorinated aromatics, and endocrine disruptors (Xu 1996; Amitai et al. 1998; Kües 2015; Yadav et al. 2016).

Table 13.1 Oxidoreductases: key catalysts in bioremediation

Feature	Unspecific peroxygenase (UPO)	Cytochrome P450 monooxygenase (P450)	Manganese peroxidase (MnP)	Lignin peroxidase (LiP)
Location	Extracellular (secreted)	Intracellular (membrane-bound)	Extracellular (secreted)	Extracellular (secreted)
Cofactor	Hydrogen peroxide (H_2O_2)	NADPH + O_2	H_2O_2 + Mn^{2+} (with chelators)	H_2O_2 + mediators (e.g., veratryl alcohol)
Mechanism	Direct oxyfunctionalization (peroxygenation) of substrates	Monooxygenation (inserts one O atom into substrate, reduces other to water)	Oxidizes $Mn^{2+} \rightarrow Mn^{3+}$, which oxidizes substrates indirectly	Direct oxidation of non-phenolic aromatics via radical formation
Substrate range	>300 substrates including PAHs, phenols, toluene, phthalate esters, nitroarenes, polychlorinated benzenes	PAHs, alkylphenols, pesticides, pharmaceuticals, PCDDs	Phenols, aromatic amines, PAHs, dyes, quinones	Lignin model compounds, PAHs, aromatic hydrocarbons
Optimal pH	Mildly acidic (pH 4–6)	Near neutral (intracellular)	Slightly acidic (pH 4–7)	Acidic (pH 2–5)
Specificity	Broad (unspecific) but affected by steric hindrance & electron-withdrawing groups	Moderate (broad but enzyme-specific isoforms)	Broad due to the indirect oxidation mechanism	Broad, but more substrate-selective than MnP
Key advantages	Simple cofactor requirement, extracellular activity, targets substrates previously thought to be P450-specific	Powerful intracellular detoxification, multiple isoforms allow redundancy	Stable under adverse conditions, capable of complete mineralization of PAHs	High redox potential, able to degrade recalcitrant lignin-like structures
Biotechnological relevance	Ideal for extracellular biocatalysis & pollutant mitigation	Target for metabolic engineering & biotransformation of pharmaceuticals	Useful in enzyme cocktails for soil and wastewater treatment	Valuable in lignin valorization & degradation of persistent pollutants

Applications of Laccases in Xenobiotic Detoxification

Laccases in Environmental and Industrial Applications

1. *PCB Detoxification*:
   (a) Laccases from *Trametes versicolor* and *Pleurotus ostreatus* effectively detoxify polychlorinated biphenyls (PCBs), previously used as electrical insulators but banned due to toxicity (Keum and Li 2004).
   (b) The enzymes catalyze PCB oligomerization and dechlorination, reducing toxicity.
2. *PAH Degradation*:
   (a) Laccases initiate oxidation of polycyclic aromatic hydrocarbons (PAHs), facilitating extracellular breakdown.
   (b) *Coriolopsis gallica* laccase oxidizes PAHs such as carbazole, N-ethylcarbazole, and dibenzothiophene in the presence of mediators like 1-hydroxybenzotriazole (Viswanath et al. 2014).
   (c) Free laccases applied to contaminated soils degrade up to 15 PAHs, including anthracene and benzopyrene (Wu et al. 2008).
   (d) Covalently immobilized laccases efficiently degrade anthracene, naphthalene, and phenanthrene (Bautista et al. 2015).
3. *Endocrine-Disrupting Chemicals (EDCs) Removal*:
   (a) Laccases degrade BPA and alkylphenols via oligomerization or complete mineralization to CO_2.
   (b) Efficiency is enhanced by mediator supplementation (Macellaro et al. 2014; Margot et al. 2013; Zhang et al. 2015).
4. *Dye Degradation*:
   (a) Laccases from *Pycnoporus sanguineus*, *Myceliophthora thermophila*, and *Ceriporiopsis subvermispora* degrade recalcitrant dyes like bromophenol blue, methyl violet, and malachite green, aiding textile effluent treatment (Antošová et al. 2018; Chmelová and Ondrejovič 2016).
5. *Industrial Applications*:
   (a) Laccases are employed in paper and pulp bleaching, textile processing, and pharmaceutical residue detoxification, including diclofenac and mefenamic acid in municipal wastewater (Margot et al. 2013).
6. *Key Benefits*:
   (a) Broad substrate specificity, extracellular activity, and enhancement by redox mediators make laccases versatile for environmental remediation and industrial bioprocesses.

Enhancing Laccase Efficiency: Mediators and Engineering Approaches

Redox mediators play a critical role in enhancing laccase activity, especially for substrates that cannot directly interact with the enzyme's active site. Research on

Coriolopsis gallica laccases has demonstrated that mediators, such as acetosyringone and syringaldehyde, significantly improve the transformation of halogenated pesticides, expanding the enzyme's substrate range (Torres-Duarte et al. 2009). Beyond naturally occurring isoforms, protein engineering and directed evolution strategies have been employed to optimize laccase performance. These approaches enhance enzyme stability, catalytic efficiency, and substrate specificity, enabling engineered laccases to effectively degrade recalcitrant xenobiotics. Such advancements hold considerable potential for industrial and environmental applications, providing scalable solutions for the bioremediation of persistent pollutants (Camarero et al. 2012; Gu et al. 2014; Mate and Alcalde 2015).

13.4.1.2 Peroxidases in Xenobiotic Bioremediation

Peroxidases embody one of the most critical classes of fungal enzymes intricate in xenobiotic detoxification. These ligninolytic enzymes possess exceptionally high redox potentials (often exceeding 1.4 V) and catalyze oxidative breakdown of lignin and a wide variety of aromatic pollutants, using hydrogen peroxide as a co-substrate (Sharma et al. 2018; Piontek et al. 2001). By generating reactive phenolic radicals, peroxidases initiate cleavage and depolymerization of recalcitrant structures, facilitating either complete mineralization to CO_2 and water or their transformation into less bioavailable forms.

Structurally, peroxidases are glycosylated secreted proteins containing a heme prosthetic group (iron protoporphyrin IX) at the catalytic center. The heme cofactor is essential for redox cycling and enables one-electron oxidation of diverse aromatic and nonaromatic substrates. In the existence of mediators, such as veratryl alcohol, the oxidative potential of peroxidases is expressively improved, intensifying their ability to act on compounds with high ionization potentials.

Fungal peroxidases are classified into lignin peroxidase (LiP), manganese peroxidase (MnP), and versatile peroxidase (VP). These enzymes underpin white-rot fungal ligninolysis, playing key roles in degrading persistent organic pollutants, PAHs, halogenated pesticides, and industrial dyes.

Lignin Peroxidase (LiP)

Lignin Peroxidase (LiP): Overview and Mechanism

Lignin peroxidase (LiP) is among the most potent fungal oxidoreductases, notable for its ability to oxidize non-phenolic aromatic substrates that are generally resistant to degradation. LiPs function optimally in acidic conditions, with pH ranges between 2 and 5, and have molecular weights of 38–43 kDa. Multiple isoforms have been identified across basidiomycete fungi, reflecting their genetic and functional diversity (Shrivastava et al. 2005). The identification of LiP in the extracellular medium of *Phanerochaete chrysosporium* in 1983 was a milestone in understanding fungal ligninolysis, emphasizing the role of extracellular enzymes in lignin degradation. (Falade et al. 2017).

LiPs display remarkably broad substrate specificity, capable of oxidizing both methylated and non-methylated lignin model compounds. They also degrade complex xenobiotics, including polycyclic aromatic hydrocarbons (PAHs), such as benzo(a)pyrene and anthracene (Kadri et al. 2017). Substrates with ionization potentials below 8 eV are particularly susceptible to LiP-catalyzed oxidation. The presence of veratryl alcohol, a redox mediator, further extends the range of substrates that LiPs can target.

The catalytic mechanism of LiP contains activation by hydrogen peroxide, which oxidizes the enzyme's heme group to form an intermediate known as Compound I. This high-energy intermediate abstracts an electron from the aromatic substrate, generating cation radicals. These radicals then undergo rearrangement, bond cleavage, or polymerization reactions, ultimately resulting in detoxification and increased solubility of the target compounds. This combination of broad substrate specificity, potent oxidizing ability, and extracellular activity underlines the pivotal role of LiPs in lignin degradation and fungal-mediated bioremediation.

Some key components to understand the mechanism involve:

- *Enzyme type*: Fungal oxidoreductase
- *Optimal pH*: 2–5
- *Molecular weight*: 38–43 kDa
- *Isoforms*: Multiple, in basidiomycetes
- *Substrate specificity*: Non-phenolic aromatics, lignin models, PAHs
- *Cofactor/activator*: Hydrogen peroxide, veratryl alcohol (mediator)
- *Mechanism*: Heme activation → compound I → cation radical formation → substrate cleavage or polymerization

Bioremediation Significance

Lignin peroxidases (LiPs) can oxidize up to 90% of lignin structures, efficiently degrading pollutants like benzo(a)pyrene, anthracene, fluoranthene, and dibenzothiophene. Their unique ability to cleave ester bonds in non-phenolic aromatics enhances xenobiotic breakdown (Falade et al. 2017; Pozdnyakova 2012).

Manganese Peroxidase (MnP)

Manganese peroxidase (MnP) is a crucial ligninolytic enzyme that operates optimally under slightly less acidic conditions (pH 4–7) likened to lignin peroxidases (LiPs) (Asgher et al. 2008). Similar to LiPs, MnPs are heme-containing extracellular enzymes predominantly secreted by *Basidiomycota* fungi (Harms et al. 2011). In distinction to laccases, MnPs are usually encoded by a limited number of genes and are not found in widespread isoenzyme families (Torres-Farradá et al. 2017). Their catalytic mechanism is characteristic, relying on the oxidation of Mn^{2+} to Mn^{3+}, which acts as a diffusible oxidant. This Mn^{3+} species can oxidize a broad spectrum of phenolic and non-phenolic compounds, enabling lignin breakdown and the degradation of various recalcitrant environmental pollutants. MnPs thus play a central role in fungal lignin degradation and bioremediation processes. The Mn^{3+} product must be stabilized by complexation with organic acid chelators, such as oxalate or malonate, to prevent disproportionation, thus enabling indirect oxidation of recalcitrant xenobiotics.

MnPs play a crucial role in ring-cleavage reactions, leading to the breakdown of aromatic pollutants. They have been shown to oxidize monoamino-dinitrotoluene, chlorophenol derivatives, phenols, aromatic amines, azo dyes, and quinones (Harms et al. 2011). Importantly, MnPs can mineralize quinones to CO_2 via ring-fission transformations, thus achieving near-complete detoxification.

Bioremediation Applications

MnPs isolated from *Anthracophyllum discolor* oxidize pyrene, anthracene, fluoranthene, and phenanthrene, while MnPs from *Trametes* spp. degrade azo and indigo dyes as well as PAHs (Zhang et al. 2016). Cross-linked MnP aggregates from *Ganoderma lucidum* have been reported to efficiently degrade endocrine disruptors, such as nonylphenol and triclosan (Bilal et al. 2017b).

MnPs exhibit considerable stability under adverse environmental conditions, which makes them promising for field applications. However, their biotechnological deployment is complicated by the requirement for organic acid chelators and reliance on redox mediators, such as Tween 80, to achieve higher catalytic efficiency. Interestingly, MnP activity is also dependent on the presence of laccases to initiate Mn^{3+} complex formation. In vitro studies have shown that Mn^{3+} complexing fails to occur in the absence of laccase and hydrogen peroxide but is restored upon their addition (Schlosser and Höfer 2002).

Environmental induction of MnP secretion is significant: for example, *Fomes* spp. produce MnP at concentrations up to 1299 U/L after 21 days of growth in PAH-polluted media, highlighting its role as a major enzymatic response to xenobiotic stress (Godoy et al. 2016).

13.4.1.3 Versatile Peroxidase (VP): Hybrid Peroxidases in Bioremediation

Versatile peroxidases (VPs) are hybrid enzymes that combine the substrate specificities and mechanisms of both LiP and MnP. Like other peroxidases, VPs are heme proteins capable of using hydrogen peroxide as an electron acceptor. They can directly oxidize phenolic and non-phenolic aromatic compounds, as well as oxidize Mn^{2+} to Mn^{3+}, thereby coupling the two catalytic pathways into one enzyme system (Kües 2015).

The first VP (PS1) was isolated from *Pleurotus eryngii* and shown to possess both a Mn^{2+} oxidation site and a high-redox-potential aromatic substrate oxidation site (Camarero et al. 1999). Notably, VPs can retain catalytic activity in conditions that would normally inactivate LiPs or MnPs, making them potentially more robust enzymes for environmental applications.

Bioremediation Significance

VPs are inducible enzymes whose production increases in response to PAH contamination (Pozdnyakova et al. 2018a, b) They have been demonstrated to degrade polyhalogenated pesticides, such as dichlorophen, bromoxynil, and pentachlorophenol (PCP), which are otherwise difficult to remediate (Davila-Vazquez et al. 2005). Because they cartel two catalytic domains, VPs have wider substrate ranges than LiP or MnP independently and represent a versatile selection for treating chemically varied xenobiotic mixtures.

13.4.1.4 Cytochrome P450 Monooxygenases

Fungal bioremediation extends beyond extracellular enzymes like laccases and peroxidases, encompassing intracellular systems, such as cytochrome P450 monooxygenases (CYP450s) (Syed and Yadav 2012). These heme-containing oxygenases catalyze oxygen incorporation into hydrophobic compounds, destabilizing aromatic rings and promoting solubilization and degradation of recalcitrant pollutants, making them vital for aerobic xenobiotic transformation (Watanabe 2001).

Mechanism of Action and Cofactor Dependence

Cytochrome P450 (CYP450) enzymes perform as monooxygenases, incorporating one atom of molecular oxygen into a substrate while reducing the second atom to water at a time. This process depends on electrons supplied by the cofactor NADPH, which drives the enzyme's redox cycle (Doddapaneni et al. 2005). Dissimilar to CYP450s, dioxygenases introduce both oxygen atoms into the substrate at the same time. Through the CYP450 catalytic cycle, xenobiotic molecules undergo reactions, such as epoxidation, demethylation, or hydroxylation. These modifications increase the compound's polarity, improving their solubility and making them more responsive to subsequent metabolic transformations, degradation, or excretion, thereby facilitating detoxification within the cell.

Diversity and Genetic Abundance

CYP450s form one of the largest fungal enzyme superfamilies, exhibiting extensive genetic diversity. *Phanerochaete chrysosporium* encodes at least 149 CYP isozymes across multiple subfamilies (Doddapaneni et al. 2005; Syed et al. 2011), with similar families in *Ascomycota*, *Basidiomycota*, *Mucoromycota*, and *Chytridiomycota*, highlighting their ecological and biochemical significance.

Substrate Range and Detoxification Potential

A notable feature of fungal CYP450s is their little substrate specificity, allowing them to metabolize a broad collection of pollutants. Several fungicides, halogenated herbicides, and pesticides are detoxified by CYPs, as are many aromatic hydrocarbons and endocrine-disrupting chemicals. For instance, CYP enzymes from *P. chrysosporium* are capable of degrading polycyclic aromatic hydrocarbons (PAHs) like anthracene, as well as alkylphenols such as nonylphenol (Hirosue et al. 2011). Class II P450 subfamilies are particularly important in these detoxification processes:

(i) *CYP53* enzymes degrade benzoate derivatives.
(ii) *CYP504* acts on phenylacetate derivatives.
(iii) *CYP57* detoxifies pisatin, a plant phytoalexin produced by peas.

These metabolic capabilities highlight CYP450s as central players in fungal xenobiotic degradation.

Relevance in Non-Ligninolytic Fungi

While ligninolytic fungi rely heavily on extracellular oxidative enzymes (LiP, MnP, LAC), CYP450 enzymes enable non-ligninolytic fungi to perform comparable detoxification. For example, *Scopulariopsis brevicaulis* completely transforms anthracene into 9,10-anthraquinone, a major metabolite identical to that produced by ligninolytic basidiomycetes (Godoy et al. 2016). This observation demonstrates that CYP450-dependent intracellular pathways can substitute for extracellular oxidative systems in bioremediation.

Detoxification of Pharmaceuticals

CYP450s are also involved in the biotransformation of pharmaceuticals, which have become emerging pollutants of concern. Antibiotics, β-blockers, and nonsteroidal anti-inflammatory drugs (NSAIDs) are susceptible to CYP-mediated hydroxylation and demethylation (Olicón-Hernández et al. 2017). A notable example is the detoxification of *naproxen*, a widely used NSAID that has been detected in drinking water supplies. Fungal CYPs convert naproxen into less harmful metabolites via demethylation and hydroxylation reactions (Aracagök et al. 2017).

Integration with Other Intracellular Enzymes

Detoxification of xenobiotics by CYP450s is rarely a stand-alone process. Shotgun proteomics studies in *Penicillium oxalicum* revealed that exposure to anthracene regulated hundreds of proteins, reflecting a coordinated metabolic network (Camacho-Morales et al. 2018). Intracellular xenobiotic metabolism generally occurs in two separate phases.:

1. *Phase I Reactions*—CYP450 enzymes add reactive or polar groups to xenobiotics, such as through hydroxylation or epoxidation.
2. *Phase II Reactions*—Transferases (e.g., glutathione-S-transferases, glucuronosyl-transferases) and epoxide hydrolases conjugate or modify Phase I metabolites, increasing their solubility for excretion or facilitating immobilization.

Limitations and Biotechnological Challenges

Despite their remarkable biotechnological potential, CYP450 enzymes face several limitations for direct application. Being intracellular, they are difficult to extract, purify, and immobilize compared to secreted extracellular enzymes. Their activity typically depends on accessory proteins, such as cytochrome P450 reductases, which facilitate efficient electron transfer. This requirement complicates the development of cell-free systems for bioremediation, as maintaining functional CYP450 activity outside the cellular environment remains technically challenging (Haroune et al. 2017).

13.4.1.5 Unspecific Peroxygenases (UPOs)

Unspecific peroxygenases (UPOs) are an interesting class of secreted hybrid enzymes that associate the catalytic characteristics of heme peroxidases and cytochrome P450 monooxygenases (Karich et al. 2017). Architecturally, UPOs occur in two key forms:

(i) Short UPOs (~29 kDa), which are extensively distributed across all fungal phyla.
(ii) Long UPOs (~44 kDa), which are found exclusively in ascomycetes and basidiomycetes.

Functionally, UPOs offer significant advantages: in xenobiotic transformation because they

- Utilize hydrogen peroxide as a cofactor.
- Simplify the reaction requirements compared to MnPs that rely on chelators and redox mediators.
- UPOs have demonstrated an extraordinary substrate versatility, with the ability to transform 41 of the EPA-listed priority pollutants and over 300 additional substrates, counting aromatic, aliphatic, polycyclic, and heterocyclic compounds.

The precise physiological role of UPOs: remains to be fully elucidated; it has been proposed that they

- Act synergistically with intracellular P450 enzymes.
- In this model, UPOs conduct crude extracellular oxidation to reduce the toxicity of xenobiotics, after which P450s carry out intracellular "fine-tuning" transformations that ultimately render the compounds inert or metabolizable as carbon sources (Olicón-Hernández et al. 2017).

Despite some limitations—such as (i) steric hindrance, (ii) substrate solubility, and (iii) the inhibitory effects of electron-withdrawing substituents—UPOs have proven capable of degrading a wide range of pollutants. For example, UPO from *Agrocybe aegerita* transformed naphthalene, phenol, anisole, toluene, ethylbenzene, acenaphthene, fluorene, phenanthrene, anthracene, pyrene, benzo(a)anthracene, and several phthalate esters, nitroarenes, and polychlorinated benzenes.

The ability of UPOs to act on substrates previously thought to be exclusive to P450s, MnPs, or LiPs underscores their biotechnological promise as versatile agents for extracellular xenobiotic degradation and environmental remediation.

13.4.2 Significance and Future Prospects

Collectively, LiPs, MnPs, and VPs form a robust enzymatic arsenal for degrading persistent environmental pollutants. Their broad substrate specificity, ability to catalyze one-electron oxidations, and capacity to initiate ring fission reactions make them indispensable tools in mycoremediation strategies.

Future research is focusing on enhancing enzyme yields, engineering variants with improved pH and temperature stability, and optimizing mediator systems for large-scale biotechnological applications. Immobilization strategies (e.g., cross-linked enzyme aggregates, encapsulation) are also being explored to improve reusability and operational stability under field conditions.

Given the rising levels of xenobiotic pollution in soil and water ecosystems, fungal peroxidases—particularly the hybrid versatile peroxidases—are poised to play an increasingly important role in green remediation technologies aimed at restoring contaminated environments in an eco-friendly and cost-effective manner.

Cytochrome P450 monooxygenases represent a versatile, genetically diverse, and biochemically powerful class of enzymes in fungal xenobiotic metabolism. Their ability to hydroxylate, demethylate, and destabilize a wide variety of pollutants—including PAHs, halogenated pesticides, and even pharmaceutical contaminants—makes them indispensable to fungal bioremediation strategies. Moreover, their presence in both ligninolytic and non-ligninolytic fungi expands the ecological reach of fungal detoxification. While technical limitations exist for industrial application, ongoing research into heterologous expression, cofactor regeneration systems, and metabolic engineering promises to unlock the full potential of fungal CYP450s in green biotechnological solutions for environmental remediation.

13.5 Mechanisms of Xenobiotic Detoxification by Fungal Enzymes

Fungi possess sophisticated biochemical systems to transform and detoxify a wide variety of xenobiotics. This ability makes them vital players in natural bioremediation and highly promising candidates for biotechnological applications in pollutant degradation. Fungal extracellular xenobiotic degradation generally occurs in two conserved steps. The first step involves the hydrolytic system, in which hydrolases break down complex macromolecules into smaller, more accessible fragments. The second step involves the ligninolytic system, which employs a battery of nonspecific oxidative enzymes to further oxidize and transform these smaller compounds into less toxic, more bioavailable, or completely mineralized forms. (Rao et al. 2010).

The chief classes of oxidoreductases comprise laccases, peroxidases (lignin peroxidase and manganese peroxidase), and cytochrome P450 monooxygenases.

These enzymes are remarkable for their ability to act on a wide range of substrates, including recalcitrant xenobiotics, often without requiring specific inducers. The following sections summarize the structural characteristics, catalytic mechanisms, and biotechnological significance of these major enzymatic systems (Table 13.2).

Table 13.2 Key strategies and considerations in fungal xenobiotic bioremediation

Strategy	Mechanism/approach	Key advantages	Example(s)
Isolation of pollutant-adapted strains	Collect and culture fungi from contaminated environments	Naturally tolerant to pollutants; produce biodegradative enzymes	*Aspergillus*, *Trametes* spp. isolated from polluted soils
Heterologous expression & genetic engineering	Clone and overexpress fungal enzyme genes in suitable hosts; apply mutagenesis (DNA shuffling, error-prone PCR)	Large-scale enzyme production, enhanced activity, improved pH/ temperature stability	Overexpressed laccase with higher dye degradation at acidic pH (77)
Enzyme immobilization	Covalent linkage, adsorption, or encapsulation of enzymes in matrices (glass beads, chitosan, alginate)	Increased half-life, stability, resistance to proteases, reusability	Laccases immobilized on glass beads retained 90% activity (81)
Use of cofactors and mediators	Add mediators/ cosubstrates to enhance catalytic cycles	Improves efficiency against recalcitrant pollutants	Veratryl alcohol boosts LiP activity; acetohydroxamic acid for laccases
Exploitation of extremophilic/marine fungi	Use fungi adapted to harsh conditions (salinity, hydrocarbons)	Enzymes function under extreme pH, salinity, temperature	*Pestalotiopsis palmarum* degrading crude oil fractions
Nanobiotechnology (nanozymes)	Employ nanoparticles with enzyme-like activity	High stability, low cost, reusable	Fe_3O_4 nanoparticles mimicking peroxidases for dye degradation
Metagenomics & bioinformatics	Identify detoxification genes from unculturable microbes; use molecular docking to predict enzyme-substrate interactions	Access to unculturable diversity; guides enzyme engineering	Screening laccase structures against toxic compound databases (88)

13.5.1 Laccases

Laccases (LACs) are multicopper oxidases with exceptionally low substrate specificity, making them key players in fungal xenobiotic degradation. In basidiomycetes, laccase families can include between 5 and 17 distinct isozymes (Yang et al. 2017). Despite this diversity, their catalytic core is highly conserved and relies on four copper atoms, all in the +2 oxidation state in the resting enzyme. These copper centers are classified into three types:

I. *T1 copper*—located near the enzyme surface, facilitates substrate oxidation.
II. *T2 copper* and *T3 binuclear copper pair*—form the trinuclear cluster (TNC), which stores electrons and catalyzes the four-electron reduction of oxygen to water.

The catalytic cycle begins when T1 copper abstracts electrons from a suitable substrate. Electrons are then relayed through a conserved His-Cys-His sequence to the TNC, reducing molecular oxygen bound at this site to water and regenerating the resting oxidized (RO) enzyme state (Jones and Solomon 2015).

13.5.1.1 Substrate Scope and Limitations

Laccases preferentially act on substrates with relatively low redox potentials, such as phenolic compounds, which are oxidized to phenoxyl radicals. These radicals can undergo coupling reactions or rearrange into quinones. However, laccases generally cannot directly oxidize compounds with high redox potentials or bulky substituents due to steric constraints.

13.5.1.2 Laccase-Mediator Systems (LMSs)

The versatility of laccases is dramatically enhanced by laccase-mediator systems (LMSs), in which low-molecular-weight mediators (e.g., ABTS, HBT, TEMPO) are first oxidized by laccase, and then, these mediator radicals oxidize otherwise recalcitrant substrates. LMSs broaden the range of accessible substrates, enhance reaction kinetics, and are widely used in xenobiotic remediation, pulp bleaching, and dye decolorization (Wong et al. 2013).

Key Advantages

(a) Broad substrate range (especially with LMSs)
(b) Mild reaction conditions (pH 4–6)
(c) Direct use of O_2 as a terminal electron acceptor
(d) Significant role in the degradation of phenolic pollutants, endocrine disruptors, and dyes

13.5.2 Peroxidases

Peroxidases represent another major class of fungal oxidoreductases involved in xenobiotic degradation. They are heme-containing enzymes that use hydrogen peroxide (H_2O_2) as a cosubstrate to catalyze powerful one- or two-electron oxidation reactions. Among these, the most studied are lignin peroxidases (LiPs) and manganese peroxidases (MnPs), both secreted by white-rot fungi.

General Catalytic Mechanism

Both LiP and MnP share a common reaction pathway. The native Fe(III) peroxidase is first oxidized by H_2O_2 to form compound I (an oxoiron(IV) porphyrin π-cation radical). Compound I then undergoes two successive one-electron reductions, each oxidizing a substrate molecule, to return to the resting Fe(III) state (Mougin et al. 2009). This cycle enables them to perform highly oxidative transformations even on substrates with high redox potentials.

13.5.2.1 Lignin Peroxidases (LiPs)

LiPs are specialized enzymes secreted primarily by white-rot basidiomycetes and are named for their ability to depolymerize lignin, a highly recalcitrant aromatic polymer. The Fe(III) center in LiP is embedded within a globular protein of ~50 × 40 × 40 Å, positioned in a heme pocket accessible via two narrow channels.

Mechanisms of Action

LiPs can act through three complementary mechanisms:

1. *Direct Oxidation*—Phenolic and non-phenolic lignin model compounds can directly access the heme center and undergo oxidative cleavage of Cα–Cβ bonds, leading to structural depolymerization.
2. *Mediator-Dependent Oxidation*—When direct access is not possible, LiPs oxidize mediators, such as veratryl alcohol (VA) to the VA^+ radical, which in turn oxidizes non-accessible substrates (Fig. 13.2).
3. *Fenton-like Reactions*—VA^+ radicals can further react with organic acids to generate superoxide anions. These anions reduce Fe^{3+} to Fe^{2+}, which then participates in a *Fenton reaction* with H_2O_2 to produce hydroxyl radicals, enabling nonenzymatic oxidation of highly recalcitrant compounds (Christian et al. 2005a, b).

Distinctive Features

(a) *Optimal pH*: ~3.0 (much lower than most peroxidases)
(b) *High Redox Potential*: Capable of oxidizing compounds other peroxidases cannot
(c) *Key Role*: Complete degradation of lignin, PAHs, and industrial pollutants

13.5.2.2 Manganese Peroxidases (MnPs)

MnPs are another group of heme peroxidases that primarily catalyze the oxidation of Cheng et al. (2007)

$$Mn^{2+} \rightarrow Mn^{3+},$$

which acts as a diffusible redox mediator. The enzyme undergoes two sequential one-electron transfers, forming compounds I and II, which oxidize Mn^{2+}(Pozdnyakova et al. 2018a, b).

The resulting Mn^{3+} is stabilized by chelation with carboxylic acids (e.g., oxalate, malonate) and diffuses into the extracellular environment, oxidizing phenolic and non-phenolic substrates indirectly.

Additional Pathways

1. Mn^{3+} can react with carboxylic acids to produce organic radicals, further broadening the substrate range.
2. MnP can also oxidize hydroquinones to semiquinone radicals, which participate in redox cycling and reduce oxidized pollutants.

Key Advantages

(a) Functions at moderately acidic pH (4–7)
(b) Capable of oxidizing a wide range of xenobiotics via small, diffusible mediators
(c) Suitable for soil and wastewater treatment where direct enzyme-substrate contact is limited

13.5.3 Cytochrome P450 Monooxygenases

Cytochrome P450 (CYP450) enzymes represent one of the most diverse enzymatic superfamilies in biology. They are intracellular, heme-thiolate monooxygenases that catalyze a wide range of regio- and stereospecific oxidations.

13.5.3.1 Mechanism of Catalysis

CYP450 enzymes catalyze the general reaction:

$$RH + O_2 + NAD(P)H \rightarrow ROH + H_2O + NAD(P)^+$$

The catalytic cycle proceeds through the following steps:

1. Substrate (RH) binds to Fe^{3+}-heme in the resting enzyme.
2. NADPH-cytochrome P450 reductase transfers an electron, reducing Fe^{3+} to Fe^{2+}.
3. O_2 binds to the reduced heme, forming a ferrous–dioxygen complex.
4. A second electron and protonation steps lead to O–O bond cleavage, releasing one oxygen as water.
5. The remaining oxygen atom forms a Fe = O (oxoferryl) species that performs a highly selective hydroxylation, epoxidation, or heteroatom oxygenation on the substrate.

An alternative "peroxide shunt" pathway allows CYP450s to use peroxides directly as oxygen donors, bypassing the need for NADPH.

13.5.3.2 Substrate Range

Fungal P450s catalyze the hydroxylation of aromatic hydrocarbons, alkylphenols, pesticides, and even heteroatom oxidations (S, N, P). Many products undergo further metabolism to dihydrodiols or phenols, priming them for ring-cleavage and mineralization.

13.5.3.3 Key Advantages

(i) Intracellular detoxification system that "fine-tunes" metabolites generated by extracellular enzymes (Rao et al. 2010).
(ii) High regio- and stereoselectivity makes them valuable for biocatalysis and synthesis of pharmaceuticals.
(iii) Critical for complete xenobiotic detoxification and assimilation as carbon sources.

13.5.3.4 Biotechnological and Ecological Implications

Together, laccases, peroxidases, and P450s represent a complementary enzymatic network. Laccases and peroxidases initiate extracellular oxidation, depolymerization, or transformation of xenobiotics, producing metabolites of lower toxicity or higher solubility. These intermediates are then taken up and subjected to intracellular P450-mediated transformations, often resulting in complete detoxification and assimilation.

Such coordinated enzymatic action has major implications for:

(i) *Bioremediation*: Treatment of soils, sediments, and effluents contaminated with PAHs, dyes, pesticides, and industrial chemicals.
(ii) *Green Chemistry*: Biocatalytic synthesis of value-added compounds, replacing harsh chemical oxidants.
(iii) *Waste Valorization*: Conversion of lignin and aromatic waste into bio-based chemicals.

The combination of broad substrate range, high redox potential, and extracellular secretion makes fungal oxidoreductases uniquely suited for biotechnological applications in xenobiotic mitigation strategies.

13.6 Techniques and Key Considerations for Pollutant Detoxification

Bioremediation using fungi or fungal-derived enzymes offers an effective approach for detoxifying xenobiotic compounds. Fungi are often superior to bacteria because several fungal phyla possess versatile enzyme systems capable of degrading a wide range of recalcitrant pollutants (Pozdnyakova et al. 2010). However, successful application of fungal bioremediation requires addressing several key limitations, including selecting pollutant-adapted strains and using biotechnological strategies to enhance enzyme performance, stability, and reusability (Fig. 13.4).

13.6.1 Limitations of Isolated Fungal Enzymes

Although fungal enzymes, such as laccases, lignin peroxidases (LiPs), and manganese peroxidases (MnPs), are powerful oxidative catalysts, their practical use faces several constraints.

(i) *Stability and Activity*: Enzymes often lose activity under field conditions due to suboptimal pH, temperature fluctuations, and the presence of denaturing agents. Laccases, for instance, exhibit an acidic pH optimum, which may not match the pH of effluents or contaminated soils (Sharma et al. 2018).
(ii) *Sensitivity to Cosubstrates*: Enzymes that rely on hydrogen peroxide, such as peroxidases, may become inactivated at high cosubstrate levels.

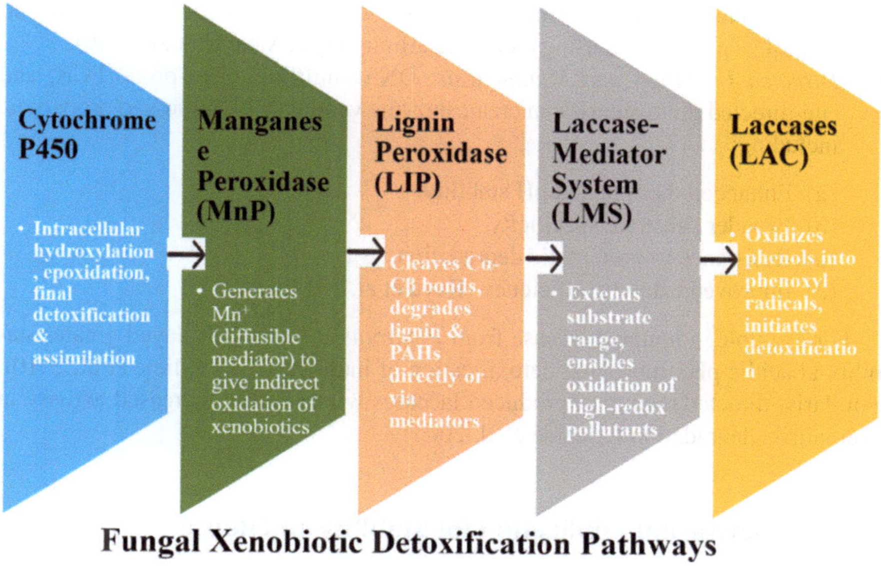

Fig. 13.4 Fungal xenobiotic detoxification pathways

(iii) *Environmental Susceptibility*: Applied enzymes face degradation by microbial proteases and destruction by physical and chemical factors. Heme-containing proteins are particularly vulnerable to oxidative inactivation (Valderrama et al. 2002).

(iv) *Economic Factors*: High costs of enzyme production, purification, and repeated application have slowed large-scale field deployment (Nicell and Wright 1997).

13.6.2 Isolation of Pollutant-Adapted Fungal Strains

One effective strategy is to isolate fungal populations from xenobiotic-polluted environments (Godoy et al. 2016). These strains are often preadapted to local pollutants and produce biodegradative enzymes naturally. However, native production is usually insufficient for direct application, necessitating strategies for scaling up enzyme yield.

13.6.3 Genetic and Enzyme Engineering

Biotechnological advances enable large-scale production and enhancement of fungal enzymes:

(i) *Heterologous Expression*: Genes encoding detoxification enzymes can be cloned and expressed in fungi, plants, or microbial hosts capable of colonizing contaminated substrates. This allows large-scale, cost-effective production and simplifies purification using cleavable affinity tags (Alcalde et al. 2006).

(ii) *Directed Evolution and Mutagenesis*: DNA shuffling, error-prone PCR, and site-directed mutagenesis can create enzyme variants with improved properties, including:

   (a) Enhanced thermal and pH stability.
   (b) Broader substrate specificity.
   (c) Resistance to proteolytic degradation.
   (d) Improved catalytic efficiency (Rayu et al. 2012).

For example, a mutated laccase from *Pleurotus ostreatus* displayed greater stability at acidic pH, improving detoxification of industrial dyes (Miele et al. 2010). Similarly, directed evolution produced laccases with 3.5-fold increased activity in acetonitrile degradation (Alcalde et al. 2005).

13.6.4 Enzyme Immobilization for Stability and Reuse

Immobilization is a crucial technique for preserving enzyme activity in adverse environments and enabling recovery and reuse.

- *Immobilization Matrices*: Enzymes can be covalently linked or adsorbed onto solid supports (glass beads, nylon membranes), entrapped in polymer gels (chitosan, gelatin), or encapsulated in matrices like sodium alginate.
- *Advantages*:

   1. Prolonged half-life and catalytic activity
   2. Protection against proteases and denaturation
   3. Increased resistance to pH, temperature, and chemical stress
   4. Cost-effectiveness through reusability (Ma et al. 2010; Su et al. 2022)

Studies show that laccases immobilized on glass beads retained 90% activity and exhibited enhanced protease resistance (Bilal et al. 2017a, b). Similar benefits were observed for MnPs (Bilal and Asgher 2016).

13.6.5 Use of Cofactors, Cosubstrates, and Mediators

Addition of mediators and cosubstrates can significantly boost enzymatic bioremediation, especially for recalcitrant pollutants.

- *Examples*:

   (a) Veratryl alcohol and 2-chloro-1,4-dimethoxybenzene enhance LiP activity (Asgher et al. 2008).
   (b) Acetohydroxamic acid acts as a mediator for laccases (Minussi et al. 2007).

(c) Supplementation with glucose as a carbon source has been shown to accelerate fungal detoxification (Asgher et al. 2008).

13.6.6 Role of Extremophilic and Marine Fungi

Marine and extremophilic fungi produce robust enzymes capable of functioning under high salinity, pressure, and pollutant load. For example, *Pestalotiopsis palmarum* thrives in extra-heavy crude oil, producing oxidative enzymes that degrade hydrocarbons (Naranjo-Briceño et al. 2013). Enzyme isolation from such organisms provides biocatalysts with intrinsic tolerance to environmental stress (Lima and Porto 2016).

13.6.7 Emerging Tools: Nanozymes and Bioinformatics

Recent innovations are replacing or complementing natural enzymes:

(i) *Nanozymes*: Engineered nanoparticles with enzyme-like activity offer high stability and low cost. Examples include Fe_3O_4 nanoparticles mimicking peroxidases for dye degradation (Wu et al. 2015) and graphene oxide nanomaterials with peroxidase-like activity (Ma et al. 2017).
(ii) *Bioinformatics and Metagenomics*: These approaches identify detoxification genes from unculturable microbes and allow heterologous expression. Molecular docking tools predict enzyme-substrate interactions and help screen laccases and other enzymes against pollutant databases (Suresh et al. 2008).

Successful fungal bioremediation requires a combination of strategies: identifying pollutant-adapted strains, genetically engineering enzymes for higher stability and broader specificity, immobilizing them for field use, supplementing cofactors and mediators, and leveraging extremophilic sources for robust biocatalysts. Cutting-edge approaches, such as nanozymes, metagenomics, and computational substrate prediction further expand the toolkit for effective xenobiotic detoxification, paving the way for cost-efficient, scalable, and sustainable remediation technologies.

13.7 Conclusion

This chapter comprehensively reviews the application of fungi for the degradation and removal of xenobiotic organic compounds (XOCs) from wastewater and aqueous solutions. Mycoremediation—which encompasses myco-enzymes, myco-degradation, and myco-sorption—has emerged as a promising, eco-friendly strategy for addressing pollutants that are otherwise resistant to conventional biodegradation.

Traditional chemical and physical treatments for removing XOCs, particularly from textile wastewater, have been widely explored. However, these processes often involve high operational costs or generate secondary pollutants. Consequently, attention has shifted toward biological approaches, particularly fungal oxidative enzymes such as manganese peroxidase (MnP), lignin peroxidase (LiP), and laccase, as well as myco-adsorption techniques. These enzymes catalyze the breakdown of complex and recalcitrant pollutants into simpler, less toxic compounds, thereby enhancing their bioavailability for further microbial degradation.

Although advanced oxidation processes (AOPs) are considered effective for XOC removal, their high cost has motivated research into low-cost alternatives. A major focus of current studies is the use of agricultural wastes as substrates for the production of fungal oxidative enzymes. These lignocellulosic residues are abundant, cost-effective, and highly inducible, making them ideal for large-scale enzyme production. The literature survey of 80 recent studies demonstrates the exceptional capacity of myco-enzymes to remove diverse XOCs, including pesticides, hydrocarbons, heavy metals, and industrial dyes.

Myco-sorption, particularly for azo dyes, has also shown promising results, although the exact mechanisms remain underexplored. It is hypothesized that functional groups on fungal cell walls facilitate dye uptake, especially for cationic dyes. Further research is needed to model sorption processes, regenerate fungal biomass, and recover dyes for potential recycling applications.

Despite its promise, mycoremediation faces challenges, including variability in efficiency depending on environmental conditions, pollutant type, and fungal species. Optimization and scaling up of these processes are required for widespread industrial applications. Nevertheless, the adaptability and resilience of fungi under diverse environmental conditions make them excellent candidates for bioremediation.

In conclusion, mycoremediation represents an innovative, sustainable, and cost-effective approach for environmental clean-up. With further research, development, and practical implementation, fungal systems—especially those utilizing agricultural waste for enzyme production—could significantly mitigate environmental pollution and contribute to healthier, more balanced ecosystems.

Acknowledgments Dr. N Saxena is grateful to MRM College for providing resources and all sorts of help for constructing the chapter.

References

Alcalde M, Bulter T, Zumárraga M, García-Arellano H, Mencía M, Plou FJ, Ballesteros A (2005) Screening mutant libraries of fungal laccases in the presence of organic solvents. J Biomol Screen 10(6):624–631

Alcalde M, Ferrer M, Plou FJ, Ballesteros A (2006) Environmental biocatalysis: from remediation with enzymes to novel green processes. Trends Biotechnol 24(6):281–287

Amitai G, Adani R, Sod-Moriah G, Rabinovitz I, Vincze A, Leader H, Chefetz B, Leibovitz-Persky L, Friesem D, Hadar Y (1998) Oxidative biodegradation of phosphorothiolates by fungal laccase. FEBS Lett 438(3):195–200

Ang EL, Zhao H, Obbard JP (2005) Recent advances in the bioremediation of persistent organic pollutants via biomolecular engineering. Enzym Microb Technol 37(5):487–496

Antošová Z, Herkommerová K, Pichová I, Sychrová H (2018) Efficient secretion of three fungal laccases from Saccharomyces cerevisiae and their potential for decolorization of textile industry effluent—a comparative study. Biotechnol Prog 34(1):69–80

Aracagök YD, Göker H, Cihangir N (2017) Biodegradation of micropollutant naproxen with a selected fungal strain and identification of metabolites. Z Naturforsch C 72(5–6):173–179

Asgher M, Bhatti HN, Ashraf M, Legge RL (2008) Recent developments in biodegradation of industrial pollutants by white rot fungi and their enzyme system. Biodegradation 19(6):771–783

Baker P, Tiroumalechetty A, Mohan R (2019) Fungal enzymes for bioremediation of xenobiotic compounds. In: Recent advancement in white biotechnology through fungi: volume 3: perspective for sustainable environments. Springer, Cham, pp 463–489

Bautista LF, Morales G, Sanz R (2015) Biodegradation of polycyclic aromatic hydrocarbons (PAHs) by laccase from *Trametes versicolor* covalently immobilized on amino-functionalized SBA-15. Chemosphere 136:273–280

Bilal M, Asgher M (2016) Enhanced catalytic potentiality of Ganoderma lucidum IBL-05 manganese peroxidase immobilized on sol-gel matrix. J Mol Catal B Enzym 128:82–93

Bilal M, Asgher M, Parra-Saldivar R, Hu H, Wang W, Zhang X, Iqbal HM (2017a) Immobilized ligninolytic enzymes: an innovative and environmental responsive technology to tackle dye-based industrial pollutants–a review. Sci Total Environ 576:646–659

Bilal M, Asgher M, Iqbal HM, Hu H, Zhang X (2017b) Bio-based degradation of emerging endocrine-disrupting and dye-based pollutants using cross-linked enzyme aggregates. Environ Sci Pollut Res 24(8):7035–7041

Bollag JM (1992) Decontaminating soil with enzymes. Environ Sci Technol 26(10):1876–1881

Camacho-Morales RL, García-Fontana C, Fernández-Irigoyen J, Santamaría E, González-López J, Manzanera M, Aranda E (2018) Anthracene drives sub-cellular proteome-wide alterations in the degradative system of Penicillium oxalicum. Ecotoxicol Environ Saf 159:127–135

Camarero S, Sarkar S, Ruiz-Duenas FJ, Martınez MJ, Martınez AT (1999) Description of a versatile peroxidase involved in the natural degradation of lignin that has both manganese peroxidase and lignin peroxidase substrate interaction sites. J Biol Chem 274(15):10324–10330

Camarero S, Pardo I, Cañas AI, Molina P, Record E, Martínez AT, Martínez MJ, Alcalde M (2012) Engineering platforms for directed evolution of laccase from Pycnoporus cinnabarinus. Appl Environ Microbiol 78(5):1370–1384

Cheng XB, Rong JIA, Ping-Sheng LI, Qin ZHU, Shi-Qian TU, Wen-Zhong TANG (2007) Studies on the properties and co-immobilization of manganese peroxidase. Chin J Biotechnol 23(1):90–96

Chmelová D, Ondrejovič M (2016) Purification and characterization of extracellular laccase produced by Ceriporiopsis subvermispora and decolorization of triphenylmethane dyes. J Basic Microbiol 56(11):1173–1182

Christian V, Shrivastava R, Shukla D, Modi H, Vyas BRM (2005a) Mediator role of veratryl alcohol in the lignin peroxidase-catalyzed oxidative decolorization of Remazol Brilliant Blue R. Enzym Microb Technol 36(2–3):327–332

Christian V, Shrivastava R, Shukla D, Modi HA, Vyas BR (2005b) Degradation of xenobiotic compounds by lignin-degrading white-rot fungi: enzymology and mechanisms involved. Indian J Exp Biol 43(4):301–312

Davila-Vazquez G, Tinoco R, Pickard MA, Vazquez-Duhalt R (2005) Transformation of halogenated pesticides by versatile peroxidase from Bjerkandera adusta. Enzym Microb Technol 36(2–3):223–231

Doddapaneni H, Chakraborty R, Yadav JS (2005) Genome-wide structural and evolutionary analysis of the P450 monooxygenase genes (P450ome) in the white rot fungus Phanerochaete chrysosporium: evidence for gene duplications and extensive gene clustering. BMC Genomics 6(1):92

Falade AO, Nwodo UU, Iweriebor BC, Green E, Mabinya LV, Okoh AI (2017) Lignin peroxidase functionalities and prospective applications. Microbiologyopen 6(1):e00394
Godoy P, Reina R, Calderón A, Wittich RM, García-Romera I, Aranda E (2016) Exploring the potential of fungi isolated from PAH-polluted soil as a source of xenobiotics-degrading fungi. Environ Sci Pollut Res 23(20):20985–20996
Gu C, Zheng F, Long L, Wang J, Ding S (2014) Engineering the expression and characterization of two novel Laccase Isoenzymes from Coprinus comatus in Pichia pastoris by fusing an additional ten amino acids tag at N-Terminus. PLoS One 9(4):e93912
Harms H, Schlosser D, Wick LY (2011) Untapped potential: exploiting fungi in bioremediation of hazardous chemicals. Nat Rev Microbiol 9(3):177–192
Haroune L, Saibi S, Cabana H, Bellenger JP (2017) Intracellular enzymes contribution to the biocatalytic removal of pharmaceuticals by *Trametes hirsuta*. Environ Sci Technol 51(2):897–904
Hirosue S, Tazaki M, Hiratsuka N, Yanai S, Kabumoto H, Shinkyo R, Arisawa A, Sakaki T, Tsunekawa H, Johdo O, Ichinose H (2011) Insight into functional diversity of cytochrome P450 in the white-rot basidiomycete Phanerochaete chrysosporium: involvement of versatile monooxygenase. Biochem Biophys Res Commun 407(1):118–123
Jones SM, Solomon EI (2015) Electron transfer and reaction mechanism of laccases. Cell Mol Life Sci 72(5):869–883
Junghanns C, Moeder M, Krauss G, Martin C, Schlosser D (2005) Degradation of the xenoestrogen nonylphenol by aquatic fungi and their laccases. Microbiology 151(1):45–57
Kadri T, Rouissi T, Brar SK, Cledon M, Sarma S, Verma M (2017) Biodegradation of polycyclic aromatic hydrocarbons (PAHs) by fungal enzymes: a review. J Environ Sci 51:52–74
Karich A, Ullrich R, Scheibner K, Hofrichter M (2017) Fungal unspecific peroxygenases oxidize the majority of organic EPA priority pollutants. Front Microbiol 8:1463
Keum YS, Li QX (2004) Fungal laccase-catalyzed degradation of hydroxy polychlorinated biphenyls. Chemosphere 56(1):23–30
Khan MF, Hof C, Niemcová P, Murphy CD (2023) Recent advances in fungal xenobiotic metabolism: enzymes and applications. World J Microbiol Biotechnol 39(11):296
Kües U (2015) Fungal enzymes for environmental management. Curr Opin Biotechnol 33:268–278
Kumari R, Singh A, Yadav AN (2021) Fungal enzymes: degradation and detoxification of organic and inorganic pollutants. In: Recent trends in mycological research: volume 2: environmental and industrial perspective. Springer, Cham, pp 99–125
Laurichesse S, Avérous L (2014) Chemical modification of lignins: towards biobased polymers. Prog Polym Sci 39(7):1266–1290
Levin L, Forchiassin F, Viale A (2005) Ligninolytic enzyme production and dye decolorization by *Trametes trogii*: application of the Plackett–Burman experimental design to evaluate nutritional requirements. Process Biochem 40(3–4):1381–1387
Lima RN, Porto ALM (2016) Recent advances in marine enzymes for biotechnological processes. Adv Food Nutr Res 78:153–192
Ma B, He Y, Chen HH, Xu JM, Rengel Z (2010) Dissipation of polycyclic aromatic hydrocarbons (PAHs) in the rhizosphere: synthesis through meta-analysis. Environmental Pollution 158(3):855–861
Ma X, Zhang L, Xia M, Li S, Zhang X, Zhang Y (2017) Mimicking the active sites of organophosphorus hydrolase on the backbone of graphene oxide to destroy nerve agent simulants. ACS Appl Mater Interfaces 9(25):21089–21093
Macellaro G, Pezzella C, Cicatiello P, Sannia G, Piscitelli A (2014) Fungal laccases degradation of endocrine disrupting compounds. Biomed Res Int 2014(1):614038
Margot J, Bennati-Granier C, Maillard J, Blánquez P, Barry DA, Holliger C (2013) Bacterial versus fungal laccase: potential for micropollutant degradation. AMB Express 3(1):63
Mate DM, Alcalde M (2015) Laccase engineering: from rational design to directed evolution. Biotechnol Adv 33(1):25–40
Miele A, Giardina P, Sannia G, Faraco V (2010) Random mutants of a Pleurotus ostreatus laccase as new biocatalysts for industrial effluents bioremediation. J Appl Microbiol 108(3):998–1006

Minussi RC, Pastore GM, Durán N (2007) Laccase induction in fungi and laccase/N–OH mediator systems applied in paper mill effluent. Bioresour Technol 98(1):158–164

Morel M, Meux E, Mathieu Y, Thuillier A, Chibani K, Harvengt L, Jacquot JP, Gelhaye E (2013) Xenomic networks variability and adaptation traits in wood decaying fungi. Microb Biotechnol 6(3):248–263

Mougin C, Boukcim H, Jolivalt C (2009) Soil bioremediation strategies based on the use of fungal enzymes. In: Advances in applied bioremediation. Springer, Berlin, pp 123–149

Naranjo-Briceño L, Pernía B, Guerra M, Demey JR, De Sisto Á, Inojosa Y, González M, Fusella E, Freites M, Yegres F (2013) Potential role of oxidative exoenzymes of the extremophilic fungus P estalotiopsis palmarum BM-04 in biotransformation of extra-heavy crude oil. Microb Biotechnol 6(6):720–730

Nicell JA, Wright H (1997) A model of peroxidase activity with inhibition by hydrogen peroxide. Enzym Microb Technol 21(4):302–310

Nousiainen P, Kontro J, Maijala P, Uzan E, Hatakka A, Lomascolo A, Sipilä J (2012) Lignin model compound studies to elucidate the effect of "natural" mediators on oxidoreductase-catalyzed degradation of lignocellulosic materials. In: Functional materials from renewable sources. American Chemical Society, pp 229–242

Olicón-Hernández DR, González-López J, Aranda E (2017) Overview on the biochemical potential of filamentous fungi to degrade pharmaceutical compounds. Front Microbiol 8:1792

Piontek K, Smith AT, Blodig W (2001) Lignin peroxidase structure and function. Biochem Soc Trans 29(2):111–116

Pozdnyakova NN, Nikiforova SV, Turkovskaya OV (2010) Influence of PAHs on ligninolytic enzymes of the fungus Pleurotus ostreatus D1. Central European Journal of Biology 5(1):83–94

Pozdnyakova NN (2012) Involvement of the ligninolytic system of white-rot and litter-decomposing fungi in the degradation of polycyclic aromatic hydrocarbons. Biotechnol Res Int 2012(1):243217

Pozdnyakova NN, Balandina SA, Dubrovskaya EV, Golubev CN, Turkovskaya OV (2018a) Ligninolytic basidiomycetes as promising organisms for the mycoremediation of PAH-contaminated Environments. In: IOP conference series: earth and environmental science, vol 107. IOP Publishing, p 012071

Pozdnyakova N, Dubrovskaya E, Chernyshova M, Makarov O, Golubev S, Balandina S, Turkovskaya O (2018b) The degradation of three-ringed polycyclic aromatic hydrocarbons by wood-inhabiting fungus Pleurotus ostreatus and soil-inhabiting fungus Agaricus bisporus. Fungal Biol 122(5):363–372

Prajapati D, Bhatt A, Gupte A, Gupte S (2022) Fungi: a sustainable and versatile tool for transformation, detoxification, and degradation of environmental pollutants. In: Progress in mycology: biology and biotechnological applications. Singapore, Springer Nature Singapore, pp 593–619

Rao MA, Scelza R, Scotti R, Gianfreda L (2010) Role of enzymes in the remediation of polluted environments. J Soil Sci Plant Nutr 10(3):333–353

Rastegari AA, Yadav AN, Gupta A (eds) (2019) Prospects of renewable bioprocessing in future energy systems, vol 10. Springer, Cham, pp 337–357

Rayu S, Karpouzas DG, Singh BK (2012) Emerging technologies in bioremediation: constraints and opportunities. Biodegradation 23(6):917–926

Salony, Mishra S, Bisaria VS (2006) Production and characterization of laccase from Cyathus bulleri and its use in decolourization of recalcitrant textile dyes. Appl Microbiol Biotechnol 71(5):646–653

Schlosser D, Höfer C (2002) Laccase-catalyzed oxidation of Mn2+ in the presence of natural Mn3+ chelators as a novel source of extracellular H2O2 production and its impact on manganese peroxidase. Appl Environ Microbiol 68(7):3514–3521

Sharma ND (2025) Reaction mechanism and redox potential of laccase. In: Laccase and polyphenol oxidase. Academic, pp 111–132

Sharma B, Dangi AK, Shukla P (2018) Contemporary enzyme based technologies for bioremediation: a review. J Environ Manag 210:10–22

Shrivastava R, Christian V, Vyas BRM (2005) Enzymatic decolorization of sulfonphthalein dyes. Enzym Microb Technol 36(2–3):333–337

Šimášková E (2020) Chemical and biochemical transformation of bioactive compounds. Univerzita Karlova

Suresh PS, Kumar A, Kumar R, Singh VP (2008) An Insilco approach to bioremediation: laccase as a case study. J Mol Graph Model 26(5):845–849

Su Y, Sun S, Liu Q, Zhao C, Li L, Chen S, Tang F (2022) Characterization of the simultaneous degradation of pyrene and removal of Cr (VI) by a bacteria consortium YH. Science of The Total Environment 853:158388

Syed K, Yadav JS (2012) P450 monooxygenases (P450ome) of the model white rot fungus Phanerochaete chrysosporium. Crit Rev Microbiol 38(4):339–363

Syed K, Porollo A, Lam YW, Yadav JS (2011) A fungal P450 (CYP5136A3) capable of oxidizing polycyclic aromatic hydrocarbons and endocrine disrupting alkylphenols: role of Trp129 and Leu324. PLoS One 6(12):e28286

Torres-Duarte C, Roman R, Tinoco R, Vazquez-Duhalt R (2009) Halogenated pesticide transformation by a laccase–mediator system. Chemosphere 77(5):687–692

Torres-Farradá G, Manzano Leon AM, Rineau F, Ledo Alonso LL, Sánchez-López MI, Thijs S, Colpaert J, Ramos-Leal M, Guerra G, Vangronsveld J (2017) Diversity of ligninolytic enzymes and their genes in strains of the genus Ganoderma: applicable for biodegradation of xenobiotic compounds? Front Microbiol 8:898

Tsutsumi Y, Haneda T, Nishida T (2001) Removal of estrogenic activities of bisphenol A and nonylphenol by oxidative enzymes from lignin-degrading basidiomycetes. Chemosphere 42(3):271–276

Valderrama B, Ayala M, Vazquez-Duhalt R (2002) Suicide inactivation of peroxidases and the challenge of engineering more robust enzymes. Chem Biol 9(5):555–565

Verma P, Madamwar D (2002) Production of ligninolytic enzymes for dye decolorization by cocultivation of white-rot fungi Pleurotus ostreatus and Phanerochaete chrysosporium under solid-state fermentation. Appl Biochem Biotechnol 102(1):109–118

Viswanath B, Rajesh B, Janardhan A, Kumar AP, Narasimha G (2014) Fungal laccases and their applications in bioremediation. Enzyme Res 2014(1):163242

Watanabe K (2001) Microorganisms relevant to bioremediation. Curr Opin Biotechnol 12(3):237–241

Wong KS, Cheung MK, Au CH, Kwan HS (2013) A novel Lentinula edodes laccase and its comparative enzymology suggest guaiacol-based laccase engineering for bioremediation. PLoS One 8(6):e66426

Wu Y, Teng Y, Li Z, Liao X, Luo Y (2008) Potential role of polycyclic aromatic hydrocarbons (PAHs) oxidation by fungal laccase in the remediation of an aged contaminated soil. Soil Biology and Biochemistry 40(3):789–796

Wu J, Xiao D, Zhao H, He H, Peng J, Wang C, Zhang C, He J (2015) A nanocomposite consisting of graphene oxide and Fe3O4 magnetic nanoparticles for the extraction of flavonoids from tea, wine and urine samples. Microchim Acta 182(13):2299–2306

Xu F (1996) Oxidation of phenols, anilines, and benzenethiols by fungal laccases: correlation between activity and redox potentials as well as halide inhibition. Biochemistry 35(23):7608–7614

Yadav AN, Sachan SG, Verma P, Kaushik R, Saxena AK (2016) Cold active hydrolytic enzymes production by psychrotrophic Bacilli isolated from three sub-glacial lakes of NW Indian Himalayas. J Basic Microbiol 56(3):294–307

Yang J, Li W, Ng TB, Deng X, Lin J, Ye X (2017) Laccases: production, expression regulation, and applications in pharmaceutical biodegradation. Front Microbiol 8:832

Zhang C, Li M, Chen X, Li M (2015) Edible fungus degrade bisphenol A with no harmful effect on its fatty acid composition. Ecotoxicol Environ Saf 118:126–132

Zhang H, Zhang S, He F, Qin X, Zhang X, Yang Y (2016) Characterization of a manganese peroxidase from white-rot fungus *Trametes* sp. 48424 with strong ability of degrading different types of dyes and polycyclic aromatic hydrocarbons. J Hazard Mater 320:265–277

MIX
Papier aus verantwortungsvollen Quellen
Paper from responsible sources
FSC® C105338
FSC
www.fsc.org

If you have any concerns about our products,
you can contact us on
ProductSafety@springernature.com

In case Publisher is established outside the EU,
the EU authorized representative is:
Springer Nature Customer Service Center GmbH
Europaplatz 3, 69115 Heidelberg, Germany

Printed by Libri Plureos GmbH
in Hamburg, Germany